AF713738

Quantitative Problems in the Biochemical Sciences

SECOND EDITION

Quantitative Problems in the Biochemical Sciences

Rex Montgomery / Charles A. Swenson

THE UNIVERSITY OF IOWA

W. H. FREEMAN AND COMPANY
New York

Library of Congress Cataloging in Publication Data

Montgomery, Rex.
Quantitative problems in the biochemical sciences.

Includes bibliographical references and index.
1. Biological chemistry—Problems, exercises, etc.
I. Swenson, Charles Allyn, 1933– joint author.
II. Title.
QD415.3.M6 1976 574.1'92'076 75–42234
ISBN 0–7167–0178–2

Copyright © 1969, 1976 by W. H. Freeman and Company

No part of this book may be reproduced
by any mechanical, photographic, or electronic process,
or in the form of a phonographic recording,
nor may it be stored in a retrieval system, transmitted,
or otherwise copied for public or private use,
without written permission from the publisher.

Printed in the United States of America

5 6 7 8 9 MP 0 8 9 8 7 6 5 4

Contents

Preface

The purpose of the second edition of *Quantitative Problems in the Biochemical Sciences,* like the first, is to illustrate ways of solving common quantitative problems in biochemistry. As the science of living systems has become more precise, it has become increasingly incumbent on the beginning student to grasp the quantitative aspects of this science as well as the less difficult descriptive elements. He is asked to use real experimental values in the calculation of some property, utilizing quantitative relationships that have been derived in general terms and with certain assumptions, approximations, and limitations. The problem is not only one of reducing an experimental situation or word problem to mathematical terms; it is also one of knowing the implications of the underlying assumptions and approximations. It is not surprising, therefore, that there is a continued request from students for study questions.

Students in biochemistry are no different from students in other sciences except that many of the basic quantitative aspects of the subject have been studied earlier in introductory chemistry courses, which have the elementary mathematical concepts as a prerequisite. However, this knowledge frequently needs restudy, particularly in the framework of problems drawn from the biochemical areas. Any encouragement to the student to think of biochemistry in these terms leads inevitably to a better understanding of the subject, but for this improvement to occur he must be given the chance to gain confidence in the quantitative approach, which requires practice. This text offers such an opportunity within the framework of experimental biochemistry.

The book is directed primarily toward the beginning student in biochemistry and many of the problems have been used for the teaching of medical, dental, and pharmacy students and, at times, of graduate students in biochemistry, chemistry, or any of the life sciences. However, the first edition of *Quantitative Problems in the Biochemical Sciences* was used by a broader spectrum of students than was envisaged at its inception. As a result it did not include enough questions and problems at the depth necessary for beginning graduate students. The new edition includes more such questions, without excluding those for the undergraduate student in the biochemical sciences. These additional problems have been accompanied by revisions of the text to include binding equilibria of macromolecules, two-substrate and sigmoidal enzyme kinetics, and smaller revisions throughout the text.

To differing extents, the discussions and problems in this book have been offered as supplementary material in the courses that have used the texts *Biochemistry* by Stryer (W. H. Freeman and Company), *Principles of Biochemistry* by White, Handler, and Smith (McGraw-Hill), *Outlines of Enzyme Chemistry* by Neilands and Stumpf (Wiley), *Dynamic Aspects of Biochemistry* by Baldwin (Cambridge University Press), *Textbook of Biochemistry* by West et al. (Macmillan), *Outlines of Biochemistry* by Conn and Stumpf (Wiley), *Biochemistry* by McGilvery (Saunders), *Biological Chemistry* by Mahler and Cordes (Harper and Row), *Biochemistry* by Lehninger (Worth), *Biochemistry: A Case-Oriented Approach* by Montgomery et al. (Mosby), or *Modern Concepts in Biochemistry* by Bohinski (Allyn and Bacon).

Only too frequently the understanding of a calculation is marred by misconceived manipulations of logarithms, the improper conversion of units, or the incorrect algebraic solution. Therefore, we review in Part 1 arithmetic and algebraic solutions, dimensions, units, and the statistical analysis of measurements. As noted earlier, this material is most likely to have been studied previously. It is included only for those students who wish to review their facility for manipulating simple mathematical expressions and chemical units.

The basic principles of quantitative biochemistry are treated in Part 2. Reversible reactions are introduced in a general sense and then applied to the dissociation of weak acids and bases, titration curves, buffer calculations, isoelectric points of amphoteric molecules, and the stability constants of complex ions. The important aspects of coupled reactions are dealt with in terms of chemical equilibrium, oxidation-reduction systems and free energy. Attempts have been made to clarify the confusion in the sign of the half-reaction potentials in oxidation-reduction systems. The various quantitative considerations of the different reversible reactions are brought together in the applications of biochemical energetics and the interconversion of energy. Simple enzyme kinetics are discussed with particular emphasis on the meaning and calculation by graphical solution of K_m, K_I, and V_{max}.

The principles discussed in Part 2 find their way into all areas of experimental biochemistry, such as the various studies of the hydrodynamic properties of biopolymers, analytical procedures such as chromatography, and more advanced kinetic analysis. Two of the basic experimental techniques, spectrophotometry and radiochemistry, are treated in some detail in Part 3. They were selected because of their broad application in following reactions without disturbing the system.

This text is not a collection of laboratory procedures or a handbook of critical tables, both of which should be consulted to learn of the experimental details and to obtain the precise physical data required to solve the problems of research. Nor is it our purpose to repeat the derivation of biochemical expressions, which would duplicate in part many excellent books. It has been necessary in certain cases, however, to consider in some detail certain of the steps in these derivations in order to indicate clearly the approximations and limitations of the final expression. Examples of related problems are worked out by steps at appropriate places so that the principles and their application to practical situations can be fully illustrated. It is difficult to separate the collection of data in the laboratory from the solving of problems, since the errors in the data must be considered and, therefore, the approximations in the experimental technique taken into account. Wherever appropriate, and in particular in the sections on spectrophotometry and radiochemistry, these experimental errors are discussed briefly so that the student will appreciate these techniques better, both in the laboratory and during the solving of problems.

The problems that are detailed in the text vary in difficulty depending upon the point to be illustrated. Similarly, the problems at the end of each section are graded into an A-type, having most direct application to mathematical expressions, and a B-type, which is frequently drawn from actual experimental observations and requires varying degrees of manipulation before the eventual calculation of the unknown. In Chapters 8 and 12 some problems particularly germane to the students in health sciences have been collected separately.

We acknowledge again the constructive suggestions of our colleagues, who also provided a number of problems from their research notebooks. In particular, we would like to thank the instructors from many institutions who sent constructive criticisms, and particularly Bryce V. Plapp and Rex Lovrein who read parts or all of the manuscript. Many of the suggestions form part of this edition; some could not have been included without appreciable expansion of the text. It is hoped that this edition will continue to stimulate the student to approach quantitative problems with confidence.

January 1976

Rex Montgomery
Charles A. Swenson

Quantitative Problems in the Biochemical Sciences

PART ONE

MATHEMATICS AND CHEMICAL UNITS

1
Problem Solving and Experimentation

To the experimental scientist, the solving of problems and the collection of data from an experiment are all part of the same process. Science begins with a detailed observation of some part of nature, but the raw data so obtained rarely represent the desired result directly. Unlike the stopwatch reading of a 100-yard dash or the body temperature of an animal, the data obtained from an experiment are usually treated in a mathematical way to give a derived value, which may be a universal constant such as a molecular weight or the speed of light, or a property of the material under study such as the concentration of a component or the enzymic activity of the preparation.

Many observations are made in the collection of the raw data, but not all are eventually useful. For example, the temperature may be recorded but the property being studied may not show any temperature dependence. Furthermore, the actual readings taken from the various instruments during the experiment may not have the same degree of accuracy and/or precision. The reliability of the result may be such that duplicate, triplicate, or multiple observations are needed, and each observation must be properly weighted.

Briefly, these are some of the general considerations that are taken in the design and performance of an experiment. After the collection of the raw data the problem solving commences. All of the problems that are given in the following pages could be a part of a large experiment and in general it is constructive to consider them as such. The problems of type A can usually be looked upon as refined data for direct inclusion in a mathematical expression. The problems of type B are closer to the true experimental observations.

The stepwise solution of problems could proceed in the following order.

1. Abstract from the data the information that is relevant for the required solution. As mentioned, temperature may be given but may not be needed. To solve a problem obviously requires a clear knowledge of its purpose and of the mathematical basis for its solution.

2. The experimental data may need to be statistically analyzed to find a reliable average. Some of the observations may be so far out of line that they should be disregarded in a particular set of data. For example, in a series of identical titrations the required milliliters of titrant may be 10.1, 10.2, 10.0, 9.9, 13.3, 10.1, 10.2, 9.9. The single value of 13.3 is clearly out of line in this series and may have been due to an unclean flask or observational error. It would not be averaged with the other values.

3. The experimentalist is usually required to make an error analysis of his results. He needs to know to what significant figures he can justifiably express the readings. There would be no point in averaging the above titration values to three decimal places.

4. The mathematical treatment of data requires that the dimensions be consistent. If the answer required is a volume, then the dimensions of the components in the mathematical formula must cancel out to give a volume. An appreciation of this type of analysis will enable a quick check to be made to see, for example, that when using the various volume units the number and not its reciprocal be introduced.

5. The mathematical manipulations must be correct. So many times the careless omission of a negative sign, the wrong collection of like terms, or the incorrect use of logarithms or slide rule may give impossible answers. One must get into the habit of checking the answers by rounding off figures to such simple numbers that the order of magnitude can at least be obtained. With a little practice it is often surprising how close approximations made in this way approach the correct answer.

6. Biochemistry is becoming more and more precise. Numerical values continue to be obtained with greater precision. It should be little consolation to you to argue that the method of approach to the problem was correct and that the actual answer is not as important. In real life the actual result is all important, and in the health sciences life itself may be prolonged only if critical calculations are correct.

In the calculations of many of the problems in the text, a slide rule was used. If, however, the accuracy of the data justified it, then logarithms or a desk calculator may have been used.

2
Exponents and Logarithms

The mathematics used in introductory biochemistry, for the most part, requires only arithmetic and simple algebra; however, a clear understanding of these concepts is essential. In Chapters 2, 3, and 4 the following topics are discussed: exponents, logarithms, algebraic equations, graphing of functions and graphic solutions of equations, dimensions and units, and error in experimental measurements. Any parts of these chapters in which the student already possesses competence should be bypassed in favor of succeeding sections.

EXPONENTS

When performing mathematical operations, exponents are frequently used. An *exponent* is a number which indicates how many times the *base* appears as a factor. Thus 5×5 can be indicated by 5^2; the exponent, 2, appears as a superscript on the base 5. The base 5 may also be said to be raised to the second power. We can generalize by calling a^m the mth power of the base a and calling m the exponent of the power.

For scientific work the properties of exponents are useful, and the representation of very large or very small numbers can be written conveniently. Consider, for example, having to write out Avogadro's number: 6.02322×10^{23} molecules per gram-mole. Scientists also often use exponential numbers for an approximate check on the result before proceeding with a detailed calculation, or to locate the decimal point when using a slide rule.

Any number can be represented in an exponential form. Although many bases are possible, ten (decimal system) is the most convenient base to use. The number 1000 can then be written as 1×10^3, 200 as 2×10^2, 0.003 as 3×10^{-3}, and so on. Numbers which are integral (whole) as well as non-integral powers of 10 can be represented. Conversion to exponential form is accomplished by shifting the decimal point of the number and multiplying the new number by the appropriate power of ten to account for the shift. Generally the exponential form is chosen such that it is the product of a number between 1 and 10 and 10 to some power. This particular exponential form is called scientific notation. Two steps are required for writing a number in scientific notation.

1. Shift the decimal point so that the resulting number is between one and ten.

2. Determine the power of ten. This is done by counting the number of positions the decimal point needs to be shifted to restore it to the original position. If this shift is to the right the exponent will be positive, and if it is to the left it will be negative.

Problem 2.1: Convert the number 7200 to scientific notation.

Solution:
Step 1. Place the decimal point so that the resulting number is between one and ten:

$$7.200$$

Step 2. The decimal point can be restored to the original position by shifting it three places to the right. The exponent on 10 is three and the number in scientific notation is 7.2×10^3.

Problem 2.2: Convert 0.000173 to scientific notation.

Solution:
Step 1. Write as a number between 1 and 10:

$$1.73$$

Step 2. The decimal point can be restored by shifting 4 places to the left. The exponent on 10 is -4, and thus the number in scientific notation is 1.73×10^{-4}.

To use these exponential forms in the mathematical operations of addition, subtraction, multiplication, and division, a few simple rules must be followed.

1. For the operations of addition or subtraction, all the terms must have the same exponent. It is convenient for our purposes to use an additional

operational rule that states: convert all terms to the largest positive exponent in the set. This gives the sum in scientific notation directly.

Problem 2.3: Perform the following addition:

$$5.0 \times 10^{-2} + 3.5 \times 10^{-1} + 7.1 \times 10^{-3}$$

Solution: Convert all terms to the largest positive exponent, in this case −1, and add

$$0.5 \times 10^{-1} + 3.5 \times 10^{-1} + 0.071 \times 10^{-1}$$

Remove the common factor, 10^{-1}, to give

$$(0.5 + 3.5 + 0.071) \times 10^{-1} = 4.071 \times 10^{-1}$$

Problem 2.4: Perform the following subtraction:

$$(5.02 \times 10^{7}) - (3.01 \times 10^{6})$$

Solution: Convert to the largest positive exponent, which is 7:

$$5.02 \times 10^{7} - 0.301 \times 10^{7}$$

Remove the common factor, 10^{7}, to give

$$(5.02 - 0.301) \times 10^{7} = 4.719 \times 10^{7}$$

---

2. For multiplication of exponential numbers the numerical portion is treated as usual and the exponential part uses the law for multiplication of exponents. This is

$$a^{m} \times a^{n} = a^{m+n} \tag{2.1}$$

and for the base 10 (decimal) it becomes

$$10^{m} \times 10^{n} = 10^{m+n} \tag{2.2}$$

The exponents are added algebraically to obtain the product in exponential form.

Problem 2.5: Multiply 500 × 3000.

Solution: Convert to exponential numbers:

$$(5 \times 10^{2}) \times (3 \times 10^{3})$$

Regroup the product:

$$5 \times 3 \times 10^{2} \times 10^{3}$$

Multiply the parts,

$$15 \times 10^{5}$$

and convert to scientific notation:

$$1.5 \times 10^6$$

Problem 2.6: Multiply 0.0013×5000.

Solution: Convert to scientific notation:

$$1.3 \times 10^{-3} \times 5 \times 10^3$$

Regroup the product:

$$1.3 \times 5 \times 10^{-3} \times 10^3$$

$$6.5 \times 10^0 = 6.5$$

---

3. Any number a to the zero power is 1:

$$a^0 = 1 \tag{2.3}$$

4. For division of exponential numbers the quotient for the numerical part is formed in the usual manner. The quotient for the exponential part is obtained by the rule for division of exponents:

$$\frac{a^m}{a^n} = a^{m-n} \tag{2.4}$$

Problem 2.7: Divide 25500 by 0.015.

Solution: Convert all factors to exponential numbers:

$$\frac{2.55 \times 10^4}{1.5 \times 10^{-2}}$$

Form quotient for parts

$$\frac{2.55}{1.5} \times 10^{4-(-2)}$$

Perform division to give

$$1.7 \times 10^6 \text{ (or 1,700,000)}$$

Problem 2.8: $140 \div 0.0002$.

Solution: Convert to exponential numbers:

$$1.40 \times 10^2 \div 2 \times 10^{-4}$$

Solution I: Form the quotient:

$$\frac{1.40 \times 10^2}{2 \times 10^{-4}}$$

Divide the parts:

$$0.7 \times 10^{2-(-4)}$$

Add the exponents:

$$0.7 \times 10^{+6}$$

Convert to scientific notation to give

$$7 \times 10^{5} \text{ (or 700,000)}$$

Solution II: Form the quotient after changing the decimal point in the numerator, so that the answer is in scientific notation directly:

$$\frac{14 \times 10^{1}}{2 \times 10^{-4}}$$

Divide the parts and obtain the quotient:

$$7 \times 10^{1-(-4)}$$

Add the exponents to obtain answer:

$$7 \times 10^{5}$$

Problem 2.9: Solve, using the rules for exponential numbers:

$$\frac{6.0 \times 10^{-27}(3.0 \times 10^{10})6.0 \times 10^{23}}{8.0 \times 10^{-5}}$$

Solution: Regroup the factors

$$\frac{108}{8} = 10^{-27+10+23-(-5)}$$

Perform multiplication and division to obtain answer

$$13.5 \times 10^{11}$$

Convert to scientific notation

$$1.35 \times 10^{12} \text{ (or 1,350,000,000,000)}$$

---

5. Two other secondary laws for exponents are useful in calculations. The first is the law for reciprocals of exponential numbers, which is a special case of the division of exponents:

$$\frac{1}{a^n} = a^{-n} \tag{2.5}$$

The second derives from the fact that powers of a number can be fractional and are expressed as

$$a^{1/n} = \sqrt[n]{a} \tag{2.6}$$

LOGARITHMS

The properties of exponents as used in the preceding section obviously facilitate the operations of multiplication and division for integral (whole) exponents. These rules, however, are applicable to nonintegral exponents as well. Thus, if we can express any number x as $x = 10^y$, we can apply the laws for exponents. This introduces the use of logarithms. The logarithm of a number is the power to which the base must be raised to give the number.

In our notation y is therefore the logarithm of x written as

$$\log_{10} x = y \tag{2.7}$$

It is the power to which the base 10 needs to be raised to give the number x. The log of 100 is 2

$$\log_{10} 100 = 2 \tag{2.8}$$

Using our definition of the logarithm, we can write

$$100 = 10^2 \tag{2.9}$$

which of course is true.

The general definition for a logarithm can be written as

$$x = a^y \tag{2.10}$$

where a is the base and y is the logarithm of x, since it is the power to which the base is raised to give x, or

$$\log_a x = y \tag{2.11}$$

It is read "the logarithm of x to the base a is equal to y."

Problem 2.10: Evaluate $\log_5 25$, $\log_{10} 10000$, $\log_3 27$, and $\log_2 32$.

Solution: $\log_5 25 = 2$, since by definition of the logarithm

$$5^2 = 25$$

The $\log_{10} 10000 = 4$, since by definition of the logarithm

$$10^4 = 10000$$

Similarly, $\log_3 27 = 3$ because

$$3^3 = 27$$

And the $\log_2 32 = 5$ because

$$2^5 = 32$$

Base 10 logarithms (common) and base e logarithms (natural) are the most useful. Logarithms to the base e, an irrational number 2.718281828 . . . , are used in theoretical calculations. To distinguish them from common logs we write the natural log as $\ln x$. Tables are available for these logarithms; however, it is as convenient to convert from base e to base 10, using the formula

$$\ln x = 2.303 \log_{10} x \tag{2.12}$$

The logarithms for numbers which have integral exponents for the various bases can be written with no difficulty. Now let us restrict ourselves to the base 10, which in the future will be understood, unless another base is indicated. Consider the logarithm of two hundred, log 200. This logarithm is divided into two parts, called the *characteristic* and the *mantissa*. The characteristic is a whole number which indicates the position of the decimal point; thus it is related to the power to which 10 needs to be raised. For any number written in scientific notation the characteristic is the exponent on the base 10. It may be positive or negative and in the latter case is indicated by a bar over the number. For our example log 200 or $\log 2 \times 10^2$, the characteristic is 2; for log 0.02 the characteristic is $\bar{2}$. The mantissa indicates the relative position of the number between any two powers of 10. It is obtained by looking up the numerical portion of the exponential number in a logarithm table. For the number two the mantissa to four places is 3010. Mantissas, since they are a fraction of an exponent, are understood to have a decimal point in front of them. Thus the logarithm of 2×10^2 is 2.3010. By the definition of logarithms

$$200 = 10^{2.3010} \tag{2.13}$$

Determining the Log of a Number. This may be easily achieved in the following way.

Step 1. Write the number in exponential form and obtain the characteristic.
Step 2. Determine the mantissa for the numerical part from the log table.
Step 3. Sum the parts to give the logarithm.

Problem 2.11: Find the log of 20.

Solution:
Step 1. Write in exponential form as 2.0×10.
Step 2. The mantissa for the numerical part (from the table) is 3010. Note that for determining logs for numbers in scientific notation, the numerical part is always a number between 1 and 10. Therefore the mantissa from the log table is its logarithm. The characteristic is zero, since these numbers in scientific notation are expressed as the number times 10^0.

Step 3. The $\log 2.0 \times 10^1 = \log 2.0 + \log 10^1$ by the definition of logs and the law for exponents.
The $\log 10^1$ is one by the definition of logarithms so substituting

$$\log 20 = 0.3010 + 1$$
$$= 1.3010$$

or

$$20 = 10^{1.3010}$$

It is apparent that the log of 2000 is 3.301, log 20,000 is 4.301, and so on. The mantissa is the same for all; only the characteristic is different.

Negative logarithms are obtained for numbers which are less than one. Two schemes are illustrated for expressing them.

Problem 2.12: Find log 0.002.

Solution:
Step 1. Write in exponential form:

$$2.0 \times 10^{-3}$$

Step 2. Find log 2.0 from the log table:

$$\log 2.0 = 0.3010$$

Step 3. The $\log 0.002 = \log 2.0 + \log 10^{-3}$, so add logs to obtain

$$0.3010 + (-3) = -2.6990$$

or

$$\log 0.002 = -2.6990$$

or

$$0.002 = 10^{-2.6990}$$

In Problem 2.12 both the characteristic and the mantissa of the logarithm are negative. It is also possible to represent this logarithm as

$$\log 2.0 \times 10^{-3} = \bar{3}.3010$$

as stated in the discussion of characteristics. Here the characteristic is negative and the mantissa is positive; that is, the statement still is $-3 + 0.3010$. The negative characteristic with a positive mantissa is represented by the bar. Another common way of representing this logarithm is $7.3010 - 10$.

The use of a negative characteristic with a positive mantissa is more convenient for most calculations; however, for calculations involving pH (see Chapter 7) it is necessary to carry out the subtraction.

Problem 2.13: Find the log of 0.0083.

Solution: Write the log of the number in exponential form:

$$\log 8.30 \times 10^{-3}$$

Find log 8.30 in the table:

$$\log 8.30 = 0.9190$$

Since $\log 8.30 \times 10^{-3} = \log 8.30 + \log 10^{-3}$, then

$$\log 8.30 \times 10^{-3} = 0.9190 + (-3)$$

$$= -2.0810$$

This can also be written as $\bar{3}.9190$ and $7.9190 - 10$.

Antilogs. Finding the antilog is the reverse of finding the log. The antilog is the number which corresponds to the logarithm. Do not make the mistake of including the characteristic in the figures that are converted in the antilog tables.

Problem 2.14: Find the antilog of 2.4771.

Solution: Look up the mantissa, 4771, in the log table and find the number which corresponds to it; in this case 3000. The characteristic is 2, and thus the antilog of 2.4771 is 3.000×10^2 or 300.0. Note that the number corresponding to the mantissa is always given as a number between 1 and 10, since looking up 4771 is finding the antilog of 0.4771.

Problem 2.15: Find the antilog of $\bar{3}.6020$.

Solution: Look up the mantissa, 6020, in the log table, and find the number which corresponds to it: 4000. Since the characteristic is -3, the number must be 4.000×10^{-3} or 0.004000.

We have already noted that numbers that are less than one have negative logarithms and furthermore that in one form both the characteristic and the mantissa are negative. Since the mantissas in logarithm tables are always positive we must convert to a form which has a positive mantissa before proceeding to the antilog.

Problem 2.16: Find the antilog of -2.337.

Solution: Convert to a positive mantissa:

$$-2.337 = -3. + 0.663$$

This is written as $\bar{3}.663$. The mantissa corresponds to 460. Since the characteristic is -3, the number is 4.60×10^{-3}, or 0.00460.

It is possible, then, to express any number in terms of the base 10 to some power. Logarithms are these exponents; thus we can use the laws of exponents for calculations. These laws for exponents enable the following equations to be written.

For multiplication,

$$\log A \cdot B = \log A + \log B \tag{2.14}$$

For division,

$$\log \frac{A}{B} = \log A - \log B \tag{2.15}$$

For raising to a power,

$$\log A^n = n \log A \text{ (n can be integral or nonintegral)} \tag{2.16}$$

Multiplication. The logs of the numbers are found, added together, and the antilog of the sum obtained.

Problem 2.17: Find the product, by Equation (2.14), of

$$23 \times 0.0031$$

Solution:

$$\begin{array}{rcrcr} \log 23 & = & 1.3617 & & 1.3617 \\ + \log 0.0031 & = & +\bar{3}.4914 & \text{or} & +7.4914 - 10 \\ \hline & & \bar{2}.8531 & \text{or} & 8.8531 - 10 \end{array}$$

Antilog:

$$0.0713$$

Division. The logs of the dividend and divisor are formed and the log of the quotient obtained by subtracting the log of the divisor from the log of the dividend.

Problem 2.18:

Divide 960 by 0.003, using Equation (2.15).

Solution 1:

$$\begin{aligned} \log 960 &= 2.9823 \\ -\log 0.003 &= \underline{-\bar{3}.4771} \\ &5.5052 \end{aligned}$$

$$\log \frac{960}{0.003} = 5.5052$$

Antilog:

$$\frac{960}{0.003} = 320{,}000$$

Solution 2:

$$\log 960 = 12.9823 - 10 \text{ (add and subtract 10)}$$

$$\log 0.003 = \frac{-(7.4771 - 10)}{5.5052}$$

$$\log \frac{960}{0.003} = 5.5052$$

Antilog:

$$\frac{960}{0.003} = 320{,}000$$

Powers. The formula for raising to a power indicates that the log of the number is multiplied by the power. If logs are used for this operation a log log term results, and it is necessary to take the antilog twice.

Problem 2.19: Evaluate $(5.75)^{3.13}$, using Equation (2.16).

Solution: Substitute in Equation (2.16) to obtain

$$\log (5.75)^{3.13} = 3.13 \log 5.75 \quad \text{(i)}$$

Find log:

$$\log 5.75 = 0.760$$

Substitute in (i):

$$\log (5.75)^{3.13} = 3.13\ (0.760) \quad \text{(ii)}$$

Find logs to evaluate the product in (ii):

$$\log 3.13 = 0.496$$

$$\log 0.760 = 9.881 - 10$$

Writing (ii) in log form we obtain

$$\log \log (5.75)^{3.13} = \log 3.13 + \log 0.760 \quad \text{(iii)}$$

Substituting in (iii), we obtain

$$\log \log (5.75)^{3.13} = 0.496 + 9.881 - 10$$

When we sum,

$$\log \log (5.75)^{3.13} = 0.377$$

Antilog to get

$$\log (5.75)^{3.13} = 2.378$$

Antilog a second time to obtain

$$(5.75)^{3.13} = 239$$

Problem 2.20: Evaluate $(0.78)^{0.32}$.

Solution: $\log (0.78)^{0.32} = 0.32 \log 0.78$

Find log:

$$\log 0.78 = \bar{1}.892$$
$$= -0.108$$

Substitute (note here that log 0.78 must be a complete negative number):

$$\log (0.78)^{(0.32)} = 0.32 \log 0.78$$
$$= -(0.32)(0.108)$$

Find logs to evaluate product:

$$\log 0.32 = \bar{1}.505$$
$$\log 0.108 = \bar{1}.032$$

Write in log form:

$$\log \log (0.78)^{0.32} = -\log 0.32 + \log 0.108$$

Substitute logs:

$$\log \log (0.78)^{0.32} = -\bar{1}.505 + \bar{1}.032$$
$$= -\bar{2}.537$$

Antilog:

$$\log (0.78)^{0.32} = -0.034$$

Convert to positive mantissa:

$$\log (0.78)^{0.32} = \bar{1}.966$$

Antilog:

$$(0.78)^{0.32} = 9.22 \times 10^{-1}$$
$$= 0.924$$

Interpolation. Logarithm tables may have different numbers of significant digits; 3, 4, 5, 6, 7, or more. If one is using a 4-place table and 4 significant figures are required, interpolation will be necessary to obtain the last decimal place.

Problem 2.21: Determine the log of 37.23.

Solution: The characteristic is 1 from the exponential form 3.723×10^1. The mantissa for 372 is 5705 and for 373 is 5717. The difference is 12; since 3723 is 0.3 of the interval between 372 and 373, we find 0.3 of $12 = 3.6 \approx 4$. This is added to the mantissa for 372 to form the mantissa for 3723 $(5705 + 4) = 5709$. The log of 37.23 is then 1.5709 to four places.

---

2. PROBLEMS A

Answers on page 332

1. Write the following numbers in exponential form.

(a) 7,231	(i) 0.00127
(b) 0.00289	(j) 3,824,434
(c) 0.00047	(k) 0.189
(d) 434	(l) 4,387
(e) 0.01456	(m) 0.0821
(f) 5,368,000	(n) 0.000073
(g) 227	(o) 620
(h) 17,450	(p) 713,000

2. Solve, using a slide rule and exponential numbers.
   (a) $(4.1 \times 10^{-7})(3.4 \times 10^{8})$
   (b) $(5.2 \times 10^{3}) + (6.7 \times 10^{-2})$
   (c) $(8.1 \times 10^{2})(0.13) + (2.9 \times 10^{7})(0.0012)$
   (d) $\dfrac{(2.2 \times 10^{-7})(1.3 \times 10^{8})}{3.0 \times 10^{-3}}$
   (e) $(8.5 \times 10^{4})(6.9 \times 10^{2}) - (6.1 \times 10^{2})(9.2 \times 10^{1})$
   (f) $\dfrac{5.6 \times 10^{-14}}{3.1 \times 10^{2}}$
   (g) $\sqrt{(1.8 \times 10^{-5})^2 + 4(3.7 \times 10^{-7})}$
   (h) $\dfrac{(2.5 \times 10^{-3})}{(9.0 \times 10^{5}) - (8.6 \times 10^{4})}$
   (i) $(1.0 \times 10^{-4})(3.8 \times 10^{3}) + (4.1 \times 10^{-5})(9.1 \times 10^{4}) - (3.6 \times 10^{-2})$
   (j) $(4.9 \times 10^{9})(6.2 \times 10^{-5})$
   (k) $[(4.3 \times 10^{2})(6.1)(3.9 \times 10^{-4})]^{1/4}$
   (l) $\dfrac{(4.7 \times 10^{3}) - (3.8 \times 10^{2})(0.24)}{4.1 \times 10^{5}}$

3. Find the logarithm of the following numbers.

(a) 271
(b) 0.00358
(c) 0.000051
(d) 59
(e) 7.2×10^{-3}
(f) 0.0415
(g) 8910
(h) 1.6×10^{-6}
(i) 0.082
(j) 1.24
(k) 57.8
(l) 343
(m) 4.20×10^{-4}
(n) 329
(o) 0.00321
(p) 0.578
(q) 47.84
(r) 0.000421
(s) 3.4×10^{-3}
(t) 5.6×10^{2}
(u) 0.000021
(v) 655
(w) 2.7×10^{-5}
(x) 78

4. Evaluate the following, using logarithms.

(a) $(6.3 \times 10^{-5}) \times (0.0415)$
(b) $3.1 \times 10^{-2}(128)$
(c) $\dfrac{59}{1.6 \times 10^{-6}}$
(d) $\dfrac{271}{3.14 \times 10^{-3}}$
(e) $4.7 \times 10^{-3}(7.14)$
(f) $3.51 \times 10^{-4}(776)$
(g) $\sqrt[2]{776}$
(h) $4.2 \times 10^{-1}(6.14)$
(i) $\dfrac{0.0080}{4.1 \times 10^{-2}}$
(j) $(12.8)^{5.10}$
(k) $\sqrt[4]{12.8}$
(l) $\dfrac{140}{4.60 \times 10^{-3}}$
(m) $\dfrac{5.10 \times 10^{-5}}{3.10 \times 10^{-3}}$
(n) $0.000051 \times 3.1 \times 10^{-3}$
(o) $4{,}420 \times 0.0011$
(p) $\dfrac{28}{0.0029}$
(q) $589 \times 5.6 \times 10^{-4}$
(r) $\dfrac{0.000358}{57}$
(s) $\sqrt[3]{0.00358}$
(t) $(0.59)^{1.6}$
(u) $(2.18)^{3.4}$
(v) $\dfrac{628}{0.00346}$
(w) $(6.18)^{6.410}$
(x) $\sqrt[3]{210}$
(y) $(0.00315)271$
(z) 4285×0.00115
(a′) $(5.21)^{0.0713}$
(b′) $\dfrac{0.0061}{525}$
(c′) $\sqrt[4]{268}$
(d′) $(3.58 \times 10^{-4})(1.6 \times 10^{-6})$
(e′) $\dfrac{0.00315}{59}$
(f′) $618 \times 4.10 \times 10^{-2}$
(g′) $\sqrt[4]{128}$
(h′) $\dfrac{329}{3.7 \times 10^{-3}}$
(i′) $52(4.20 \times 10^{-4})$
(j′) $\dfrac{0.00315}{227}$
(k′) $(0.0329)^{6.1}$
(l′) $(5.89)^{2.9}$
(m′) $\dfrac{343}{6.3 \times 10^{-5}}$
(n′) $\dfrac{0.00385}{60}$
(o′) $24(7.2 \times 10^{-3})$
(p′) $\sqrt[3]{589}$

(q′) $\dfrac{343}{0.0417}$

(r′) $7{,}130 \times 0.00358$

(s′) $647 \times 6.3 \times 10^{-5}$

(t′) $(798)^{4.80}$

(u′) $(5.89)^{3.4}$

(v′) $\dfrac{0.0415}{115}$

---

2. PROBLEMS B   No answers given

1. An estimate of the heat of combustion of coal in BTU per pound can be calculated from the data of a carbon, hydrogen, oxygen, and sulfur analysis, using the equation

$$\text{BTU pound}^{-1} = 14{,}540\text{C} + 62{,}028\left(\text{H} - \frac{\text{O}}{8}\right) + 4050\text{S}$$

where C, H, O, and S are the weight fractions of carbon, hydrogen, oxygen, and sulfur. Calculate the heat of combustion of a sample of coal which gave the following analysis:

C 54.8%, H 6.3%, O 33.8%, S 0.3%

2. The viscosity of water, η, in centipoises, as a function of temperature is given by the equation

$$\eta = 2.1482[(t - 8.435) + \sqrt{8.078.4 + (t - 8.435)^2}] - 120$$

where t is the temperature in degrees Celsius. Calculate the viscosity of water at 37°C.

3. The heat capacity, C_p, in calories per degree per gram for carbon, as a function of temperature, can be calculated using the equation

$$C_p = a + (b \times 10^{-3})T + (c \times 10^{-6})T^2 + \frac{d \times 10^5}{T^2}$$

where a, b, c, and d are constants and T is the absolute temperature. The constants for carbon are as follows: a, 4.10; b, 1.02; c, 0; and d, −2.10. Calculate the heat capacity of carbon at 30°C.

4. Carbon dioxide gas shows significant deviations from ideality and as a result van der Waal's equation is often used as its equation of state:

$$\left[P + \frac{n^2 a}{V^2}\right]\left[V - nb\right] = nRT$$

In this equation, P is the pressure in atmospheres, V the volume in liters, R is the gas constant 0.08205 in liter-atmospheres per degree per mole, T is the temperature in degrees Kelvin, n is the number of moles, and a and b are constants, which for carbon dioxide are 3.592 and 0.04267, respectively. Calculate the volume of one mole of carbon dioxide at 0°C and 1 atm pressure.

5. The vapor pressure of most substances over a limited temperature range can be given by an equation of the form

$$\log_{10} p = -\frac{0.05223a}{T} + b$$

where p is the pressure in millimeters, T is the temperature in degrees Kelvin, and a and b are constants characteristic of the compound. Calculate the vapor pressure of mercury at 25°C if a and b are respectively 73,000 and 10.383.

6. For liquid deuterium oxide, D_2O, the dielectric constant, ϵ, varies with temperature according to the empirical equation

$$\epsilon = 78.25[1 - 4.617(10^{-3})(t - 25) + 1.22(10^{-5})(t - 25)^2 - 2.7(10^{-8})(t - 25)^3]$$

where t is the temperature in degrees Celsius. Calculate the dielectric constant of D_2O at 50°C.

7. At a temperature, t, in °C and a pressure, p, in centimeters of Hg the density of air in grams per milliliter is given by the equation

$$d = \frac{0.001293}{1 + 0.00367t} \cdot \frac{P}{76}$$

Calculate the density of air at 37°C and a pressure of 74 cm of Hg.

8. The reflection of light at the interface of a transparent medium and air is given by Fresnel's equation

$$R = \left[\frac{(n_2 - 1)}{(n_2 + 1)}\right]^2$$

where n_2 is the refractive index of the medium. Calculate the fraction of the light reflected at an air-quartz interface if quartz has a refractive index of 1.567.

9. The free energy of formation, ΔG_f°, of liquid water in calories per mole as a function of temperature can be calculated from the equation

$$\Delta G_f^\circ = -70{,}600 + 2.303aT \log T + b \times 10^{-3}T^2 + c \times 10^5 T^{-1} + IT$$

where a, b, c and I are constants and T is the temperature in degrees Kelvin. If a, b, c, and I are, respectively, −18.26, +0.64, −0.04 and 91.67, calculate ΔG_f° at 298°K.

10. The observed net rotation, α, for a 10.414% solution (w/v) of the amino acid L-serine in water at 20°C in a 2-decimeter cell is −1.42°. The specific rotation

$$[\alpha]_D^{20} = \frac{\alpha \times 100}{c \times l}$$

where c is the concentration in grams per 100 ml and l is the path length in decimeters. Calculate the specific rotation at 20°C.

11. The mean square displacement, Δ^2, of a sphere of radius r in a medium of viscosity η is

$$\Delta^2 = \frac{kTt}{3\pi r\eta}$$

where k is Boltzmann's constant 1.38×10^{-16} erg deg^{-1}, T is the absolute temperature and t is the time allowed for the displacement. Calculate the 1-second displacement in μ of a subcellular particle of 100-Å radius in a fluid, viscosity 1 poise, at 27°C.

12. The total lung capacity (VC) in liters of a man is given by

$$\text{VC} = 0.125[27.63 - (0.112 \times \text{age(years)})(\text{height in meters})]$$

Calculate this for a 43-year-old man, 6 ft 1 in. tall, and 70 kg in weight.

3

Algebraic and Graphic Solutions

ALGEBRAIC EQUATIONS

Algebraic equations often use symbols for which numerical values can be substituted. A variable is a symbol which can be assigned to any numerical value. The symbols x, y, and z are commonly used to denote the variable in algebraic expressions, except where physical properties of compounds are involved when various symbols are used such as c, concentration, T, temperature, and V, volume. A constant on the other hand has a specific value, such as the number 3.58, and the letters a, b, and c are examples of those used. In the equation $ax + by + c = 0$, for example, the symbols x and y are generally considered to be variables while the symbols a, b, and c are thought of as constants. The symbol x is also used to denote the unknown in an algebraic equation and a symbol such as a is called the coefficient of x.

Linear Equations. An equation is a mathematical statement that two expressions are equal; the two expressions are usually referred to as sides of the equation. In the application of mathematics to physical problems, equations arise which are true for only certain numerical values. These are called conditional equations. The equation is linear when the unknown appears only to the first degree (power).

To illustrate, consider the equation $2x - 3 = 5$. A conditional equation such as this may be thought of as asking the question: For what value of x is the equation true? The letter whose value is being requested is called an unknown. Values of the unknown which satisfy the equation are called solutions or roots. The equation presented here is true only when x is equal to 4, and thus 4 is the root. Linear equations have only one root.

Several operations can be performed on an equation which leave it unchanged, yet facilitate the solution.

1. Addition of the same number to both sides.
2. Multiplication (or division) of both sides by the same number provided it does not involve the unknown or is not zero.

Problem 3.1: Solve the equation $2x - 3 = 5$ for x.

Solution: Adding 3 to both sides gives

$$+3 + 2x - 3 = 5 + 3$$

When we collect terms,

$$2x = 8$$

Thus

$$x = 4$$

The steps for solving equations in one unknown can be summarized as follows.

1. Any fractions which appear should be cleared by multiplication by the lowest common denominator (L.C.D.).
2. Transpose all terms involving the unknown to one side of the equation and all other terms to the opposite side.
3. Combine terms.
4. Divide both sides by the coefficient of the unknown.
5. The solution can be checked by substitution in the original equation.

Problem 3.2: Solve for x in the equation

$$\frac{3x-2}{4} + \frac{10x-8}{12} = \frac{13}{4} + \frac{x-2}{3}$$

Solution: 12 is the L.C.D., so multiply by it:

$$\frac{12(3x-2)}{4} + \frac{12(10x-8)}{12} = \frac{12(13)}{4} + \frac{12(x-2)}{3}$$

Reduce:

$$3(3x-2) + 10x - 8 = 3(13) + 4(x-2)$$

Expand factors:

$$9x - 6 + 10x - 8 = 39 + 4x - 8$$

Transpose by adding +6, +8, and $-4x$ to both sides:

$$9x + 10x - 4x = 39 - 8 + 8 + 6$$

Collect terms:

$$15x = 45$$

Divide by the coefficient of x:

$$x = 3$$

Problem 3.3: Solve for n in the equation $1 = a + (n - 1)d$.

Solution: Expand:

$$1 = a + nd - d$$

Transpose:

$$1 + d - a = nd$$

Divide by the coefficient of the unknown:

$$\frac{1 + d - a}{d} = n$$

Problem 3.4: If water increases its volume by one-tenth upon freezing, how much water is required to produce 220 ft^3 of ice?

Solution: To translate this problem into mathematical terms we assign x to the unknown quantity of water. Then it is reasoned that for any volume of water the volume of ice produced is the volume of water plus one-tenth of its volume. The equation is

$$x + \frac{1}{10}x = 220 \text{ ft}^3$$

(Note that x will necessarily be in cubic feet.)

Solve for x by collecting terms to give

$$1.1x = 220 \text{ ft}^3$$

Divide by the coefficient of x to obtain the solution:

$$\frac{1.1}{1.1}x = \frac{220}{1.1}\text{ft}^3$$

$$x = 200 \text{ ft}^3$$

Problem 3.5: The equation which relates degrees Fahrenheit to degrees Celsius is $F = \frac{9}{5}C + 32$. (a) Solve for C in terms of F. (b) Find C if F = 50°.

Solution: (a) Given

$$F = \tfrac{9}{5}C + 32$$

Transpose terms to isolate C:

$$F - 32 = \tfrac{9}{5}C$$

Divide by the coefficient of C to obtain

$$C = \frac{F - 32}{\frac{9}{5}}$$

Multiply the numerator and denominator by 5 to clear the fraction:

$$C = \frac{5(F - 32)}{9}$$

(b) Substitute 50 for F in the equation derived in part (a):

$$C = \frac{5(50 - 32)}{9}$$

Subtract and reduce the fraction to give

$$C = \frac{5(18)}{9}$$

Obtain the product:

$$C = 5 \times 2 = 10$$

Thus a temperature of 10°C is equivalent to 50°F.

Quadratic Equations. A quadratic equation is an equation of the second degree, which means that the highest power in the variable is two. The standard form for the quadratic equation is

$$ax^2 + bx + c = 0 \tag{3.1}$$

Two roots are found for quadratic equations, which in some cases may be equal. Two methods will be discussed for the solution of quadratic equations – solution by factoring and solution by quadratic formula.

1. Solution by factoring. The basis for this method is that the product of two factors is zero, when and only when at least one of the factors is zero. The two steps are the following.

(a) Transpose the equation into the standard form and factor.

(b) Equate each factor to zero and solve for x.

Problem 3.6: Solve $x^2 - 3x = 10$.

Solution: Transpose to standard form:

$$x^2 - 3x - 10 = 0$$

Factoring gives

$$(x + 2)(x - 5) = 0$$

Thus the equation is satisfied if

$$x + 2 = 0$$

or

$$x - 5 = 0$$

The roots are then $x = -2$ and $x = 5$.

Problem 3.7: $16x^2 = 24x - 9$.

Solution: Transpose to standard form:

$$16x^2 - 24x + 9 = 0$$

Factor:

$$(4x - 3)(4x - 3) = 0$$

The equation is satisfied if

$$4x - 3 = 0$$

Thus

$$x = \tfrac{3}{4} \text{ (two identical roots)}$$

---

2. Solution by quadratic formula. The solution of a quadratic equation can be obtained by use of the formula

$$x = \frac{-b \pm \sqrt{b^2 - 4ac}}{2a} \tag{3.2}$$

where the notation refers to the standard form $ax^2 + bx + c = 0$. The two steps are the following.

(a) Transpose the equation to standard quadratic form.

(b) Substitute numerical values of a, b, and c in Equation (3.2) and solve.

Problem 3.8: Solve $x - 1 = -6x^2$ for x.

Solution: Transpose to standard form:

$$6x^2 + x - 1 = 0$$

where

$$a = 6, \quad b = 1, \quad c = -1$$

Substitute in Equation (3.2):

$$x = \frac{-1 \pm \sqrt{1 - 4 \cdot 6(-1)}}{12} = \frac{-1 \pm \sqrt{1 + 24}}{12} = \frac{-1 \pm \sqrt{25}}{12}$$

This simplifies to

$$x = \frac{-1 \pm 5}{12}$$

The two roots are

$$x = \frac{-1 + 5}{12} = \frac{1}{3}$$

$$x = \frac{-1 - 5}{12} = -\frac{1}{2}$$

Problem 3.9: Solve $4x^2 + 8x = -9$ for x.

Solution: Transpose to standard form:

$$4x^2 + 8x + 9 = 0$$

where

$$a = 4, \quad b = 8, \quad c = 9$$

Substitute in Equation (3.2):

$$x = \frac{-8 \pm \sqrt{64 - (4 \cdot 4 \cdot 9)}}{2 \times 4} = \frac{-8 \pm \sqrt{64 - 144}}{8} = \frac{-8 \pm \sqrt{-80}}{8}$$

Note here the square root of a negative number; thus the solutions will be imaginary. That is, they involve i where $i = \sqrt{-1}$. Thus

$$x = \frac{-8 \pm i\sqrt{80}}{8}$$

Remove $\sqrt{16}$ from radical

$$x = \frac{-8 \pm i4\sqrt{5}}{8}$$

Divide all terms by 4

$$x = \frac{-2 \pm i\sqrt{5}}{2}$$

The two roots are $x = \dfrac{-2 + i\sqrt{5}}{2}$ and $x = \dfrac{-2 - i\sqrt{5}}{2}$.

In experimental problems involving finite quantities, only one root will be physically possible (see, for example, Problem 6.7, page 90). Imaginary solutions are not possible. Thus, if such a solution is obtained, one should question the form of the equation or the mathematics involved by substituting in the quadratic formula.

GRAPHING OF FUNCTIONS

Functions. Often an equation is used to relate two variables—for example, x and y. In such an equation, when a value is assigned to x, the value of y is fixed and can be evaluated. Then x is called the independent variable and y is called the dependent variable, or y is said to be a function of x, which means that its value depends on x. Schematically it is written $y = f(x)$, which is read: y is a function of x.

Problem 3.10: $y = 7x + 2 = f(x)$. Evaluate $f(x)$ for $x = 2$.

Solution: $y = f(x) = 7x + 2$. Substitute $x = 2$:

$$y = f(2) = 7 \cdot 2 + 2 = 16$$

For this function, $y = 16$ when $x = 2$. In other words, x is the independent variable, y is the dependent variable.

Problem 3.11: Evaluate $f(x)$ for $x = 1$ in the equation $y = x^2 + 3x + 1 = f(x)$.

Solution: Substitute $x = 1$ in $f(x)$:

$$y = f(1) = 1^2 + 3 \cdot 1 + 1 = 5$$

Thus $y = 5$ when $x = 1$.

Problem 3.12: The titration of iodine with thiosulfate shows a linear relationship between moles of iodine added and moles of thiosulfate titrated. Determine the functional relationship.

Solution: Let x equal the amount of iodine titrated (moles) and let y be the moles of thiosulfate used as titrant. Experimental data show that two moles of thiosulfate are required to react with one mole of iodine. Thus the function is

$$y = 2x$$

or

$$[S_2O_3^{-2}] = 2[I_2]$$

This is an oxidation-reduction reaction wherein the molar ratio of the reactants is 2/1.

Graphing. To graph a function is to find the set of points whose coordinates form pairs of corresponding values of x and y. A linear function of x has the general form

$$y = mx + b \tag{3.3}$$

where m is the slope and b is the intercept. Its graph is a straight line.

Problem 3.13: Graph the function $f(x) = 2x + 1$.

Solution: Calculate a table of values.

x	0	1	2	3
y	1	3	5	7

The graph is shown in Figure 3.1.

Figure 3.1

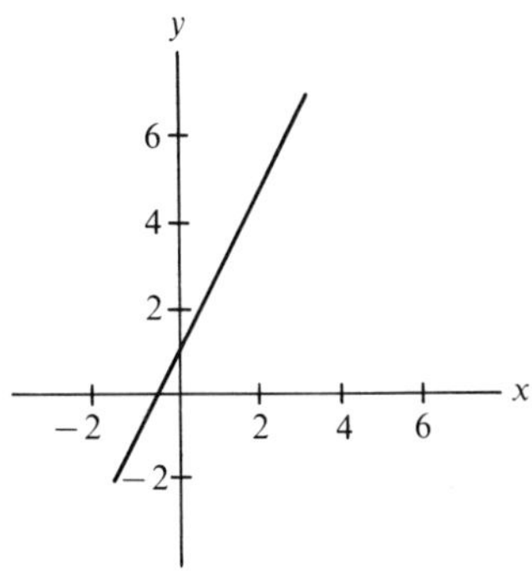

The slope m is the number of units change in y per unit change in x. For the function in Problem 3.13 the slope is 2. The intercept b, the value of y where the graph intercepts the y-axis, is 1.

Problem 3.14: Graph the function $f(x) = x^2 + 3$.

Solution: Calculate a table of values.

x	−2	−1	0	1	2
y	7	4	3	4	7

The graph is shown in Figure 3.2.

Figure 3.2

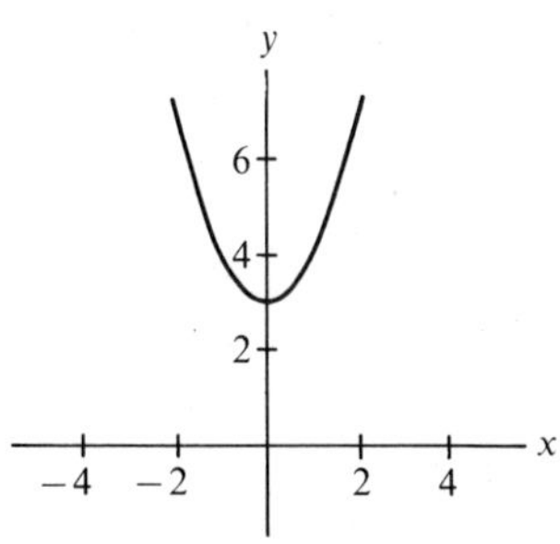

It is possible to linearize a function of higher degree by rearrangement or substitution to obtain the general form for a linear function, given in Equation (3.3).

Problem 3.15: Show what function can be plotted so that a straight line graph is obtained from the function of $y = x^2 + 3$.

Solution: This is a linear form if we let x^2 be the variable rather than x. Plot y versus x^2.

x	-2	-1	0	1	2
x^2	4	1	0	1	4
y	7	4	3	4	7

The graph is shown in Figure 3.3.

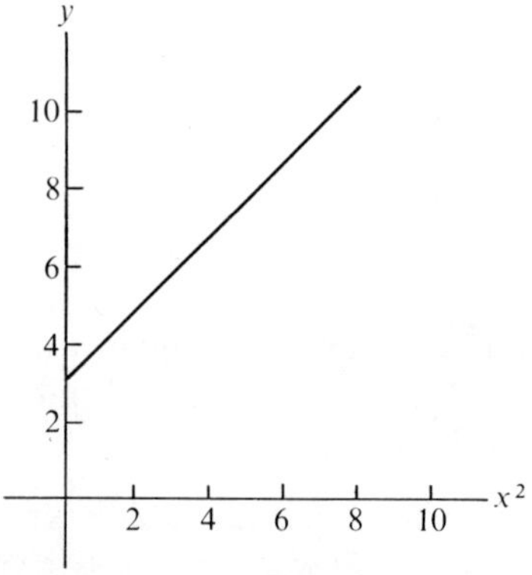

Figure 3.3

Problem 3.16: Given the equation

$$v = \frac{V_{max} S}{K_m + S}$$

where v and S are variables and V_{max} and K_m are constants. Obtain a linear form.

Solution: Rearrange to separate the variables v and S

$$v(K_m + S) = V_{max} S$$

Obtain the reciprocal

$$\frac{1}{v(K_m + S)} = \frac{1}{V_{max} S}$$

Transpose to give

$$\frac{1}{v} = \frac{K_m + S}{V_{max} S} = \frac{K_m}{V_{max} S} + \frac{S}{V_{max} S}$$

Thus
$$\frac{1}{v} = \frac{K_m}{V_{max}} \cdot \frac{1}{S} + \frac{1}{V_{max}}$$

(The form is $y = mx + b$)

Plot $1/v$ versus $1/S$ to obtain a linear graph.

Problem 3.17: Given the equation $v = Ae^{-E_a/RT}$, where v and T are the variables, and A, E_a, and R are constants. Obtain a linear form for the equation.

Solution: Convert the equation to log form:

$$\ln v = \ln A + \ln e^{-E_a/RT}$$

Rewrite, using Equation (2.16) and the definition of logs,

$$\ln v = \ln A - \frac{E_a}{RT}$$

and convert to base 10 logs:

$$2.303 \log v = 2.303 \log A - \frac{E_a}{RT}$$

or
$$\log v = \log A - \frac{E_a}{2.303R} \cdot \frac{1}{T}$$

The form is $y = b + (-m \cdot x)$.

This equation will give a linear graph if $\log v$ is plotted versus $1/T$.

In the two previous problems the equations were converted to linear form so that the constants in the equation could be evaluated from corresponding values of the variables. The function itself is not known, as is generally the case with experimental problems. To obtain the constants we graph the data and draw the best straight line through the points. The slope of the line which contains the desired constant can then be readily calculated from the graph, with the use of the formula

$$m = \frac{Y_2 - Y_1}{X_2 - X_1}$$

where X_1, Y_1 and X_2, Y_2 are the values of the variables at any two points on the line.

Problem 3.18: Determine the velocity in micromoles per minute at zero time for an enzyme-catalyzed reaction from the following data.

Micromoles reacted	6	12	17.5	22.5	27
Time (min)	1	2	3	4	5

Solution: Graph the data.

The graph is shown in Figure 3.4.

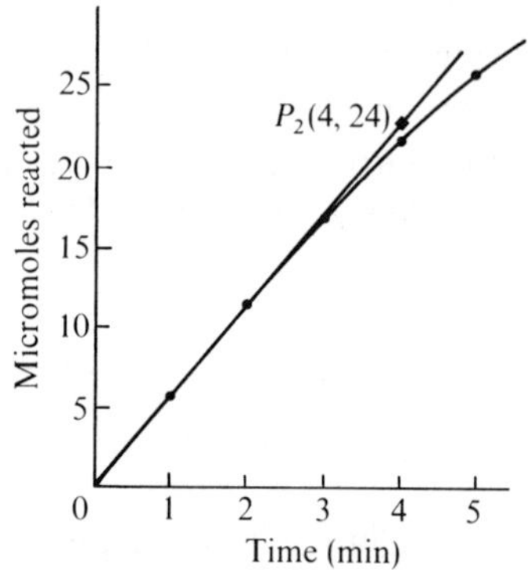

Figure 3.4

The velocity of the reaction at zero time will be the slope of this curve at this point. A line is therefore drawn tangent to the curve at $t = 0$ and the slope of the line determined. Two points are chosen. In this case it is most convenient to use the origin as one point. The slope is then

$$m = \frac{(24 - 0) \text{ micromoles}}{(4 - 0) \text{ minutes}}$$

The velocity of the reaction at zero time is then 6 micromoles per minute.

LEAST-SQUARES FIT OF DATA FOR LINEAR FUNCTIONS

In the previous section the slope was calculated from the line drawn through the experimental points. This method is not without considerable error as a result of subjective positioning of the line through the points; however, it should be apparent that for such a construction a linear function is subject to less error than a nonlinear one. In this method individuals may differ by several percent when using the same data. These errors in plotting are eliminated by using the least-squares method, which is one of the most reliable procedures for obtaining the function directly from the experimental data.

In the method of least-squares the coefficients m and b in the equation of a line $y = mx + b$ are chosen so that the average sum of the deviations squared is a minimum. The deviation d is defined by

$$b + mx - y = d \qquad (3.4)$$

where the exact equation of the line in this form is

$$b + mx - y = 0 \qquad (3.5)$$

From these principles the coefficient, m, the slope, and b, the intercept, are defined by the equations

$$b = \frac{\Sigma(x)\Sigma(xy) - \Sigma(x^2)\Sigma(y)}{[\Sigma(x)]^2 - n\Sigma(x^2)} \tag{3.6}$$

$$m = \frac{\Sigma(x)\Sigma(y) - n\Sigma(xy)}{[\Sigma(x)]^2 - n\Sigma(x^2)} \tag{3.7}$$

The symbol sigma, Σ, indicates that these terms are sums of n data points.

Problem 3.19: Calculate the value of the slope and intercept by the least-squares method, given the experimental data for x and y shown under **Solution.**

Solution: Since it is best to tabulate the data for this type of problem the given experimental information, as well as the calculated terms, are included in the following table.

x	y	x^2	xy
1	3.1	1	3.1
2	6.0	4	12.0
3	8.7	9	26.1
4	12.9	16	51.6
5	15.3	25	76.5
6	17.9	36	107.4
7	22.0	49	154.0
8	23.7	64	189.6
$\Sigma x = 36$	$\Sigma y = 109.6$	$\Sigma x^2 = 204$	$\Sigma xy = 620.3$

The calculated values are now substituted into Equations (3.6) and (3.7) to obtain the intercept and slope.

Intercept:

$$b = \frac{\Sigma(x)\Sigma(xy) - \Sigma(x^2)\Sigma(y)}{[\Sigma(x)]^2 - n\Sigma(x^2)}$$

When we substitute numbers,

$$b = \frac{(36)(620.3) - (204)(109.6)}{(36)^2 - (8)(204)} = 0.082$$

Slope:

$$m = \frac{\Sigma(x)\Sigma(y) - n\Sigma(xy)}{[\Sigma(x)]^2 - n\Sigma(x^2)}$$

When we substitute numbers,

$$m = \frac{(36)(109.6) - (8)(620.3)}{(36)^2 - (8)(204)} = 3.026$$

Thus the equation which fits this data is

$$y = 3.026x + 0.082$$

GRAPHICAL SOLUTION OF AN EQUATION

Problems in complex chemical equilibria often require the solution of equations of higher degree than quadratic in the general form $y = f(x)$.

The roots of an equation are the values of x for which $f(x)$ is zero. If we write $y = f(x)$ the root is the value of x for which y equals zero. An equation of degree n will have n roots. The steps for obtaining approximate roots by graphing are the following.

1. Simplify and transpose to the form $f(x) = y$.
2. Obtain a table of values and graph to obtain values of x for which the function crosses the x-axis. The values are the roots of $f(x)$.

Problem 3.20: Solve for x by graphing the equation $x^3 + 8x^2 - 4x = 1$.

Solution: Transpose to the form $y = f(x)$:

$$y = x^3 + 8x^2 - 4x - 1$$

Obtain a table of values:

x	y
1	4
0	−1
−1	10
−2	31
−7	76
−8	31
−9	−45

The three approximate roots are 0.7, −0.2, −8.4 from graph 3.5.

Figure 3.5

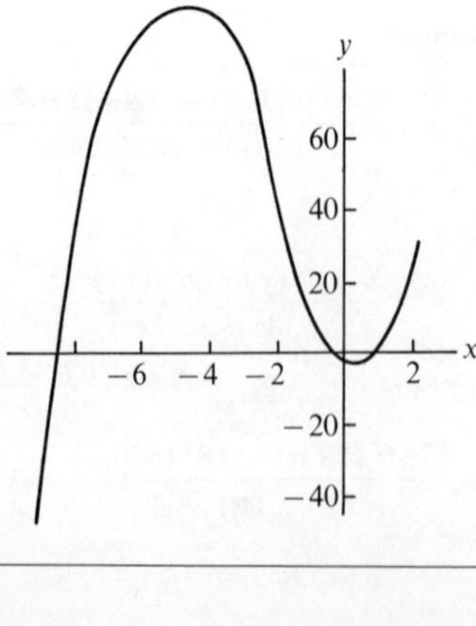

If the roots obtained from the graph are not sufficiently accurate, they can be adjusted by trial and error to best satisfy the equation

$$f(x) = 0 \tag{3.8}$$

Should the function be complicated this is not an economical procedure. Instead a numerical method which converges to the required root should be used.*

---

3. PROBLEMS A

Answers on page 333

1. Solve the following equations.

(a) $f = m\alpha$ for α
(b) $5x^2 = 15$ for x
(c) $4x^2 + 9x = 9$ for x
(d) $S = \frac{1}{2}gt^2$ for t
(e) $A = \dfrac{HB}{2}$ for H
(f) $\dfrac{1}{\mu} = \dfrac{1}{m_1} + \dfrac{1}{m_2}$ for m_1
(g) $\epsilon = \dfrac{hc}{\lambda}$ for h
(h) $4az - a^2 = 4z - 1$ for z
(i) $3x^2 - 7x = 0$ for x
(j) $\dfrac{4}{5x} - \dfrac{2}{6x} = \dfrac{1}{15}$ for x
(k) $6x = 3 + c$ for x
(l) $4x^2 = c$ for x
(m) $8x^2 + 6x + 1 = 0$ for x
(n) $\dfrac{PV}{n} = RT$ for V
(o) $\dfrac{P_1V_1}{T_1} = \dfrac{P_2V_2}{T_2}$ for T_2
(p) $\dfrac{A^2}{3} = 7$ for A
(q) $S = \pi r^2 d$ for d
(r) $\frac{1}{2}x^2 - 1 = \frac{1}{3}x^2$ for x
(s) $x^2 - ex = 10$ for x
(t) $V = \frac{4}{3}\pi r^3$ for r
(u) $M = \dfrac{3\lambda}{n^2 + 2}$ for n
(v) $S = 7r^2 + 1$ for r
(w) $A = \epsilon_m bc$ for ϵ_m
(x) $6y + \frac{1}{4} = 7y + \frac{1}{2}$ for y
(y) $fr = mv^2$ for v
(z) $8a + 2bx^2 = 4cd$ for x
(a′) $8x^2 - 2x - 6 = 0$ for x
(b′) $2x^2 = 3$ for x
(c′) $\dfrac{5x}{8} - \dfrac{x}{9} = \dfrac{1}{6}$ for x
(d′) $v = \dfrac{V_{\max}S}{K_m + S}$ for K_m
(e′) $4x^2 + 8x = -3$ for x
(f′) $7x^2 = 5 - 3x^2$ for x
(g′) $P = \dfrac{I}{rt}$ for r
(h′) $5x^2 - 3 = 0$ for x
(i′) $25x^2 + 4 = 20x$ for x
(j′) $PV = nRT$ for R
(k′) $4x^2 - 2 = 6x$ for x
(l′) $c(x-4) = 5x$ for x
(m′) $4x = ax + 8$ for x
(n′) $3 - 3x = -7 + 5x$ for x
(o′) $\dfrac{x^2}{4} = 25$ for x
(p′) $7x^2 + 3x - 4 = 0$ for x

---

*See R. A. Buckingham, *Numerical Methods*. London: Pitman, 1957.

(q′) $3x^2 = 12$ for x
(r′) $x^2 = 2x + 7$ for x
(s′) $9x^2 - 4 = 5x$ for x
(t′) $S = vt - \frac{1}{2}gt^2$ for v
(u′) $3ax^2 = h$ for x
(v′) $4x^2 + 3x = 8$ for x

2. Graph the functions. What are their slopes? And what are their intercepts?
   (a) $y = 4x$
   (b) $y = \frac{x}{4} + 4$
   (c) $x = 8y - 3$
   (d) $4x - 3y + 2 = 0$
   (e) $6y + x = 1$
   (f) $3x + y = 0$
   (g) $2y = 5x + 4$
   (h) $ay = 7x + 1$
   (i) $y = 6x + 3$
   (j) $4y = x - 1$
   (k) $\frac{y}{2} = 3x + 7$
   (l) $3x + 7y = 0$
   (m) $3y = 4x + 1$
   (n) $\frac{1}{y} = \frac{1}{x}$
   (o) $2y + 6x = 1$
   (p) $3y = 6x + 12$

3. Indicate what should be plotted to yield a linear graph for the following functions.
   (a) $2x^2 - y + 7 = 0$
   (b) $2a - y + \frac{b}{x^2} = 0$, where a and b are constants
   (c) $y - 9 = 4x^2$
   (d) $\frac{4y}{a} = x^2$, where a is a constant
   (e) $-4x^2 = 21 - y$
   (f) $y + \frac{a}{x^2} = 1$, where a is a constant
   (g) $S - 8r^2 = 1$
   (h) $y = \frac{4x^3}{3} + 7$

4. Find the real roots by graphing the functions.
   (a) $1.74 \times 10^{-4} = \frac{x^2}{(0.1 - x)}$
   (b) $5x - 5 = x^3 + 2x^2$
   (c) $2x^4 - 6x^2 + 7 = 0$
   (d) $12x = x^3 + 3$
   (e) $8 + x^3 = 4x^2 + 2x$

---

3. PROBLEMS B No answers given

1. The values of the viscosity coefficient η of glycerol obey the equation $\ln \eta = A + BT^{-3}$, where A and B are constants and T is the absolute temperature. From the following data evaluate A and B. Calculate by extrapolation, the viscosity of glycerol at -50°C. Calculate, by interpolation, the viscosity at 10°C.

T°C	$\eta \times 10^{-1}$
50	0.123
25	0.692
0	7.95
−25	200.0

[From D. Gegiou, K. A. Muszkat, and E. Fischer, *J. Amer. Chem. Soc.* **90:**12 (1968).]

2. The activation energy E_a for the hydrolysis of sucrose with the enzyme invertase can be determined by measuring the velocity of the reaction as a function of temperature. A plot of log v versus $1/T$ is linear with the slope of $-E_a/2.3R$, where R is the gas constant. Calculate the activation energy from a least-squares fit of the following data from three separate experiments.

T°C	v_1 (mole min^{-1})	v_2 (mole min^{-1})	v_3 (mole min^{-1})
20	2.82	2.76	2.79
35	6.25	6.31	6.10
45	10.20	10.14	10.27
55	13.10	13.30	13.23

3. The hydrolysis of the model peptide, benzoyl arginine ethyl ester, is catalyzed by trypsin. Calculate the initial velocity, mmole min^{-1}, of the reaction from the following data, using the least-squares method.

Time (min)	millimoles hydrolyzed $\times 10^3$
0.4	1.02
0.9	2.25
1.4	3.55
1.9	4.82
2.4	6.12
2.9	7.28
3.4	8.52
3.9	9.72
4.4	11.0
4.9	12.2

4. The absorbance, log I_0/I, of a solution, where I_0 is the intensity of the incident light and I is the intensity of the transmitted light at a given wavelength, is proportional to C, the concentration in moles per liter, and b, the path length in centimeters. The proportionality constant, a_m, which is called the molar absorption coefficient and is characteristic of the compound, can be determined by plotting absorbance versus concentration and evaluating the slope which is equal to $a_m b$. Calculate the molar absorption coefficient for the enzyme ribonuclease at 277.5 nm from a least-squares fit of the following two sets of data.

C (mmole liter^{-1})	$(\log I_0/I)_1 (b = 1$ cm$)$	$(\log I_0/I)_2 (b = 1$ cm$)$
0.108	0.115	0.111
0.271	0.275	0.273
0.379	0.384	0.389
0.503	0.511	0.509
0.792	0.813	0.807

5. The activation energy, 10,900 cal mole^{-1} for the isomerization of nitrous acid, HNO_2, is related to the rate constant, k, by the equation

$$k = 1.33 \times 10^{13} e^{-10,900/(1.986)T}$$

where T is the absolute temperature. What is the rate constant at 50°C?

6. Radioisotopes decay according to a first-order rate equation

$$A = A_0 e^{-\lambda t}$$

where A is the activity at any time t, A_0 is the activity at zero time, and λ is the decay constant. Rearrange and indicate what should be plotted to give a linear graph.

7. The molecular weight of bovine serum albumin can be calculated by combining the following experimental data: the intrinsic viscosity, (η); the sedimentation coefficient, $S^\circ_{20,w}$; the partial specific volume, $\bar{v}$; and the density, ρ. These quantities are related by the equation

$$M = \frac{4700(S^\circ_{20,w})^{3/2}[\eta]^{1/2}}{(1 - \bar{v}\rho)^{3/2}}$$

For bovine serum albumin $[\eta]$ is 0.04, $S^\circ_{20,w}$ is 4.4, and ρ is 1.03. Calculate the molecular weight M for the following values of $\bar{v}$: 0.70, 0.71, 0.73, and 0.75. Make a plot of M versus $\bar{v}$.

8. The mathematical treatment of buffers requires the solution of the equation

$$x = y + \log \frac{[S]}{[A]}$$

where [S] is the concentration of salt and [A] is the concentration of acid. If x is 4.83 and y is 5.71, calculate the ratio [S]/[A].

9. Conductivity measurements on potassium chloride solutions at 25.0°C yielded the following results.

Concentration ($N \times 10^3$)	Resistance (ohms)
100.0	67.8
50.0	132
25.0	258
12.5	503
6.25	987
3.12	1930
1.56	3820

(a) A quantity called the specific conductance, L, is defined by the equation

$$K = LR$$

where K is the cell constant and R is the resistance in ohms. Calculate L for each of these solutions if K is 0.8780.

(b) Calculate Λ, the equivalent conductance, for each potassium chloride solution, using the equation

$$\Lambda = \frac{1000L}{N}$$

where N and L are as defined earlier.

(c) The equivalent conductance at infinite dilution, Λ_0, can be determined by plotting Λ versus $f(c)$ and extrapolating to zero concentration. What $f(c)$ gives the best extrapolation?

10. A transient radioactive equilibrium is set up under certain conditions by the decay of a parent and daughter radionuclide. The daughter activity is given by the relationship

$$A_t = \frac{\lambda_1}{\lambda_2 - \lambda_1} A_0^1 e^{-\lambda_1 t} - \frac{\lambda_1}{\lambda_2 - \lambda_1} A_0^1 e^{-\lambda_2 t} + A_0^2 e^{-\lambda_2 t}$$

where A_t is the total activity (counts per minute) at any time, t, starting from a parent radioisotope of activity A_0^1 counts per minute, decay constant, λ_1, and a daughter radioisotope of initial activity A_0^2 counts per minute, decay constant, λ_2. If one starts from pure parent with A_0^1 90,000 cpm and λ_1 0.0021 min^{-1}, then A_0^2 is zero. The daughter has λ_2 0.02 min^{-1}. Plot A_t against time, every 60 min up to 6 hr, and give the total counts per minute and slope of the line at $5\frac{1}{2}$ hr.

11. The measured optical rotation of polyglutamic acid as a function of wave length, λ, can be fitted by the equation

$$[\alpha] = \frac{K}{\lambda^2 - \lambda_c^2}$$

where $[\alpha]$ is the specific rotation, K is constant, and λ_c is the wavelength of the band center of an optically active absorption. If λ_c is a constant, what should be plotted to yield a linear graph? How can λ_c be determined?

4

Dimensions, Units, and Experimental Measurements

DIMENSIONS AND UNITS

When learning basic mathematical operations we generally use pure numbers. In scientific work these numbers often have a dimension associated with them. Time, mass and length are examples of dimensions. Furthermore, these dimensions need not appear alone. For example, the dimensions of velocity are length divided by time. A dimension such as time can be expressed as seconds, hours, days, or years. These are referred to as units of time or, more generally, units. The dimensions of velocity are length per time, and thus can be expressed as miles per hour, feet per second, or any other suitable combinations of units which are of the proper dimension. It should be noted that often the words dimension and unit are used interchangeably in spite of the difference pointed out here. This arises since, in order to have physical significance, the dimension must be given in some suitable units. Thus we say that our velocity is 45 miles/hr, not length/time.

A term in common usage is "per." This means divided by or the number or units of some substance per unit of another; for example, 5280 ft/mile or 12 in./ft. Two general rules suffice for the use of dimensions in calculations.

1. For the operations of addition and subtraction, the dimensions and the units must be the same.

2. For the operations of multiplication and division, the dimensions are multiplied and divided just as are the numbers, the result being the product or quotient of the dimensions.

Dimensions are useful quantities for several reasons. They guide the derivation of formulas and serve as valuable checks on the memory and the accuracy of equations. Factors can be conveniently derived to convert a variable from one set of units to another with the same dimension.

Dimensions should be included with numbers whenever they occur. This facilitates those calculations which use variables with multiple units.

Problem 4.1: Calculate the weight of salt one can purchase with 91 cents, if a pound of salt costs 7 cents.

Solution: Write the factor, with pounds in the numerator as this is to be determined,

$$\frac{1 \text{ pound salt}}{7 \text{ cents}}$$

The factors or their reciprocals are inserted into the equation to give the desired units. The units are canceled algebraically. Thus

$$91 \cancel{\text{cents}} \times \frac{1 \text{ pound}}{7 \cancel{\text{cents}}} = 13 \text{ pounds}$$

It is often more convenient to express units in the following way. The factor 7 cents/pound can be written as 7 cents pound^{-1}. Thus

$$\frac{91 \text{ cents}}{7 \text{ cents pound}^{-1}} = \frac{91 \text{ cents pound}}{7 \text{ cents}} = 13 \text{ pounds}$$

Problem 4.2: Calculate the number of centimeters in 1 ft if

$$1 \text{ in.} = 2.54 \text{ cm}$$

Solution:

$$1 \text{ ft} \times \frac{12 \text{ in.}}{\text{ft}} \times 2.54 \frac{\text{cm}}{\text{in.}} = 12 \times 2.54 \text{ cm} = 30.48 \text{ cm}$$

or

$$1 \text{ ft} \times 12 \text{ in. ft}^{-1} \times 2.54 \text{ cm in.}^{-1} = 30.48 \text{ cm}$$

Problem 4.3: A car traveling at 40 miles/hr has its speed increased by 38 ft/sec. What is its new speed in feet per second?

Solution: Add the two speeds to get a total speed:

$$40 \text{ miles/hr} + 38 \text{ ft/sec}$$

Check to see that the dimensions are the same. Both are length/time. Thus the addition can be performed after converting to some common unit. Since the answer is desired in feet per second, this is the common unit:

$$\frac{40 \text{ mile}}{\text{hr}} \times \frac{5280 \text{ ft}}{\text{mile}} \times \frac{\text{hr}}{3600 \text{ sec}} + \frac{38 \text{ ft}}{\text{sec}}$$

Cancel units and collect terms:

$$\frac{40 \times 5280 \text{ ft}}{3600 \text{ sec}} + \frac{38 \text{ ft}}{\text{sec}}$$

Multiply the first term and add the terms to give

$$\frac{59 \text{ ft}}{\text{sec}} + \frac{38 \text{ ft}}{\text{sec}} = \frac{97 \text{ ft}}{\text{sec}}$$

Problem 4.4: Find a conversion factor which converts gallons to milliliters given 946 ml/qt.

Solution:

$$1 \text{ gal} \times \frac{4 \text{ qt}}{\text{gal}} \times \frac{946 \text{ ml}}{\text{qt}} = 3784 \text{ ml}$$

The conversion factor is 3784 ml/gal.

Problem 4.5: The human eye can perceive objects which have a diameter of 0.1 mm. What is this diameter in inches? Given 2.54 cm/in.

Solution:

$$0.1 \text{ mm} \times \frac{1 \text{ cm}}{10 \text{ mm}} \times \frac{1 \text{ in.}}{2.54 \text{ cm}}$$

Cancel units and collect terms,

$$\frac{0.1 \text{ in.}}{25.4 \text{ cm}}$$

and divide to obtain

$$0.004 \text{ in.}$$

THE GAS CONSTANT, *R*, AND ENERGY CONVERSION FACTORS

The gas constant R, is used in a variety of expressions. In calculations involving an ideal gas it has units of liter-atmosphere degree^{-1} mole^{-1}, whereas for free energy calculations it has the units calorie degree^{-1} mole^{-1}. It is apparent that in general R has the units energy degree^{-1} mole^{-1} and the unit that changes is energy.

Problem 4.6: Calculate the gas constant in joule degree^{-1} mole^{-1}. Given $R = 0.08205$ liter-atm deg^{-1} mole^{-1} and the conversion factors: 1 calorie = 4.18 joules (J) and 1 liter-atm = 24.21 cal.

Solution: Write

$$R = 0.08205 \text{ liter-atm deg}^{-1} \text{ mole}^{-1} \times 24.21 \frac{\text{cal}}{1 \text{ liter-atm}} \times 4.18 \frac{\text{J}}{1 \text{ cal}}$$

Cancel terms and evaluate:

$$R = 8.314 \text{ J deg}^{-1} \text{ mole}^{-1}$$

Some common energy conversions are given in Table 4.1.

Table 4.1
Energy conversion factors

	Calorie	*Erg*	*Joule*
Calorie	1	4.186×10^{7}	4.186
Erg	2.39×10^{-8}	1	1×10^{-7}
Joule	2.39×10^{-1}	1×10^{-7}	1
Liter-atmosphere	2.421×10^{1}	1.013×10^{9}	1.013×10^{2}
Volt-coulomb	2.39×10^{-1}	1×10^{7}	1
Einstein (λ in Å)	$\frac{2.86 \times 10^{8}}{\lambda}$	$\frac{1.20 \times 10^{16}}{\lambda}$	$\frac{1.20 \times 10^{9}}{\lambda}$

	Liter-atmosphere	*Volt-coulomb*	*Einstein (λ in Å)*
Calorie	4.131×10^{-2}	4.186	$3.50 \times 10^{-9} \times \lambda$
Erg	9.87×10^{-10}	1×10^{-7}	$8.33 \times 10^{-17} \times \lambda$
Joule	9.87×10^{-3}	1	$8.33 \times 10^{-10} \times \lambda$
Liter-atmosphere	1	1.013×10^{2}	$8.47 \times 10^{-8} \times \lambda$
Volt-coulomb	9.87×10^{-3}	1	$8.33 \times 10^{-10} \times \lambda$
Einstein (λ in Å)	$\frac{1.18 \times 10^{7}}{\lambda}$	$\frac{1.20 \times 10^{9}}{\lambda}$	1

EXPERIMENTAL MEASUREMENTS

Significant Figures. The measurement of a physical quantity, no matter how precise, is still unreliable in the mathematical sense. It has a certain amount of error associated with it. If, for example, the weight of a sample of salt is found to be 0.733 and 0.732 g on two successive measurements, the answer is unreliable in the third decimal place. The measurement is said to contain three *significant figures;* that is, three of the digits are obtained with some reliability. The average weight obtained for the sample is 0.7325 g. Here the averaging process has introduced a fourth digit which is not proper, since only three significant digits were found by measurement. The proper answer is either 0.732 or 0.733 g.

Problem 4.7: The weight of a beaker of water on three successive weighings was found to be 10.15, 10.17, 10.21 g. Express the average weight to the proper number of significant figures.

Solution:

10.15 g
10.17 g
10.21 g
Total 30.53 g

Average 10.1766 g

The weighings disagree in the second place beyond the decimal. Seven is then the last remaining significant digit, so the answer rounded off to the proper number of significant digits is 10.18 g.

The preceding problem illustrates the principle that the results of a calculation cannot be more reliable than the least reliable number used. Thus, we must drop digits which are not reliable. This process, which is called "rounding off," is governed by three rules.

1. When the digit to be dropped is less than 5, leave the last significant digit unchanged.
2. If it is more than 5, increase the last significant digit by one.
3. If the digit to be dropped is 5 followed by zero, the last remaining significant digit is left even; for example, 0.6350 = 0.64.

Problem 4.8: Indicate the number of significant digits in the following numbers.

Solution:

Number	Significant digits	Number
1430.0	1, 4, 3, 0, 0	5
14.3	1, 4, 3	3
1.430	1, 4, 3, 0	4
0.0143	1, 4, 3	3
0.01430	1, 4, 3, 0	4

Problem 4.9: Determine the sum of $4.65 + 7.1 + 367.573$:

$$\begin{array}{r} 4.65 \\ 7.1 \\ \underline{367.573} \\ 379.323 \end{array}$$

Since the number 7.1 is significant to only tenths, the answer can be expressed only to the nearest tenth, or as 379.3.

Problem 4.10: Multiply $201. \times 7.3$. Give the answer to the proper number of significant digits.

Solution:

$$201. \times 7.3 = 1467.3$$

The minimum number of significant digits is 2, so that the product can be expressed to only 2 significant digits; thus $201. \times 7.3 = 1500$.

Problem 4.10 poses a question with respect to representing a number such as 1500 when it has only two significant figures. Writing the results in exponential numbers as $15. \times 10^2$ shows that only the 1 and 5 are significant, and thus clarifies this point. In general, however, we assume the final zeros before the decimal point to be significant unless indicated otherwise. Final zeros after the decimal point are significant digits and should not be carelessly omitted. Thus a reading of 10.0 g indicates that the weight is known to tenths of grams. Zeros appearing ahead of a number are not significant even though they may be to the right of the decimal point (see Problem 4.8).

Reliability of Measurements. The variable errors that result when an experimental measurement is repeated suggest that the laws of chance may be involved. If only a small number of measurements are made, each has the same probability of being correct, and we have no way of knowing which is more accurate. When a large number of readings are made, one notes that certain values occur with a higher frequency than others. If one were to plot the frequency of a given reading versus the reading, a curve similar to that in Figure 4.1, page 46, would be obtained. A symmetrical curve results only if the variable errors are random. The true value of the experimental quantity being measured is indicated by the solid vertical line in Figure 4.1.

The true value, which has the highest probability of occurrence, is realized only by averaging the results of an infinite number of measurements. The distribution curve shows that readings with small deviations from the true value occur more frequently than those with large deviations. Experiments done with differing degrees of precision will change the breadth of the distribution curve, but not the most probable value. The distribution of values is narrow for carefully performed experiments.

Figure 4.1

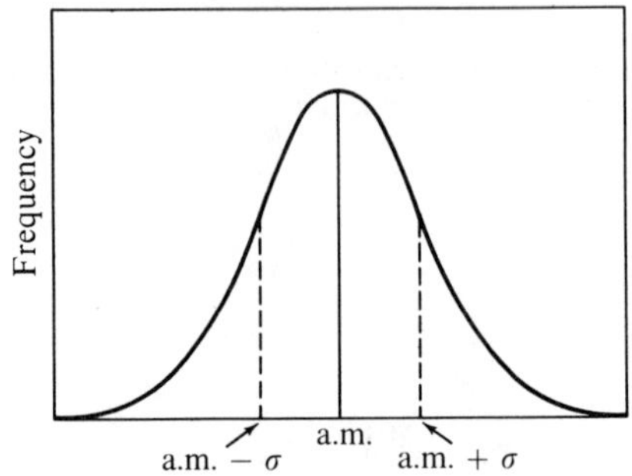

Although the true value is desired whenever a physical quantity is measured, it is not possible to perform an infinite number of experiments. Thus for the true value we use a quantity called the arithmetic mean, a.m., which is defined as

$$\text{a.m.} = \frac{\Sigma X_i}{n} \tag{4.1}$$

where X is the value of the ith measurement and n is the number of measurements. It is apparent that when only a small number of measurements are made the a.m. has a low probability of being the true value. For example, two observations of a physical quantity could with high probability lie on the same side of the true value.

A measure of the reliability should then be some function of the shape of the distribution curve and the number of observations. The average error and standard deviation (root-mean-square error) are often used as measures of the reliability of experimental observations. The type of error calculation must be indicated. These are discussed in the succeeding paragraphs.

Average Error. The average error is calculated by summing the deviations of the observations from the arithmetic mean, without regard to sign, and dividing the sum by the number of observations, n:

$$\text{average error} = \frac{\Sigma|X_i - \text{a.m.}|}{n} \tag{4.2}$$

Problem 4.11: The following data were gathered for the pH of a given buffer solution on six successive readings: 6.71, 6.75, 6.74, 6.77, 6.73, 6.74. Calculate the average error for these results. Include the average error in an expression of the result of the measurements.

Solution: Calculate the arithmetic mean, the deviations from the mean, and the average error. Be sure to retain the proper number of significant figures. An exception is made in the case of the average error, where one additional decimal place is retained for comparison with the standard deviation. These numbers are underlined.

Observations	$\lvert X_i - \text{a.m.}\rvert$
6.71	0.03
6.75	0.01
6.74	0.00
6.77	·0.03
6.73	0.01
6.74	0.00
$\Sigma X_i = 40.44$	$\Sigma\lvert X_i - \text{a.m.}\rvert = 0.08$

$$\text{a.m.} = \frac{\Sigma X_i}{n} = \frac{40.44}{6} = 6.74 \qquad \text{average error} = \frac{\Sigma\lvert X_i - \text{a.m.}\rvert}{n} = \frac{0.08}{6} = 0.013$$

The pH of the solution can now be expressed as

$$6.74 \pm 0.01\underline{3}$$

Standard Deviation. Standard deviation, or root-mean-square error, is preferred for some types of measurements. The standard deviation, σ, is defined as

$$\sigma = \pm\sqrt{\frac{\Sigma\lvert X_i - \text{a.m.}\rvert^2}{n-1}} \tag{4.3}$$

To understand the significance of standard deviation we need to examine the nature of the distribution function that determines the curve in Figure 4.1. Its mathematical representation for a large number of measurements is generally called the Gaussian Distribution Function,

$$F = \frac{1}{\sigma(2\pi)^{1/2}}\, e^{-\frac{(X_1 - \text{a.m.})^2}{2\sigma^2}} \tag{4.4}$$

where X_i, a.m. and σ are as defined earlier, and F is the frequency of occurrence of a particular observation. Every experimental measurement that is made lies somewhere on this curve. The most probable value is the true value.

For a given range of X_i about the true value the probability that a measured value will fall in this range is proportional to the area under the curve. Since the standard deviation is a range of values spanning the true value, it is thus related to the probability that a measurement lies in this range. These probabilities for the Gaussian function, using a large number of measurements, are as follows:

a.m. $\pm\ \sigma$	68.3%
a.m. $\pm\ 2\sigma$	95.4%
a.m. $\pm\ 3\sigma$	99.7%

Thus 95.4% of the measurements should be within two standard deviations, and 99.7% of the measurements should be within three standard deviations.

Problem 4.12: Calculate the standard deviation for the measurements in the preceding problem.

Solution: Evaluate the arithmetic mean and the sum of the square of the deviations from the arithmetic mean.

Observations	$\lvert X_i - \text{a.m.}\rvert$	$\lvert X_i - \text{a.m.}\rvert^2$
6.71	0.03	0.0009
6.75	0.01	0.0001
6.74	0.00	0.0000
6.77	0.03	0.0009
6.73	0.01	0.0001
6.74	0.00	0.0000
$\Sigma X_i = 40.44$	$\Sigma\lvert X_i - \text{a.m.}\rvert = 0.08$	$\Sigma\lvert X_i - \text{a.m.}\rvert^2 = 0.0020$

The arithmetic mean is

$$\text{a.m.} = \frac{\Sigma X_i}{n}$$

Substitute

$$\text{a.m.} = \frac{40.44}{6} = 6.74$$

The standard deviation is calculated by Equation (4.3):

$$\sigma = \pm\sqrt{\frac{\Sigma\lvert X_i - \text{a.m.}\rvert^2}{n-1}}$$

Substitute

$$\sigma = \pm\sqrt{\frac{0.002}{5}}$$

$$\sigma = \pm\sqrt{0.0004}$$

$$\sigma = \pm\ 2.0 \times 10^{-2}$$

The pH of the solution can then be expressed as

$$6.74 \pm 0.020$$

This means that if more measurements were made, then 68.3% would fall between a pH of 6.72 and 6.76.

Confidence Levels and Confidence Intervals. The percentage of measurements which lie within a certain interval is called a confidence level. Any desired confidence level can be chosen. Thus a 70% confidence level would indicate that 70% of the measurements lie in a given range. The range of values represented by a confidence level is called the confidence interval and in general is (a.m. $-$ $t\sigma$), where t is the fractional value of the standard deviation required to give the desired confidence level. For the 70% confidence level $t = 1.036$ for an infinite number of measurements. The values of t are obtained from a compilation such as Table 4.2.

Table 4.2
t-Values for various sample sizes and confidence levels

Sample size (n)	*Percent confidence level*						
	50	60	70	80	90	95	99
2	1.000	1.376	1.963	3.078	6.314	12.706	63.657
3	.816	1.061	1.386	1.886	2.920	4.303	9.925
4	.765	.978	1.250	1.638	2.353	3.182	5.841
5	.741	.941	1.190	1.533	2.132	2.776	4.604
6	.727	.920	1.156	1.476	2.015	2.571	4.032
7	.718	.906	1.134	1.440	1.943	2.447	3.707
8	.711	.896	1.119	1.415	1.895	2.365	3.499
9	.706	.889	1.108	1.397	1.860	2.306	3.355
10	.703	.883	1.100	1.383	1.833	2.262	3.250
20	.688	.861	1.066	1.328	1.729	2.093	2.861
30	.683	.854	1.055	1.311	1.699	2.045	2.756
40	.681	.851	1.050	1.303	1.684	2.021	2.704
50	.680	.849	1.048	1.299	1.676	2.008	2.678
60	.679	.848	1.046	1.296	1.671	2.000	2.660
120	.677	.845	1.041	1.289	1.658	1.980	2.617
∞	.674	.842	1.036	1.282	1.645	1.968	2.576

Source: From C. Pierce and R. N. Smith, *General Chemistry Workbook* (3rd ed.). San Francisco: W. H. Freeman and Company. © 1965.

Problem 4.13: For 50 measurements the arithmetic mean was found to be 673, with $\sigma = 17$. Calculate the 70% confidence interval.

Solution: From Table 4.2 we find the t value for 50 measurements at the 70% confidence level. Substituting, we obtain

$$673 \pm (1.048)17 = 673 \pm 18$$

If only a small number of measurements are made, then the Gaussian Distribution Function is not obeyed. We have already stated that the reliability of a measurement should relate to the number of observations and that the difficulty with a small number of measurements is that the average value is not the most probable, or true, value. Thus a new function is necessary—one that has a broader distribution. This in general decreases the confidence level and increases the confidence interval. This is noted in Table 4.2. For example, to maintain an 80% confidence level the number of standard deviations is 1.282 for an infinite number of measurements, while for two measurements it is 3.078. One should use the proper function for the number of observations made.

REFERENCE

Robert R. Sokal and F. James Rohlf, *Biometry*. San Francisco: W. H. Freeman and Company, 1969

4. PROBLEMS A

Answers on page 338

1. Work out a conversion factor which will convert the following.
   (a) Kilometers per hour to centimeters per second
   (b) Gallons per second to milliliters per second
   (c) Cubic feet per second to liters per day
   (d) Seconds to weeks
   (e) Inches to microns
   (f) Centimeters per second to miles per second

2. Calculate the number of molecules in 4×10^{-8} g of the following compounds.
   (a) NH_4Cl
   (b) $(NH_4)_2SO_4$
   (c) NaCl
   (d) C_2H_5OH

3. The volume of a mole of water is 18.1 ml. What is the volume of 5 kg of water in liters?

4. Calculate the volume of a molecule of glucose-1-phosphate (mol. wt. 260.14) if its density is 1.71 g/cm^3

5. Calculate the volume of 0.40 mole of glucose in cm^3 if its density is 1.56 g/cm^3 (mol. wt. 180.16)

6. The surface area of a powdered sample was reported to be 95 m^2/g. How many grams are needed to give 1000 m^2 of surface?

7. The velocity of a reaction is 3.1×10^{-2} mole/min. How many minutes will it take to react 0.40 mole?

8. Express the results of the following operations to the proper number of significant figures.

(a) $71.5 + 689.32 + 35.7$

(b) $0.000716 + 0.0004 + 0.0051$

(c) $\dfrac{4.5 \times 61 \times 35}{45}$

(d) $\dfrac{6.10 \times 322.3 \times 4.78}{126}$

(e) $0.08 + 0.032 + 0.0035$

(f) $7.291 + 9.340 + 1.7$

(g) $\dfrac{0.148 \times 48 \times 711}{81}$

(h) $0.1846 + 8.350 + 731.32$

(i) $0.0632 + 0.00171 + 0.003$

(j) $3.18 + 4.2 + 6.79 - 0.3$

(k) $4120 + 3682.4 - 121.62$

(l) $0.003127 + 0.00512 + 0.0004260$

(m) $0.0062 + 0.0190 + 0.00834$

(n) $\dfrac{4.41 \times 720 \times 430}{510}$

(o) $4.25 + 46.77 + 6.713$

(p) $\dfrac{73 + 4.1 + 67.25}{22.2}$

(q) $\dfrac{3.1 \times 35.7 \times 62.5}{14.6}$

(r) $0.0071 + 0.00362 + 0.00119$

(s) $3120.1 + 610.3 + 55$

(t) $0.00713 + 0.0027 + 0.0013$

9. The pH of an acetate buffer was measured five times on a pH meter. The readings were 4.15, 4.21, 4.18, 4.30, and 4.06. Express the average reading to the proper number of significant figures.

10. The weight of a sample of bovine serum albumin on a microbalance showed the following variation in 4 successive weighings: 0.314279, 0.314259, 0.314264, 0.314269. Express the average reading to the appropriate number of significant figures.

11. Six separate determinations of the concentration of a solution of HCl gave these results: 0.1021, 0.1017, 0.1023, 0.1013, 0.1020, and 0.1016. Calculate the average deviation, standard deviation, and 90% confidence interval.

12. Calculate the average deviation, standard deviation, and 80% confidence interval for a radioactive sample which when counted gave the following data for ten 5-min counts: 22,700, 21,650, 22,200, 23,100, 23,900, 22,000, 21,400, 22,300, 21,700, and 23,050.

---

4. PROBLEMS B No answers given

1. The molecular weight of a sample of DNA is 6.00×10^5. If the density is 1.1 g/cm³, what is the volume of one molecule of DNA in cubic angstroms? (1 Å $= 10^{-8}$ cm.)

2. Calculate the volume of a glucose molecule (mol. wt. 180.16) in cubic angstroms if its density is 1.56 g/cm³.

3. The volume of a protein molecule was found to be 300 Å³. Assuming it to be spherical, calculate its diameter.

4. The surface area of a powdered sample was reported to be 100 m^2/g. How many milligrams would be needed for an area of 1 cm^2?

5. The velocity of a chemical reaction is 0.12 mmole/sec. How many minutes would be necessary to react 0.2 mole?

6. The velocity of sound waves is 1100 ft/sec. Convert to miles per hour.

7. The molecular weight of a sample of DNA is 6.0×10^5. Calculate the length of a single molecule, using a density of 1.1 g/cm^3 and assuming it to be a cylinder with a cross-sectional area of 75 $Å^2$.

8. The enzyme catalase converts H_2O_2 to H_2O and O_2 at a velocity of 5×10^6 molecules H_2O_2 per minute per molecule of enzyme. What is this rate in moles per minute per molecule of enzyme?

9. The weight of a sample of hexokinase on six successive weighings showed this variation: 1.3146, 1.3131, 1.3137, 1.3135, 1.3141, and 1.3138 g. Express the average reading to the proper number of significant figures.

10. Four observations of the light absorption of a solution of oxyhemoglobin on a Klett spectrophotometer gave readings of 231, 243, 237, and 226 (in Klett units). Express the average reading to the appropriate number of significant figures.

11. The viscosity of 0.1% solution of cytochrome-c was measured to assess the changes in shape which the molecule had experienced. Four observations on the solution gave outflow times of 34.81, 34.03, 34.51, and 34.37 sec. Express the average reading and the average deviation to the proper number of significant figures.

12. Five determinations of the molecular weight of the enzyme ribonuclease by ultracentrifugation gave the following results: 13,100, 13,640, 13,400, 13,250, and 13,790. Calculate the average deviation in molecular weight to the appropriate number of significant digits.

13. The manometer constant, k, for a Warburg determination is given by the equation

$$k = \frac{(V_g \cdot 273/T) + V_f \cdot \alpha}{P}$$

where V_g and V_f are the volumes of gas space above the fluid and fluid phase, respectively, in ml, T is the temperature, P is the standard pressure in millimeters of manometer fluid, and α is the Bunsen coefficient. Determine the units of α and k. (See Chapter 5 for a discussion of α.)

14. The dissociation constant of 2,6-dinitrophenol has been measured in the solvent dimethyl sulfoxide. For six determinations at various concentrations the following values were obtained: 1.3, 1.3, 1.4, 1.3, 1.5, and 1.3, all times 10^{-6}. Calculate the average value, the average error, the standard deviation, and the 90% confidence interval for these results. [From I. M. Kolthoff, M. K. Chantoom, and S. Bhownik, *J. Amer. Chem. Soc.* **90**:23 (1968).]

15. The Michaelis-Menten rate equation,

$$v = \frac{V_m[S]}{K_m + [S]}$$

can be linearized by plotting $v[S]$ versus v or by plotting $[S]/v$ versus $[S]$. If v is in millimoles per minute and $[S]$ is in moles per liter, what are the units of the abscissa and ordinate and what are the units of the slopes?

16. For the Gibbs free-energy equation,

$$\Delta G = \Delta H - T\,\Delta S$$

where ΔG is in cal per mole and T is the absolute temperature, what are the units of ΔH, the enthalpy, and ΔS, the entropy?

17. The molecular weight of various macromolecules can be determined by osmotic pressure measurement. For ideal solutions, the osmotic pressure π is related to solute concentration C_2 in g/100 ml solution, the molecular weight M_2, the gas constant R, and the temperature T, according to the equation

$$\pi = \frac{C_2RT}{M_2}$$

The experimental pressure in such a measurement is given in centimeters of H_2O. What are the units of R? (Density of mercury = 13.56 g/cm^3.)

18. The rat adrenal contains 138 mg glutathione per 100 g wet tissue, with a standard deviation of ± 15 mg for 42 rats. In another set of experiments three male litter mates showed 154 mg per 100 g wet adrenal with an experimental error of ± 5 mg. To what level of confidence are these new values the same as those reported earlier?

19. It is said that one antibiotic molecule added per ribosome is sufficient to stop protein syntheses, and hence cell growth. It is desirable to use streptomycin, mol. wt. 580 g/mole, and make up 100 ml of a stock solution of it such that 0.10 ml would be sufficient to stop growth in 10 ml of a cell suspension of *E. coli*. The suspension has a density of 8×10^8 cells per milliliter and there are 1.5×10^4 ribosomes per cell. How many grams of streptomycin should be weighed out?

5

Weight, Concentration, and Stoichiometry

WEIGHT

The weight of a single molecule of any particular compound is so small, $10^{-18} - 10^{-24}$ g, that it is impossible to measure, even with our most sensitive balances. The large numbers of molecules used for experiments are expressed in terms of a gram-molecular weight, which is the weight in grams of a particular substance containing Avogadro's number of molecules (6.02×10^{23}). The basic unit for this scale was one-sixteenth of the atomic weight of oxygen until 1961, when it was changed to one-twelfth of the atomic weight of the most abundant isotope of carbon. The term gram-molecular weight is referred to as a mole.

Although the gram-molecular weight (mole) is an adequate way of expressing a quantity of a compound, it is often more convenient to use a gram-equivalent weight when discussing a chemical reaction. The gram-equivalent weight is the weight in grams which contains one gram-mole or Avogadro's number of a key ion or atom of the reaction. The equivalent weight then depends on the reaction being considered. For example, in the neutralization of an acid with a base or vice versa, the gram-equivalent weight of an acid contains one gram-molecular weight of hydrogen ion, while the gram-equivalent weight of a base contains one gram-molecular weight of hydroxyl ion. It is apparent that one gram-equivalent weight of

H^+ combines exactly with one gram-equivalent weight of OH^-. The source of the OH^- or H^+ ion is of no consequence, as far as the reaction to produce H_2O is concerned.

The gram-molecular weights of HCl and H_2SO_4 contain 1 and 2 gram-equivalent weights of hydrogen ion, respectively. The molecular weight of HCl is 36.46 g, and thus the gram-equivalent weight is 36.46 g. For H_2SO_4 the gram-equivalent weight is 49.05; that is, it is the molecular weight, 98.1, divided by two. Gram-equivalent weights for bases are determined in a similar manner. Thus the gram-equivalent weight of NaOH is equal to the molecular weight, 40, and for $Ca(OH)_2$, which has two hydroxyl groups per molecule, the gram-equivalent weight is equal to the molecular weight divided by two.

SOLUTIONS

A solution is a system of different chemical substances which is homogeneous with respect to its physical and chemical properties. Various types of solutions are possible, the most common being those in which the solvent is a liquid and the solute is a gas, liquid, or solid. These, together with the type gas dissolved in a gas (for example, the atmosphere), will be emphasized in the following discussion.

LIQUIDS AND SOLIDS DISSOLVED IN LIQUIDS

The composition of a solution can be expressed in various ways, each with an advantage for a particular application. Solution concentrations can be divided into two types, those based on volume and those based on weight. Some concentration units of each basic type are defined here.

Volume Concentrations. These are of six kinds.

1. Molar. A solution is one molar, 1 *M*, when it contains one mole of solute in a liter of solution.

2. Normal. A solution is one normal, 1 *N*, when it contains one gram-equivalent weight of solute in a liter of solution. For normal solutions it is necessary to specify the reaction being considered, for example, redox or acid-base reactions. In addition, if the reaction proceeds by steps, the stage of the reaction must be indicated.

Problem 5.1: Calculate the equivalent weight of H_3PO_4 in a neutralization reaction.

Solution: The neutralization can be carried to any one of the three stages indicated by the following equations. The equivalent weight at any stage is the molecular weight of H_3PO_4 divided by the number of moles of OH^- used in the reaction per mole of H_3PO_4.

$$H_3PO_4 + NaOH \rightarrow NaH_2PO_4 + H_2O \qquad \text{(i)}$$

$$\text{Equivalent weight} = \frac{98 \text{ g (mole } H_3PO_4)^{-1}}{1 \text{ mole } H^+ \text{ (mole } H_3PO_4)^{-1}} = 98 \text{ g (mole } H^+)^{-1}$$

$$H_3PO_4 + 2NaOH \rightarrow Na_2HPO_4 + 2H_2O \qquad \text{(ii)}$$

$$\text{Equivalent weight} = \frac{98 \text{ g (mole } H_3PO_4)^{-1}}{2 \text{ mole } H^+ \text{ (mole } H_3PO_4)^{-1}} = 49 \text{ g (mole } H^+)^{-1}$$

$$H_3PO_4 + 3NaOH \rightarrow Na_3PO_4 + 3H_2O \qquad \text{(iii)}$$

$$\text{Equivalent weight} = \frac{98 \text{ g (mole } H_3PO_4)^{-1}}{3 \text{ mole } H^+ \text{ (mole } H_3PO_4)^{-1}} = 32.6 \text{ g (mole } H^+)^{-1}$$

---

3. Osmolar. The osmolar concentration of a solute is the molar concentration multiplied by the number of particles produced per molecule in solution. For glucose, a nondissociating solute, a 1 *M* solution is 1 osmolar. A 1 *M* solution of sodium chloride, which dissociates into two particles, is 2 osmolar. Two solutions producing the same osmotic effect—for example, the plasma solution, which bathes the red blood cells, and isotonic saline—are said to be isoosmolar. The osmolar concentration of plasma is 0.308. A solution of glucose, which is 0.308 *M* or 0.308 osmolar, would thus be isoosmotic with red blood cells. An isoosmolar (isotonic) solution of sodium chloride is 0.308 osmolar or 0.154 *M*. Plasma cells which are bathed in either of these solutions will neither swell nor shrink due to osmotic effects.

4. Weight-volume percent (w/v). A solution is 1% (w/v) when it contains 1 g of solute per 100 ml of solution.

5. Milligram percent or milligrams per deciliter (w/v). A solution is 1 mg % or 1 mg/dl when it contains 1 mg of solute per 100 ml of solution. Note that in this expression of concentration it is possible to have solution concentrations that are greater than 100%. For example, a 250 mg % solution indicates 250 mg of solute in 100 ml of solution.

6. Parts per million (w/v). A solution is one part per million (ppm) when it contains 1 mg of solute per liter of solution.

Weight Concentrations. Weight concentrations have a distinct advantage over volume concentrations in that they are not influenced by temperature. Volume concentrations will vary with temperature due to changes in density. It is apparent that the density of the solution must be known to convert from a solution concentration based on weight to one based on volume.

1. Molal. A 1 molal solution, 1 *m*, contains 1 mole of solute plus 1000 g of solvent.

2. Mole fraction. The mole fraction of a substance in solution is the ratio of the number of moles of the substance to the total number of moles of all the substances in the solution. The mole fraction of the solute in a two-component system is expressed as

$$X_2 = \frac{n_2}{n_1 + n_2} \tag{5.1}$$

where n_1 and n_2 are the number of moles of solvent and solute, respectively.

3. Percent by weight (w/w). A solution is 1% (w/w) when it contains 1 g of solute plus 99 g of solvent. Any mole of solvent of crystallization must be taken into account in the calculation. For example, crystalline sodium thiosulfate ($Na_2S_2O_3 \cdot 5H_2O$) molecular weight 248.19 contains 90.08 g of water per mole of solute.

Problem 5.2: Express the concentration of a 2.08% (w/w) H_2SO_4 solution, density 1.0140 g cm^{-3}, in terms of molarity, normality, molality, and mole fraction.

Solution: A 2.08% H_2SO_4 solution contains 2.08 g of H_2SO_4 per 97.92 g of H_2O.

Molarity. We need to find the number of moles of solute in a liter of solution. The density indicates that a liter of solution weighs 1014 g. Thus we can set up the equation

$$\frac{2.08 \text{ g}}{100 \text{ g}} = \frac{x \text{ g liter}^{-1}}{1{,}014 \text{ g liter}^{-1}} = \frac{1014}{100} \times 2.08, \ x = 21.1 \text{ g liter}^{-1}$$

$$\text{Molar} = \frac{\text{moles}}{\text{liter}} = \frac{21.1 \text{ g liter}^{-1}}{98.08 \text{ g mole}^{-1}} = \frac{0.215 \text{ moles}}{\text{liter}} = 0.215\ M$$

Normality. Find equivalents per liter. Use 21.1 g liter^{-1} from the previous part and the equivalent weight 49.04 g.

$$\text{Normal} = \frac{\text{equiv}}{\text{liter}} = \frac{21.1 \text{ g liter}^{-1}}{49.04 \text{ g equiv}^{-1}} = \frac{0.430 \text{ equiv}}{\text{liter}} = 0.430\ N$$

Molality. Find the moles of solute per 1000 g of solvent.

$$\frac{2.08 \text{ g}}{97.92 \text{ g of solvent}} = \frac{x}{1000 \text{ g of solvent}}, \ x = 21.2 \text{ g}$$

$$\text{Molality} = \frac{\text{moles}}{1000 \text{ g of solvent}} = \frac{21.2 \text{ g } (1000 \text{ g solvent})^{-1}}{98.08 \text{ g mole}^{-1}}$$

$$= \frac{0.217 \text{ mole}}{1000 \text{ g of solvent}} = 0.217\ m$$

Mole Fraction. Find the number of moles of each component. Calculate the fraction of the total moles contributed by each component:

$$\frac{2.08 \text{ g}}{98.08 \text{ g mole}^{-1}} = 0.0212 \text{ mole} = n_{H_2SO_4}$$

$$\frac{97.92 \text{ g}}{18.016 \text{ g mole}^{-1}} = 5.435 \text{ mole} = n_{H_2O}$$

$$X_{H_2SO_4} = \frac{n_{H_2SO_4}}{n_{H_2O} + n_{H_2SO_4}} = \frac{0.0212}{5.435 + 0.0212} = \frac{0.0212}{5.4562} = 0.00388$$

$$X_{H_2O} = \frac{n_{H_2O}}{n_{H_2O} + n_{H_2SO_4}} = \frac{5.435}{5.435 + 0.0212} = \frac{5.435}{5.4562} = 0.996$$

OTHER METHODS OF EXPRESSING CONCENTRATION

Activity. As the concentration of solute in a solution increases, there occurs an interaction between solute molecules. This interaction is especially large in solutions that contain ions. Because of this interaction the ions behave differently from those in dilute solution. An illustration may clarify this point. For a particular reaction we may use a reactant concentration of 0.5 *M* and obtain a certain reaction velocity. If we now rerun the same reaction, but increase the concentration of reactant to 1.0 *M*, we might expect the velocity to double. This is not realized, as the interaction between the solute molecules increases with concentration and the effective concentration is less than 1.0 *M*.

We refer to these effective concentrations as activities and represent them by the symbol a. Activity, a, is related to concentration, c, by the activity coefficient γ:

$$\gamma = \frac{a}{c} \tag{5.2}$$

The activity and the activity coefficient both depend on concentration. As the concentration approaches zero, the interactions between solute molecules decrease and the activity and concentration become equal.

The advantage of using this concept of activity is then apparent. Let us use equilibrium constants to illustrate the point. An equilibrium constant varies with concentration due to interaction between solutes in the solution. It can be made a true constant over the concentration range with no deviation by using activities in place of concentrations. Thus we do not need to change the form of our equations where concentrated solutions are used; we just substitute activity for concentration.

Activities should be used in place of concentrations whenever possible. Since biochemical systems are complex, this is often difficult to realize in practice. Fortunately many in vitro enzyme systems are at low solute concentrations, where activity and concentration are interchangeable.

Concentration has been used interchangeably with activity throughout the text in spite of the obvious limitation imposed by such use.

Ionic Strength. The activity coefficients of the previous section correlate with the total ion concentration in solution. This effect, which is electrostatic in nature, depends on the number of ions and the number of charges per ion. G. N. Lewis included this electrostatic effect in a concept called ionic strength, which is defined in the equation

$$\frac{\Gamma}{2} = \mu = \tfrac{1}{2}\Sigma c_i Z_i^2 \tag{5.3}$$

where ionic strength is indicated either by $\Gamma/2$ or μ and c_i and Z_i are respectively the molar concentration and charge of the ith ionic species.

Problem 5.3: Calculate the ionic strength of 0.1 M Na_2HPO_4.

Solution: $\Gamma/2 = \mu = \tfrac{1}{2}\Sigma c_i Z_i^2$.

Assume complete ionization of the salt to give

$$Na_2HPO_4 \rightarrow 2Na^+ + HPO_4^{--}$$

Thus $Na^+ = 0.2\ M$ with $Z = +1$ and $HPO_4^{-2} = 0.1\ M$ with $Z = -2$.

Write out:

$$\frac{\Gamma}{2} = \mu = \frac{[Na^+][Z_{Na^+}]^2 + [HPO^{--}][Z_{HPO_4^{--}}]^2}{2}$$

Substitute:

$$\frac{\Gamma}{2} = \mu = \frac{[0.2][1]^2 + [0.1][-2]^2}{2}$$

$$= \frac{0.2 + 0.4}{2}$$

$$= 0.3\ M$$

The units of ionic strength are those of concentration.

Electrostatic effects are important structure-determining factors for many biochemical molecules. Thus, if one is studying a property of a biochemical system, for example, the pH optimum of an enzyme, other factors should be kept constant. Electrostatic effects can be minimized if the buffers used for the study are of constant ionic strength.

Ionic strength has an effect on simple equilibria as well. The second pK_a of phosphoric acid is markedly affected by ionic strength. At $\mu = 0$ the $pK_a = 7.16$, while at $\mu = 1.0$ the $pK_a = 6.60$. The concentration of phosphate must then be considered when performing calculations on this buffer system.

Expressing Concentrations in Logarithmic Form. In experimental studies it is not uncommon for the concentration of certain species to vary by several orders of magnitude. The hydrogen ion is one of the best known examples of such a species and for convenience Sørensen, in 1909, introduced the term pH, which expresses the $[H^+]$ in terms of a logarithmic function:

$$pH = -\log a_{H^+} \tag{5.4}$$

where a_{H^+} is the activity of the H^+ ion. Since we are making no distinction between activities and concentrations, the equation can be written as

$$pH = -\log [H^+] \tag{5.5}$$

where $[H^+]$ is the hydrogen ion concentration in moles per liter.

Problem 5.4: Calculate the pH of a 0.025 *M* solution of HCl.

Solution: HCl is completely ionized so the $[H^+]$ is 0.025 *M*. Find the log:

$$pH = -\log [H^+] = -\log 0.025$$
$$= -\log 2.5 \times 10^{-2}$$
$$\log 2.5 = 0.40$$
$$pH = -\log (10^{0.40} \times 10^{-2}) = -\log 10^{-1.60}$$
$$= -(-1.60) = 1.60$$

The general form of this function

$$pX = -\log [X] \tag{5.6}$$

has proved to be a convenient means of expressing concentration. Thus the concentration c of a metal ion is expressed as pc_m, and so on.

This logarithmic function is also used to indicate acid and base strength by the expression

$$pK_a = -\log K_a \tag{5.7}$$

where K_a is the ionization constant for the acid or base (see Chapter 6).

It is apparent that this function can give positive, negative, or zero values because the equilibrium constant can be, for example, one or any number greater or less than one.

Problem 5.5: Calculate the pK_a for acetic acid; $K_a = 1.8 \times 10^{-5}$.

Solution: $pK_a = -\log K_a$
$= -\log 1.8 \times 10^{-5}$

Find the log: $pK_a = -\log 10^{0.25} \times 10^{-5}$

Combine exponents to give

$$pK_a = -\log 10^{-4.75}$$

Therefore $pK_a = (-4.75)(-\log 10) = +4.75$

GASES DISSOLVED IN GASES

The mixing of 1 liter of O_2 with 1 liter of CO_2, at constant temperature and pressure, will result in a volume of 2 liters for the mixture, provided the gases are ideal. Even nonideal gases show only a very slight departure from additivity. Our discussion is limited to ideal gases. When these two gases are mixed at constant volume rather than constant pressure, it is apparent that n, the number of moles of each gas, is the same; however, the total pressure is doubled. This can be shown as follows. The number of moles of each gas is calculated from the ideal gas law, $PV = nRT$, where P is the pressure, V is the volume, n is the number of moles, R is the gas constant, and T is the absolute temperature. The total number of moles is $n_{tot} = n_{O_2} + n_{N_2}$. Multiplying each term by RT/V, we obtain

$$n_{tot}\frac{RT}{V} = n_{O_2}\frac{RT}{V} + n_{N_2}\frac{RT}{V} \tag{5.8}$$

which is an expression for the pressure exerted by each gas and the total pressure, where both gases occupy the same volume, V (from the ideal gas law $P = nRT/V$). The pressure at a given temperature is thus determined by the total number of gas molecules per volume, a unit of concentration, multiplied by two constants R and T. We can rewrite Equation (5.8) in terms of pressure as $P_{tot} = P_{O_2} + P_{N_2}$, which is known as Dalton's Law. P_{O_2} and P_{N_2} are called partial pressures. This law can be extended to include any number of components. $P_{tot} = P_A + P_B + P_C + \cdots$.

Problem 5.6: Calculate the total pressure and the partial pressures in atmospheres exerted in a volume of 2 liters at 25°C which contains 5 g of O_2, 7 g of N_2, 0.5 g of CO_2, and 0.1 g of H_2O.

Solution: Calculate the number of moles of each gas and use the ideal gas law to find their partial pressures:

$$P = \frac{nRT}{V}\begin{bmatrix} R \text{ in liter atm deg}^{-1} \text{ mole}^{-1} = 0.08205 \\ T = (273.16 + 25°) = 298.16° \text{ (absolute)} \end{bmatrix}$$

$$n_{O_2} = \frac{5.0}{32}, \; P_{O_2} = \frac{5}{32}\,\frac{0.08205}{2.0}(298.16) = 1.91 \text{ atm}$$

$$n_{N_2} = \frac{7.0}{28}, \; P_{N_2} = \frac{7}{28}\,\frac{0.08205}{2.0}(298.16) = 3.06 \text{ atm}$$

$$n_{CO_2} = \frac{0.5}{44}, \; P_{CO_2} = \frac{0.5}{44}\,\frac{0.08205}{2.0}(298.16) = 1.38 \text{ atm}$$

$$n_{H_2O} = \frac{0.10}{18}, \; P_{H_2O} = \frac{0.1}{18}\,\frac{0.08205}{2.0}(298.16) = 0.67 \text{ atm}$$

Sum the partial pressures to obtain the total pressure:

$$P_{tot} = P_{O_1} + P_{N_2} + P_{CO_2} + P_{H_2O}$$

$$P_{tot} = 1.91 + 3.06 + 1.38 + 0.67 = 7.02 \text{ atm}$$

The ideal gas law, $PV = nRT$, which was conveniently used in the above calculations, needs to be discussed more fully. Let us calculate the volume of 1 mole of a gas at 0°C and 1 atm pressure:

$$V = \frac{nRT}{P} = \frac{(1 \text{ mole})}{(1 \text{ atm})}\,\frac{(0.08205)(\text{liter-atm})(273.16 \text{ deg})}{(\text{deg mole})}$$

$$= 22.414 \text{ liters} \qquad (5.9)$$

The conditions used for this calculation, 0°C and 1 atm, are known as standard conditions, SC, and the volume 22.414 liters is known as the gram-molecular volume. These standard conditions serve as a point of reference for gas phase measurements, but have no further significance. Any set of temperatures and volumes could have been used.

Problem 5.7: Calculate the number of moles in 71 liters of oxygen at SC.

Solution: 22.414 liters mole^{-1} at SC. Therefore

$$\text{number of moles} = \frac{71 \text{ liters}}{22.414 \text{ liters mole}^{-1}} = \frac{71 \text{ moles}}{22.414} = 3.17 \text{ moles of } O_2$$

The calculation in Problem 5.7 could also have been performed by using the ideal gas law. However, it is more conveniently done using the known molar volume at SC and it is thus useful to be able to convert any experimental measurements of volume to volume at SC. Two relationships permit this conversion. The first is known as Boyle's Law, which states that $PV = k$ at constant temperature. For two sets of conditions of pressure and volume, A and B, then

$$P_A V_A = P_B V_B \quad \text{or} \quad V_A = \frac{P_B}{P_A} V_B \tag{5.10}$$

This mathematical equation states that an increase in pressure will decrease the volume at constant temperature. The second expression is known as Charles' Law. It states that the volume of a gas is proportional to the absolute temperature at constant pressure. $V = kT$ or $V/T = k$. For two conditions of volume and temperature, A and B,

$$\frac{V_A}{T_A} = \frac{V_B}{T_B} \quad \text{or} \quad V_A = V_B \frac{T_A}{T_B} \tag{5.11}$$

Thus the volume of a sample at a new set of conditions is calculated by multiplying the old volume by the proper ratios of pressure and temperature.

Problem 5.8: 7.1 liters of O_2 were collected at laboratory conditions, which were 741 mm Hg and 31°C. What is the volume of this sample at SC?

Solution: 7.1 liters at 741 mm and 304°K is to be converted to SC, 760 mm and 273°K. Form a product of the original volume and the pressure and temperature ratios to give the new volume:

$$7.1 \text{ liters} \times \frac{741 \text{ mm}}{760 \text{ mm}} \times \frac{273°\text{C}}{304°\text{C}} = 6.2 \text{ liters}$$

Both ratios are less than one in this case since in going from initial to final conditions we are increasing the pressure, which decreases the volume (Boyle's Law), and we are decreasing the temperature, which decreases the volume (Charles' Law).

It is also possible to write a relation between temperature and pressure at constant volume:

$$P = k \times T \tag{5.12}$$

For two sets of conditions, A and B, $P_A/T_A = P_B/T_B$. This expression is, however, less useful, since most of the calculations will involve volumes.

Problem 5.9: Calculate the number of moles of a gas which occupies 17.3 liters at 729 mm pressure and 26°C.

Solution: Convert the volume to SC and divide by 22.414 liters mole^{-1}.

$$\text{Volume at SC} = 17.3 \times \frac{729}{760} \times \frac{273}{299}$$

$$= 15.15 \text{ liters}$$

The number of moles is

$$\text{moles} = \frac{15.15 \text{ liters}}{22.414 \text{ liters mole}^{-1}}$$

$$= 0.676 \text{ moles}$$

Problem 5.10: Calculate the number of moles of O_2 in 7.3 liters of the gas collected over H_2O at 25°C and 720 mm of pressure. The vapor pressure of H_2O is 23.8 mm at 25°C.

Solution: Determine the O_2 partial pressure. 720 mm is the total pressure:

$$720 = P_{O_2} + P_{H_2O} = P_{O_2} + 23.8$$

Then P_{O_2} is 696 mm. Use this pressure to calculate the volume at SC:

$$\text{Volume at SC} = 7.3 \times \frac{696}{760} \times \frac{273}{298} = 6.1 \text{ liters}$$

The number of moles is

$$\frac{6.1 \text{ liters}}{22.414 \text{ liters mole}^{-1}} = 0.272 \text{ mole of } O_2$$

GASES DISSOLVED IN LIQUIDS

The extent to which a gas dissolves in a liquid is related to the partial pressure of the gas over the liquid. Henry's Law is a mathematical statement of this fact:

$$P_{\text{solute}} = KX_{\text{solute}} \tag{5.13}$$

In this expression P is the pressure of the gaseous solute over the solution, X is the mole fraction of the solute in the solution, and K is the Henry's Law constant at a given temperature for a particular gas and solvent. The constant K generally increases with temperature. This means that the solubility of the gas in the solution decreases with an increase in temperature.

Problem 5.11: Calculate the mole fraction of CO_2 in an aqueous solution, which is in equilibrium with vapor having a P_{CO_2} of 0.02 atm. ($K = 1.25 \times 10^6$ mm at 25°C.)

Solution:

$$X_{CO_2} = \frac{P_{CO_2}}{K}$$

If we substitute values and convert atmospheres to millimeters,

$$X_{CO_2} = \frac{0.02 \text{ atm } 760 \text{ mm}}{1.25 \times 10^6 \text{ mm atm}}$$

When we evaluate,

$$X_{CO_2} = 1.22 \times 10^{-5}$$

The concentration of gas dissolved in a liquid can also be expressed in terms of a Bunsen coefficient. A Bunsen coefficient is the number of liters of gas (reduced to standard conditions) which will dissolve in 1 liter of solvent under a pressure of 760 mm of gas at the specified temperature. It may also be expressed in milliliters of gas per milliliter of solvent, which is numerically equivalent.

The Bunsen coefficient is sometimes confusing, since the volume of the gas dissolved at a specified temperature is expressed as volume at SC. An obvious advantage of this expression of the solubility of the gas is that it is readily converted to millimolar concentration, using the millimolar volume. Bunsen coefficients vary with temperature and thus can be used only at the temperature specified.

Problem 5.12: (a) Calculate the milliliters of O_2 dissolved in 25 ml of blood plasma at 38°C and a partial pressure of 140 mm Hg. (b) Calculate the millimolar concentration of O_2 in plasma.

Solution: (a) Bunsen coefficient:

$$\frac{0.024 \text{ ml } O_2 \text{ (SC)(ml plasma 38°C)}^{-1}}{760 \text{ mm Hg}}$$

$$\text{ml } O_2 = \frac{0.024 \text{ ml } O_2 \text{ (SC)}}{\text{ml plasma } 38°} \times \frac{140 \text{ mm Hg}}{760 \text{ mm Hg}} \times 25 \text{ ml plasma 38°C}$$

$$\text{ml } O_2 \text{ (SC)} = 0.11 \text{ ml}$$

(b) 0.11 ml of O_2 at SC per 25 ml plasma at 38°C. We assume ideal gas behavior (22.414 ml mmole^{-1} of O_2 at SC), so substituting gives

$$\text{m}M = \frac{[0.11 \text{ ml } O_2 \text{ (SC)}][1000 \text{ ml plasma 38°C}] \text{ mmole}}{[25 \text{ ml plasma 38°C}][\text{liter plasma 38°C}][22.414 \text{ ml } O_2 \text{ (SC)}]}$$

$$= 0.20 \text{ mmole liter}^{-1} = 0.20$$

Carbon dioxide deviates significantly from ideal gas behavior. Thus for calculations involving CO_2 the real millimolar volume, 22.26 ml SC mmole^{-1}, is used.

Warburg respirometers, which measure gas production or uptake, contain both liquid and gaseous phases at equilibrium. To calculate the moles of gas it is necessary to use either the Henry's Law constant or the Bunsen coefficient to account for the quantity of gas dissolved in the liquid phase. The Bunsen coefficient is also essential for calculations on the bicarbonate buffer system and respiration in the living system, as there is an equilibrium between CO_2 in the gas phase (see Chapter 8) and CO_2 in plasma.

STOICHIOMETRY

Stoichiometry is the branch of chemistry concerned with the weight relationships between atoms in a molecule and/or compounds in a chemical reaction. The principles are given in the law of conservation of mass and the law of definite proportions. Chemistry is an exact science because of these basic principles. In biochemistry these same principles are applied to molecules of biological importance.

Calculation of Empirical Formulas From Chemical Analysis. For a biologically important molecule, such as a protein, a determination of the elemental composition might give the following results: C, 50%; H, 7%; O, 23%; N, 16%; and S, 4%. The elemental composition of another protein might give identical results. Clearly this does not help us to establish a true formula. A more useful technique is to obtain an empirical formula in terms of the 20 or so constituent amino acids which are present. Although the C, H, N, O, and S percentages may be nearly identical for all proteins, their amino acid compositions are not. The usual procedure is to hydrolyze a pure sample of protein of known weight and submit it to quantitative amino acid analysis. An empirical formula can then be suggested for the protein in terms of amino acids. Similar approaches are made for other biological compounds. The true formula can be deduced from an independent measurement of the molecular weight.

Problem 5.13: Calculate the empirical formula and the true formula in terms of amino acids for the following peptide. Amino acid analysis showed 0.3 g of glycine, 0.18 g of alanine, and 0.33 g of phenylalanine. An independent measurement of the molecular weight indicated it to be 1050 ± 25.

Solution: Find the number of moles of each amino acid.

$$\text{Glycine: Number of moles} = 0.3/75 = 0.004$$

$$\text{Alanine: Number of moles} = 0.18/89 = 0.002$$

$$\text{Phenylalanine: Number of moles} = 0.33/165 = 0.002$$

Divide by the smallest number of moles to obtain the empirical formula.

Glycine: 0.004/0.002 = 2

Alanine: 0.002/0.002 = 1

Phenylalanine: 0.002/0.002 = 1

Empirical Formula = Gly, 2; Ala, 1; Phe, 1. The molecular weight of a peptide involving two glycines, one alanine, and one phenylalanine is 350. Thus the true formula is three times the empirical formula, or 6 glycine, 3 alanine, and 3 phenylalanine residues.

Weight and Volume Calculations on Chemical Equations. The step-by-step method for calculations involving chemical equations is as follows.

1. Balance the equation.
2. Determine the molecular weights or mole ratios for the products and/or reactants of interest.
3. Find the weight or the mole relationships for the products and/or reactants of interest.
4. Set up a proportion taking note of the units.

Problem 5.14: Calculate the number of grams of NH_3 produced by the action of the enzyme urease on 5 g of urea:

$$H_2N-\overset{\overset{\displaystyle O}{\|}}{C}-NH_2 + H_2O \xrightarrow{\text{urease}} NH_3 + CO_2$$

Solution: Balance the equation:

$$H_2N-\overset{\overset{\displaystyle O}{\|}}{C}-NH_2 + H_2O \xrightarrow{\text{urease}} 2NH_3 + CO_2$$

The equation states that two moles of NH_3 are produced from each mole of urea, or in grams:

$$\frac{2 \times 17}{60}\ \frac{\text{g } NH_3}{\text{g urea}}$$

Thus we can set up the proportion

$$\frac{34 \text{ g } NH_3}{60 \text{ g urea}} = \frac{x \text{ g } NH_3}{5 \text{ g urea}}$$

When we solve,

$$x = \frac{170}{60} = 2.83 \text{ g of } NH_3$$

An alternate way of setting up the problem is to write the quantities with the equation:

$$\underset{(60\ \text{g})}{\overset{(5\ \text{g})}{NH_2{-}\overset{\overset{\large O}{\|}}{C}{-}NH_2}} + H_2O \rightarrow \underset{(2)(17\ \text{g})}{\overset{(x\ \text{g})}{2NH_3}} + CO_2$$

Then write the same proportion and solve as above:

$$\frac{5\ \text{g urea}}{60\ \text{g urea}} = \frac{x\ \text{g of}\ NH_3}{34\ \text{g of}\ NH_3}$$

Problem 5.15: Calculate the volume of ammonia produced in Problem 5.14 at 740 mm and 25°C.

Solution: The 2.83 g of NH_3 produced can be converted to volume at SC:

$$V = \frac{2.83\ \text{g mole}}{17\ \text{g}} \times \frac{22.414\ \text{liters}}{\text{mole}} = 3.73\ \text{liters}$$

Now convert to the conditions given in the problem.

$$V = 3.73\ \text{liter} \times \frac{298}{273} \times \frac{760}{740}$$

$$= 4.17\ \text{liters of}\ NH_3\ \text{produced.}$$

We can also work this problem in the following manner, using volume instead of weight for NH_3:

$$\underset{(60\ \text{g})}{\overset{(5\ \text{g})}{NH_2{-}\overset{\overset{\large O}{\|}}{C}{-}NH_2}} + H_2O \rightarrow \underset{(2)(22.4\ \text{liters})}{\overset{(x\ \text{liters})}{2NH_3}} + CO_2$$

Here instead of the weight of NH_3 we use the volume at SC.

$$(2)(22.4\ \text{liters})\ \frac{(5\ \text{g})}{(60\ \text{g})} = x$$

$$x = 3.73\ \text{liters}$$

From this point on, the calculation is the same as the previous method.

Problem 5.16: Calculate the moles of O_2 and alcohol required for the production of 5 g of H_2O in the complete oxidation of ethanol C_2H_5OH.

Solution: Write a balanced equation: $C_2H_5OH + 3O_2 \rightarrow 3H_2O + 2CO_2$. The equation states that 3 moles of O_2 are consumed for each 3 moles of H_2O, so the mole ratio is

$$\frac{1\ \text{mole}\ O_2}{1\ \text{mole}\ H_2O}$$

We wish to produce 5 g of H_2O which is $\frac{5}{18}$ or 0.28 mole. Thus

$$(0.28 \text{ mole } H_2O) \times \frac{1 \text{ mole alcohol}}{3 \text{ moles } H_2O} = 0.093 \text{ moles of alcohol}$$

Concentration Calculations on Chemical Equations. Chemical reactions are commonly carried out in some solvent. In a solvent we use concentration and volume to describe the amount of material present. The product of the concentration of material and the volume of solution gives the quantity of substance:

$$M \text{ moles liter}^{-1} \times V \text{ (liters)} = \text{number of moles} \quad (5.14)$$

$$N \text{ equiv liter}^{-1} \times V \text{ (liters)} = \text{number of equivalents} \quad (5.15)$$

Using these relationships for the volume concentrations, molar and normal, it is possible to convert to moles and grams or vice versa.

One application of these relationships is in volumetric analysis. Titration is a process whereby a titrant, one of the reactants in solution, is quantitatively added to another reactant until chemically equivalent amounts are present as dictated by the mole relationships in the chemical equation. The end point—when equivalent amounts are present—is generally detected by an indicator, which changes color the instant the titrant is in excess.

If no indicator is present it is still possible to mix the reactants in the mole ratios dictated by the equation and assume conversion to the products. This assumption must be checked for the particular reaction under consideration. If complete conversion of the reactants to the products occurs, we say the reaction is quantitative. The calculations in this section assume quantitative reactions.

Problem 5.17: Calculate the number of milliliters of 1 *M* HCl required to react with 17.5 ml of 1.2 *M* NaOH. How many grams of NaCl are formed?

Solution: Write a balanced equation for the neutralization reaction:

$$HCl + NaOH \rightarrow NaCl + H_2O$$

The equation shows that 1 mole of NaOH reacts with 1 mole of HCl. The number of moles of NaOH is

$$\begin{aligned}(\text{volume} \times \text{molarity})_{NaOH} &= (0.0175 \text{ liter})\ 1.2 \text{ moles liter}^{-1} \\ &= 0.021 \text{ mole}\end{aligned}$$

The number of moles of HCl must also be 0.021. Substitute:

$$(\text{volume} \times \text{molarity})_{HCl} = 0.021 \text{ mole}$$

$$(V \text{ liter} \times 1 \text{ mole liter}^{-1})_{HCl} = 0.021$$

Solve for the volume and substitute values:

$$V(\text{liter})_{\text{HCl}} = \frac{0.021 \text{ mole}}{1 \text{ mole liter}^{-1}} = 0.021 \text{ liter} = 21 \text{ ml}$$

The amount of NaCl produced is 0.021 mole. Convert this to grams by multiplying by the gram-molecular weight.

$$0.021 \text{ mole} \times 58.5 \text{ g mole}^{-1} = 1.23 \text{ g of NaCl}$$

For problems of this type, the weight of solute in a solution of given volume and concentration is

$$\text{g solute} = V \text{ (liters)} \times M \text{ moles liter}^{-1} \times \text{mol. wt. (g mole}^{-1}) \quad (5.16)$$

or,

$$\text{g solute} = V \text{ (ml)} \times M \text{ moles liter}^{-1} \times \text{mol. wt. (g mmole}^{-1}) \quad (5.17)$$

Molar can be expressed as moles per liter or millimoles per milliliter. Numerically they are equal.

Problem 5.18: Calculate the volume of 8 M NH_4OH required to produce 25 ml of 0.1 M ammoniacal silver in the reaction

$$4NH_4OH + AgNO_3 \rightarrow Ag(NH_3)_4NO_3 + 4H_2O$$

Solution: Calculate the number of mmoles of the ammoniacal silver:

$$25 \text{ ml} \times 0.1\ M = 2.5 \text{ mmole } Ag(NH_3)_4NO_3$$

The equation indicates 4 mmole NH_4OH per mmole $Ag(NH_3)_4NO_3$:

$$\text{mmole } NH_4OH = \frac{2.5 \text{ mmole } Ag(NH_3)_4NO_3 \times 4 \text{ mmole } NH_4OH}{\text{mmole } Ag(NH_3)_4NO_3}$$

$$= 10$$

Substitute in Equation (5.13), remembering that 8 mmole ml^{-1} is equivalent to 8 moles $liter^{-1}$:

$$10 \text{ mmole} = V \text{ (ml)} \times 8 \text{ mmole ml}^{-1}$$

The volume of 8 M NH_4OH required is

$$V \text{ (ml)} = \frac{10}{8} = 1.25 \text{ ml}$$

Problem 5.19: Calculate the volume of 0.5 M HCl required to neutralize 22 ml of 0.5 M $Ba(OH)_2$.

$$2HCl + Ba(OH)_2 \rightarrow 2H_2O + BaCl_2$$

Solution: Calculate the millimoles of $Ba(OH)_2$:

$$22\ (\text{ml}) \times 0.5\ \text{mmole ml}^{-1} = 11\ \text{mmole of } Ba(OH)_2$$

The equation states that the ratio is 2 mmole HCl per 1 mmole $Ba(OH)_2$, so

$$\text{mmole of HCl} = 11\ (\text{mmole of } Ba(OH)_2) \times \frac{2(\text{mmole HCl})}{(\text{mmole } Ba(OH)_2)}$$

$$= 22$$

The volume of HCl required is

$$22\ (\text{mmole}) \times \frac{(\text{ml})}{0.5\ (\text{mmole})} = 44\ \text{ml}$$

Problem 5.20: Calculate the volume of 0.5 N HCl required to neutralize 22 ml of 0.5 N $Ba(OH)_2$:

$$2HCl + Ba(OH)_2 \rightarrow 2H_2O + BaCl_2$$

Solution: Calculate the milliequivalents (meq) of $Ba(OH)_2$ in the 22 ml of solution.

$$\text{meq } Ba(OH)_2 = 22\ (\text{ml}) \times 0.5\ \text{meq ml}^{-1}$$

$$\text{meq } Ba(OH)_2 = 11$$

The equation states that the ratio is 2 mmole HCl per 1 mmole of $Ba(OH)_2$. For HCl 1 mmole equals 1 meq and for $Ba(OH)_2$ 1 mmole equals 2 meq. A meq of HCl is equal to a meq of $Ba(OH)_2$. The volume of HCl required, since its concentration is the same as $Ba(OH)_2$, is therefore 22 ml.

The advantages of using equivalent weight are now apparent. If the concentrations are given in terms of normality (equiv l^{-1}), then the mole ratio is not a concern in a calculation using volumes and we can write

$$V_1 \times N_1 = V_2 \times N_2 \tag{5.18}$$

This is not to say that the mole ratios indicated in the equation are ignored. It simply means they were taken into account when making the solutions.

Problem 5.21: What is the molar concentration of a 50-ml solution of glucose, which, when reacted with 1.5 mmole of NaOI, gives the following results?

Reaction scheme (balanced equations)

$$\text{glucose} + NaOI + NaOH \rightarrow \text{gluconic acid} + NaI + H_2O \tag{i}$$

$$NaOI + NaI + H_2SO_4 \rightarrow I_2 + Na_2SO_4 + H_2O \tag{ii}$$

$$I_2 + 2S_2O_3^{-2} \rightarrow S_4O_6^{-2} + 2I^- \tag{iii}$$

The excess NaOI is converted to I_2 (Equation (ii)) and titrated with thiosulfate (Equation (iii)). 50 ml of 0.01 *M* thiosulfate were required.

Solution: The equation indicates that 1 mole of glucose is oxidized per mole of NaOI. 1.5 mmole of sodium hypoiodite were added to the reaction mixture. The excess NaOI was titrated with thiosulfate. This excess is in millimoles of thiosulfate:

$$50 \text{ ml} \times 0.01 \text{ mmole ml}^{-1} = 0.5 \text{ mmole } S_2O_3^{-2}$$

2 moles of $S_2O_3^{-2}$ are used per mole of I_2 (Equation (iii)), so

$$\text{mmole iodine} = \frac{0.5 \text{ (mmole } S_2O_3^{-2})(\text{mmole } I_2)}{2 \text{ (mmole } S_2O_3^{-2})}$$

$$= 0.25$$

The molar ratio to convert NaOI to I_2 equals one (Equation (ii)), so mmole I_2 = mmole of NaOI. The amount of NaOI used to react with the sugar is then 1.50 mmole − 0.25 mmole = 1.25 mmole. They react on an equal molar basis (Equation (i)), so the solution contained 1.25 mmole of glucose. The concentration of glucose in 50 ml is

$$M \frac{\text{(moles)}}{\text{(liter)}} = \frac{1.25 \text{ (mmole)}}{50 \text{ (ml)}} = 0.025\ M$$

---

5. PROBLEMS A

Answers on page 339

1. How would you prepare the following solutions?
   - (a) 1 liter of 0.5 *M* $(NH_4)_2SO_4$ from solid $(NH_4)_2SO_4$
   - (b) 5 liters of 0.1 *M* NaH_2PO_4 from solid $NaH_2PO \cdot H_2O$
   - (c) 0.2 liter of 0.3% $MgCl_2$ (w/v) from solid $MgCl_2$
   - (d) 37 ml of 0.8 *N* H_2SO_4 from 17.8 *M* H_2SO_4
   - (e) 3 ml of 0.20 *N* NaOH from 6 *M* NaOH
   - (f) 50 ml of 0.5 *N* H_2SO_4 from 1 *M* H_2SO_4
   - (g) 10 ml of 0.45 *N* H^+ from 2 *N* H^+
   - (h) 200 ml of 0.2 *N* KNO_3 from solid KNO_3
   - (i) 50 ml of 0.1% NaCl (w/v) from solid NaCl
   - (j) 20 ml of 0.01 *M* NH_4Cl from solid NH_4Cl
   - (k) 50 ml of 10^{-4} *N* HCl from 10^{-1} *M* HCl
   - (l) 10 ml of 1 *N* H_2SO_4 from 2 *M* H_2SO_4
   - (m) 40 ml of 0.1 *N* Na_2CO_3 with respect to Na^+ from 2 *M* Na_2CO_3
   - (n) 5.1 ml of 0.15 *N* HCl from 0.38 *M* HCl
   - (o) 100 ml of 10^{-3} *M* H^+ from 10^{-1} *M* H^+
   - (p) 7.5 ml of 0.1 *N* H_2SO_4 from 1 *M* H_2SO_4
   - (q) 10 ml of 30% saturated $(NH_4)_2SO_4$ from saturated solution
   - (r) 1.71 ml of 0.50 *M* NH_4OH from 1 *M* NH_4OH
   - (s) 1 liter of 1 *M* Na_2HPO_4 from solid $Na_2HPO_4 \cdot 2H_2O$
   - (t) 105 ml of 0.15 *N* HCl from 11.7 *N* HCl

(u) 4.7 ml of 3.1 M H_2SO_4 from 17.8 M H_2SO_4
(v) 250 ml of 0.3 M NaCl from 1.2 M NaCl
(w) 9 liters of 0.2 M $NaC_2H_3O_2$ from solid $NaC_2H_3O_2$
(x) 20 ml of 0.3% NaCl (w/v) from 1% NaCl (w/v)
(y) 5.1 ml of 0.1 N KOH from 0.87 M KOH
(z) 20 ml of 0.1 N H_2SO_4 from 1.5 M H_2SO_4

2. Calculate the following.
   (a) Weight in grams of 1×10^{23} molecules of NaCl
   (b) Weight in grams of 0.45 mole of C_2H_5OH
   (c) The molecular weight in grams if one molecule weighs 6.644×10^{-20} g
   (d) The number of moles of KCl in 15 g
   (e) The number of moles of $CaCl_2 \cdot H_2O$ in 15 g

3. How many grams of the following compounds are contained in 250 ml of a 0.1 M solution?
   (a) sucrose (mol. wt. 342.3)
   (b) NaOH
   (c) $MgCl_2$

4. Calculate the molarity and molality of the following solutions.
   (a) 2% (w/v) NaCl which has a density of 1.012 g ml^{-1}
   (b) 1% (w/v) $MgCl_2$ which has a density of 1.007 g ml^{-1}
   (c) 0.8% (w/v) solution of sucrose which has a density of 1.002 g ml^{-1}
   (d) 1% (w/v) NaOH which has a density of 1.05 g ml^{-1}
   (e) 22.2 g of $CaCl_2$ dissolved in 800 ml of solution which has a density of 1.015 g ml^{-1}
   (f) 900 mg% NaCl which has a density of 1.009 g ml^{-1}

5. Calculate the number of milliequivalents of the following.
   (a) Cl^- in 160 ml of 0.2 M NaCl
   (b) HCl in 50 ml of 0.1 N HCl
   (c) H^+ in 1 ml of 0.9 M H_2SO_4
   (d) H^+ in 1 ml of 0.3 M HNO_3
   (e) OH^- in 10 ml of 0.5 M NaOH
   (f) A solute in 60 ml of 0.2 N solution
   (g) Na^+ in 0.5 ml of 0.2 M Na_2HPO_4
   (h) Cl^- in 50 ml of 0.2 M $AlCl_3$
   (i) OH^- in 43 ml of 0.25 M NaOH
   (j) SO_4^{-2} in 50 ml of 0.50 M $Al_2(SO_4)_3$

6. Calculate the pH of the following solutions.
   (a) 0.0173 N HCl
   (b) 200 ml of 0.1 N $AgNO_3$ mixed with 50 ml of 0.3 M HCl
   (c) 50 ml of 0.01 N HCl mixed with 100 ml of 0.0025 N KOH
   (d) 0.007 M NaOH
   (e) 0.0031% (w/v) HCl
   (f) 80 ml of 0.12 N HCl mixed with 20 ml of 0.06 M KOH
   (g) 0.02 M H_2SO_4

7. Calculate the hydrogen ion concentration in the following solutions.
   (a) 0.5 *M* HCl
   (b) 0.5 *M* HNO_3
   (c) Blood at pH 7.4
   (d) NaOH at pH 10.66
   (e) Acetate buffer which has a pH of 4.74
   (f) Urine which has a pH of 6.4

8. The average concentration of cations of a serum is given below. Express the concentration of these cations in milliequivalents per liter.
   (a) Sodium (at. wt. 23) 330 mg/100 ml
   (b) Potassium (at. wt. 39) 19 mg/100 ml
   (c) Calcium (at. wt. 40) 10 mg/100 ml
   (d) Magnesium (at. wt. 24) 2.4 mg/100 ml

9. A 0.1-g sample of a strong acid, when titrated to a phenolphthalein end point with NaOH, required 40.5 ml of 0.05 *N* base. What is the equivalent weight of the acid?

10. Assuming that all the gases behave in an ideal way, calculate the following.
    (a) The weight of CO_2 which occupies 15 liters at SC
    (b) The volume of O_2 at SC if its volume at 27°C and 740 mm Hg pressure is 11 ml
    (c) The number of millimoles of CO_2 in a 5.3-liter sample of gas at 38°C and 747 mm Hg.
    (d) The liters of O_2 at 27°C and 740 mm Hg required for the oxidation of 10 g of C_2H_5OH

11. Calculate the number of moles of each of the constituent gases in one liter of inspired air at 38°C with the following partial pressures: $O_2 = 158.2$ mm, $CO_2 = 0.5$ mm, $N_2 = 597$ mm, $H_2O = 5.0$ mm. What is the total pressure in atmospheres?

12. Calculate the moles of O_2, N_2, and CO_2 in 1 liter of expired air if the partial pressures are 116, 568, and 28.5 mm Hg, respectively.

13. Calculate the number of millimoles per liter of O_2 in plasma exposed to an oxygen partial pressure of 155 mm Hg at 38°C. The molar volume of oxygen is 22.414 liters at SC. The Bunsen coefficient is 0.024 at 38°C.

14. What is the millimolar concentration of CO_2 in plasma, which is in equilibrium with a partial pressure of CO_2 of 50 mm Hg. The Bunsen coefficient at 38°C is 0.51. The molar volume of CO_2 is actually 22.260 liters at SC.

15. Calculate the grams of sodium D-gluconate produced by oxidation of 0.15 moles of D-glucose with NaOI in the reaction

$$C_6H_{12}O_6 + NaOI + NaOH \rightarrow C_6H_{11}O_7Na + NaI + H_2O$$

16. How many moles of O_2 are required for the complete oxidation of 0.1 ml of glycerol measured at 20°C? Calculate the ml of O_2 at SC required for the reaction. The density of glycerol is 1.263 g ml^{-1} at 20°C.

17. The enzyme urease catalyzes the decomposition of urea to produce CO_2 and NH_3.

$$NH_2-\overset{\overset{\displaystyle O}{\|}}{C}-NH_2 + H_2O \rightarrow CO_2 + 2NH_3$$

Calculate the number of moles of water consumed in two minutes, if in this same time 230 ml of NH_3 were produced by the reaction at 740 mm Hg pressure and 29°C.

18. The acid number of a fat is the number of milligrams of potassium hydroxide necessary to neutralize the free fatty acids in 1 g of fat. What is the acid number of pure oleic acid? (mol. wt. 282.5.)

---

5. PROBLEMS B No answers given

1. Blood can be deproteinized by treating with an equal volume of 6% (w/v) perchloric acid at 0°C. Calculate the milliliters of 70% (w/v) perchloric acid required to make 100 ml of the 6% (w/v) solution.

2. Ringers solution is a physiological solution used for perfusing and incubating tissues. Its composition is NaCl = 8 g, KCl = 0.42 g, $CaCl_2$ = 0.24 g. $NaHCO_3$ = 0.20 g; made up to 1 liter with H_2O. What is the concentration, in milliequivalents per liter, of the individual (a) anions? (b) cations? and of (c) total cations?

3. In the determination of plasma volume, a small amount of a nontoxic dye, which is slowly cleared from the blood stream, is injected intravenously and its concentration is determined as soon as equilibrium is reached. 20 ml of an Evan's Blue solution (0.2 mg ml^{-1}) is injected and after 3 min a blood sample was withdrawn and found to contain 0.53 mg% (w/v) of the dye. What is the plasma volume?

4. An isolated strip of skeletal muscle weighing 1 g was placed in 9 ml of Ringer's solution and N_2 gas was bubbled through the system to flush out any O_2. The muscle was then stimulated electrically, causing contraction. Before stimulation, the lactic acid content of the solution was 8 mg% (w/v); after stimulation, it was 26 mg% (w/v). How many milligrams of glucose must have disappeared from the system? How many moles of ATP could have been produced from ADP and Pi by this metabolic conversion?

5. Hemoglobin reacts with O_2 to form a complex which contains 4 moles of oxygen per mole of hemoglobin.
   (a) Calculate the number of hemoglobin molecules required to carry 1 ml of O_2 gas at SC.
   (b) Calculate the number of ml of O_2/100 ml plasma at 38°C. The molecular weight of hemoglobin is 68,000 and its concentration in plasma is 15 g/100 ml.

6. The partial pressure of inspired air is 158 mm Hg and that of expired air is 116 mm Hg with respect to oxygen. How many moles of O_2 are absorbed by the lungs per liter of inspired air?

7. Cyclopropane is mixed with oxygen for use in anesthesia. A common mixture would have the following partial pressures: $P_{\text{cyclopropane}} = 170$ mm Hg, $P_{O_2} = 570$ mm Hg. Calculate the concentration of cyclopropane in volume percent.

8. Nitrous oxide is mixed with oxygen for use in anesthesia. A common mixture is 65% O_2 and 35% N_2O. Calculate the partial pressure of each gas in mm at 1 atm total pressure.

9. A diver was rapidly decompressed accidentally from 6 atm pressure to the pressure at sea level; Caisson disease (intravascular bubbles of N_2 gas and clotting) resulted. How much N_2 gas could theoretically be released from his tissues by this change in pressure? Additional data: inhaled air is $\frac{1}{5}O_2$ and $\frac{4}{5}N_2$ by volume; Bunsen coefficient of N_2 gas is 0.024 in tissue fluid; tissue fluid volume of diver = 50 liters.

10. Bromine, Br_2, or iodine, I_2, reacts quantitatively with the double bonds of fatty acids in methanol solution. A quantity called the iodine number, which is defined as the grams of iodine reacting with 100 g of fatty acid, is often calculated. Calculate the iodine number for a 2-g sample of lipid which reacted with 27 ml of 0.01 *M* iodine solution.

11. In determining the titratable acidity of urine, according to Folin, an aliquot is titrated with NaOH to a phenolphthalein end point. The result is expressed as milliliters of 0.1 *N* acid excreted per day. A 25-ml aliquot of a daily urine sample of 1570 ml requires 7.65 ml of 0.0890 *N* NaOH for neutralization. What is the titratable acidity?

12. Succinate can be determined by its quantitative conversion to fumarate by the reaction

$$\text{succinate} + 2[\text{Fe(CN)}_6]^{-3} \rightarrow \text{fumarate} + 2\text{H}^+ + 2[\text{Fe(CN)}_6]^{-4}$$

which is catalyzed by the enzyme succinic dehydrogenase. The specific activity of the enzyme solution is 120 μmoles min^{-1} ml^{-1}. Is a quantitative conversion achieved using 0.003 ml of enzyme solution and a 1.5-minute reaction time with a 4.0 μmole sample of succinate?

13. Serum chlorides are sometimes determined by precipitating chloride with a known amount of $AgNO_3$ and then back-titrating the unused $AgNO_3$ with sodium thiocyanate (NaSCN). With ferric alum as an indicator, 0.3 ml of 0.047 *M* $AgNO_3$ were added to 1 ml of serum; 2.61 ml of 0.019 *M* NaSCN were required to titrate the excess $AgNO_3$. Calculate the Cl concentration of this sample in terms of both meq ml^{-1} and mg% (w/v) of NaCl.

14. D-Glucose is quantitatively determined by using a coupled enzyme system of D-glucose oxidase and catalase. The reaction can be followed by measuring manometrically the breakdown of hydrogen peroxide formed by the oxidase, to water and oxygen. Calculate the volume of oxygen in milli-

liters produced by the breakdown of 0.1 mmole of hydrogen peroxide at 740 mm Hg and 38°C.

15. Amino acids can be quantitatively determined with ninhydrin by measuring the CO_2 produced according to the following reaction:

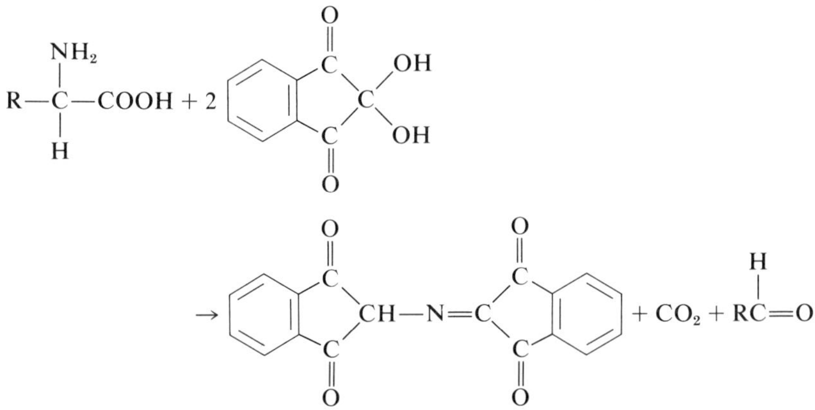

    (a) Calculate the number of moles of CO_2 produced in the reaction with 0.1 g of phenylalanine.
    (b) Calculate the volume of CO_2 produced by 10^{-4} mole of an amino acid at 25°C and at a pressure of 741 mm Hg.

16. The protein content of spinal fluid can be determined according to the method of Kjeldahl, which converts the nitrogen to $(NH_4)_2SO_4$. Ammonia is liberated by addition of concentrated alkali and distilled into a known amount of acid. The unused acid is then titrated with standard base. 2 ml of pathological spinal fluid containing 1.25% protein (6.25 g protein is assumed to contain 1 g of N) is digested and the ammonia distilled into 5 ml of 0.1009 N H_2SO_4. How much 0.052 N NaOH is required to neutralize the unused acid?

17. The oxidation of D-amino acids, which is catalyzed by the enzyme D-amino acid oxidase, occurs according to the reaction

$$\underset{\displaystyle |\atop \displaystyle NH_2}{R\text{—}CH}\text{—}COOH + O_2 \rightarrow \underset{\displaystyle \|\atop \displaystyle NH}{R\text{—}C}\text{—}COOH + H_2O_2$$

    Alanine is oxidized at a rate of 64 μl O_2 consumed per hour per milligram of enzyme dry weight at 742 mm Hg and 38°C.
    (a) How many micromoles are consumed per minute per milligram of enzyme?
    (b) The volume of gas, which is determined in a Warburg apparatus, can be measured to ± 0.1 μl. What is the uncertainty in the rate calculated in part (a)?

18. Serum sodium can be determined by precipitating sodium as the complex salt, sodium zinc uranyl acetate, $NaZn(UO_2)_3(CH_3COO)_9 \cdot 9H_2O$

(mol. wt. 1538), which can be weighed. If the precipitate from 1 ml of serum weighs 215 mg, what is the concentration in serum of Na expressed as milliequivalents per liter; as milligram percent (w/v)?

19. For fluoroscopic examination of the gastrointestinal tract, barium sulfate is frequently administered as a radio-opaque substance. Barium sulfate, being insoluble, is nontoxic whereas barium chloride and other soluble barium salts are deadly poisons. Suppose you had to prepare 200 g of barium sulfate for use in fluoroscopy and you had available solid $BaCl_2$ and $Na_2SO_4 \cdot 10\ H_2O$. How much of each would you use to prepare this material? How would you check to make certain no Ba^{++} remains in your preparation?

PART TWO

PRINCIPLES OF BIOCHEMICAL QUANTITATION

6

Chemical Equilibria

REVERSIBLE REACTIONS

Most simple chemical systems are composed of two opposing reactions in which certain molecules, ions, or elements are reacting with each other to form new chemicals, which in turn recombine to form the original materials. Carbon dioxide reacts with water to form carbonic acid,

$$CO_2 + H_2O \rightarrow H_2CO_3 \tag{6.1}$$

which then breaks down to give carbon dioxide and water:

$$H_2CO_3 \rightarrow CO_2 + H_2O \tag{6.2}$$

Such a chemical system represents a reversible reaction,

$$H_2O + CO_2 \rightleftharpoons H_2CO_3 \tag{6.3}$$

and when the two reactions are proceeding at the same rate, the proportions of the three chemical species under the conditions prevailing will remain constant. Chemical equilibrium is then said to have been attained and so long as the physical conditions such as temperature and pressure remain constant, the relative proportions of these materials will be constant. The simplest reversible reaction involves two chemical species A and B, represented by

$$A \rightleftharpoons B \tag{6.4}$$

By the *Law of Mass Action* the rate (v) at which A is converted to B is proportional to the concentration of A, which is usually represented [A], and thus

$$v_{A \to B} \propto [A] \tag{6.5}$$

and may be expressed mathematically as

$$v_{A \to B} = k_{A \to B}[A] \tag{6.6}$$

where the proportionality constant, $k_{A \to B}$, is defined as the rate constant. Similarly in reaction (6.4) the rate at which B is converted to A will be proportional to [B]; thus

$$v_{B \to A} = k_{B \to A}[B] \tag{6.7}$$

In the reversible reaction (6.4) these two rates, $v_{A \to B}$, $v_{B \to A}$, are always calculable under any condition of concentration for [A] and [B] if the two proportionality constants are known. At equilibrium, by definition,

$$v_{A \to B} = v_{B \to A} \tag{6.8}$$

and therefore from Equations (6.6) and (6.7)

$$k_{A \to B}[A] = k_{B \to A}[B]$$

or

$$\frac{k_{A \to B}}{k_{B \to A}} = \frac{[B]}{[A]}$$

Since $k_{A \to B}$ and $k_{B \to A}$ are both constants, the ratio of the two will be a constant, which is called the *equilibrium constant:*

$$K_{eq} = \frac{[B]}{[A]} \tag{6.9}$$

In more complex systems such as

$$C + D \rightleftharpoons M + N \tag{6.10}$$

$$K_{eq} = \frac{[M][N]}{[C][D]} \tag{6.11}$$

or for the general reaction

$$aA + bB + cC + \cdots \rightleftharpoons xX + yY + zZ + \cdots \tag{6.12}$$

$$K_{eq} = \frac{[X]^x[Y]^y[Z]^z \cdots}{[A]^a[B]^b[C]^c \cdots} \quad (6.13)$$

The dimensions of K_{eq} will vary depending upon the number of chemical species on each side of the reversible reaction.

Problem 6.1: Express K_{eq} with the appropriate units in examples (a) to (c).

Solution: (a) $A \rightleftharpoons B + C$:

$$K_{eq} = \frac{[B] \text{ moles/liter} \times [C] \text{ moles/liter}}{[A] \text{ moles/liter}}$$

Cancel out units and K_{eq} will be in moles per liter (mole liter^{-1}).

(b) $2A \rightleftharpoons B$

$$K_{eq} = \frac{[B] \text{ moles/liter}}{[A]^2 \text{ (moles/liter)}^2}$$

Cancel out units and K_{eq} will be in 1/(moles/liter), i.e., liter/mole or liter mole^{-1}.

(c) $A + B \rightleftharpoons 2C + 2D$

$$K_{eq} = \frac{[C]^2 \text{ (moles/liter)}^2 \times [D]^2 \text{ (moles/liter)}^2}{[A] \text{ moles/liter} \times [B] \text{ moles/liter}}$$

K_{eq} will be expressed in moles^2/liter^2 (mole^2 liter^{-2}).

Problem 6.2: The condensation of compound A with B to give compound C proceeded to equilibrium, at which point the concentrations were: A, 2.5 mM; B, 3.3 mM; and C, 0.03 mM. What is K_{eq}?

Solution: (a) Define the K_{eq} for the reversible reaction. Write the equation

$$A + B \rightleftharpoons C$$

and then

$$K_{eq} = \frac{[C]}{[A][B]}$$

(b) Substitute the experimental values, expressed in moles per liter

$$K_{eq} = \frac{0.03 \times 10^{-3} \text{ mole liter}^{-1}}{2.5 \times 10^{-3} \text{ mole liter}^{-1} \times 3.3 \times 10^{-3} \text{ mole liter}^{-1}}$$

Cancel the numbers and the units

$$K_{eq} = \frac{3 \times 10^{-2}}{2.5 \times 3.3 \times 10^{-3} \text{ mole liter}^{-1}}$$

$$= 3.6 \text{ liter mole}^{-1}$$

Calculations involving chemical equilibria must fulfill the following prerequisites.

1. The concentrations of all the components in the system are expressed in moles per liter since tables of equilibrium constants are by convention derived using these units.

2. By convention, the product of the concentrations of components on the right side, as the reversible reaction is written, is divided by the product of the concentration of components on the left side to calculate the equilibrium constant. If Equation (6.4) were written

$$B \rightleftharpoons A$$

then

$$K'_{eq} = \frac{[A]}{[B]} \tag{6.14}$$

and clearly

$$K_{eq} = \frac{1}{K'_{eq}} \tag{6.15}$$

Both K'_{eq} and K_{eq} are proper equilibrium constants, but one must know which is product and reactant in each case. The unqualified statement that the equilibrium constant for the reversible isomerization reaction involving glyceraldehyde-3-phosphate and dihydroxyacetone phosphate is 25 does not inform the reader as to which of the two compounds is in excess at equilibrium. Actually the equilibrium constant 25 refers to the reaction written as

$$\underset{\text{(Gly-3-P)}}{\text{glyceraldehyde-3-P}} \rightleftharpoons \underset{\text{(DHAP)}}{\text{dihydroxyacetone phosphate}} \tag{6.16}$$

where

$$K_{eq} = \frac{[\text{DHAP}]}{[\text{Gly-3-P}]} \tag{6.17}$$

3. Do not include solids or solvent concentrations in the calculation, since these remain essentially constant. We may consider their concentration as unity.

4. Equilibrium constants for reactions of gases may be calculated from measurements of their partial pressures (usually given in millimeters of mercury in biological systems). In solution, the partial pressures of gaseous reactants are converted to molar concentrations (see Chapters 5 and 8).

5. In equilibrium reactions involving the growing chain of a biopolymer, the concentration of this component does not change.

With the above points in mind we may approach problems involving equilibria in the following orders of steps.

1. Write the reversible reaction so that the K_{eq} can be defined:

$$A + B \rightleftharpoons C + D \tag{6.18}$$

2. Write the expression for K_{eq} (see 2 above), putting solid reactants and solvent concentrations equal to 1 (see 3 above):

$$K_{eq} = \frac{[C][D]}{[A][B]} \tag{6.19}$$

3. Convert all given concentrations into moles per liter.

4. Initial concentrations of the reactants and products are changed to equilibrium concentrations by allowing x moles of the reactants to be converted to x moles of the products. The stoichiometry of the reaction will permit this relationship to be drawn, and as a check it is clear that a material balance between the initial and equilibrium conditions will be maintained.

For example, in the above equation the initial concentrations in mole liter^{-1} are $[A]_i$, $[B]_i$, $[C]_i$, and $[D]_i$. Let the change in the concentration of [A] be x moles; then at equilibrium

$$[A]_{eq} = [A]_i - [x]$$

If x moles of A react then from the equation x moles of B must react with them. Similarly, x moles of C and x moles of D must be formed. The equilibrium concentrations will therefore be

$$[A]_{eq} = [A]_i - [x] \qquad [C]_{eq} = [C]_i + [x]$$

$$[B]_{eq} = [B]_i - [x] \qquad [D]_{eq} = [D]_i + [x]$$

and

$$K_{eq} = \frac{([C]_i + [x])([D]_i + [x])}{([A]_i - (x])([B]_i - [x])} \tag{6.20}$$

This will result in a quadratic equation for which either a positive or negative value for x might satisfy the conditions. A negative value for x would indicate that the reaction moved to the left in going to equilibrium. The initial conditions of the experiment will tell if this is possible or not.

5. Check the result for material balance and the K_{eq} value.

Problem 6.3: In the reaction $A \rightleftharpoons B$ there is three times as much A as B at equilibrium. What is K_{eq}?

Solution: Since the reaction is an isomerization with no change of molecular weight, the molarity of 1 liter of reactants would be

$$[\mathrm{A}] = \frac{3x}{M} \text{ mole liter}^{-1} \qquad [\mathrm{B}] = \frac{x}{M} \text{ mole liter}^{-1}$$

where x is the weight of B, and M is the molecular weight of A and B. Then

$$K_{eq} = \frac{[\mathrm{B}]}{[\mathrm{A}]}$$

$$= \frac{\dfrac{x}{M} \text{ mole liter}^{-1}}{\dfrac{3x}{M} \text{ mole liter}^{-1}}$$

$$= \frac{x}{M} \cdot \frac{M}{3x}$$

or

$$K_{eq} = \tfrac{1}{3}$$

Problem 6.4: Isocitrate at pH 7.0 splits into two products, glyoxylate and succinate, in a biochemical system. The reaction is reversible and K_{eq} is 0.07 mole liter^{-1}. How much glyoxylate and succinate will be present at equilibrium, starting from a one-molar solution of pure isocitrate?

Solution: isocitrate $\rightleftharpoons$ glyoxylate + succinate

$$K_{eq} = \frac{[\text{glyoxylate}][\text{succinate}]}{[\text{isocitrate}]}$$

Since equimolar amounts of the two products will be present [(glyoxylate] = [succinate]),

$$K_{eq} = \frac{[\text{glyoxylate}]^2}{[\text{isocitrate}]}$$

When we substitute values,

$$\frac{[\text{glyoxylate}]^2}{[1.0\text{-glyoxylate}]} = 0.07 \text{ mole liter}^{-1} \qquad \text{(i)}$$

As a first approximation assume that the amount of glyoxylate formed is small; therefore [1.0-glyoxylate] $\approx$ 1.0 and

$$[\text{glyoxylate}] = \pm\sqrt{0.07}\ M$$

$$= \pm\sqrt{7 \times 10^{-2}}\ M$$

$$= +0.265\ \mathrm{m}M$$

The value of [glyoxylate] cannot be negative because there was none present in the initial reaction solution.

Without the above approximation, cross-multiplying (i) and collecting like terms gives

$$[\text{glyoxylate}]^2 + 0.07[\text{glyoxylate}] - 0.07 = 0$$

Solving this quadratic equation for [glyoxylate]:

$$[\text{glyoxylate}] = \frac{-0.07 \pm \sqrt{(0.07)^2 + 4 \times 0.07}}{2} M$$

$$= \frac{-0.07 \pm 0.534}{2} M$$

$$= \frac{+0.464}{2} \quad \text{or} \quad \frac{-0.604}{2} M$$

Rejecting the negative root for reasons given before, (see page 27)

$$[\text{glyoxylate}] = [\text{succinate}] = 0.232\ M$$

Problem 6.5: Equilibrium between monosodium glucose-1-phosphate (G-1-P) and monosodium glucose-6-phosphate (G-6-P) is attained when there is 5% by weight of G-1-P and, by difference, 95% by weight of G-6-P. What is the equilibrium constant?

Solution: Set up an equation so that K_{eq} will be defined. We will choose

$$\text{G-1-P} \rightleftharpoons \text{G-6-P}$$

and therefore

$$K_{eq} = \frac{[\text{G-6-P}]}{[\text{G-1-P}]}$$

Convert the chemical components into the same concentration units, moles per liter.

It will be noted that the two components are chemical isomers and therefore have the same molecular weight. In such simple cases as this K_{eq} is calculated directly from weight units, so long as these are the same for each component. Furthermore, the unit volume need not be taken as the liter so we may conveniently take the relative weight percentages as our concentrations. Therefore

$$K_{eq} = \frac{95}{5} = 19$$

Problem 6.6: In the phosphorolysis of starch $(C_6H_{10}O_5)_x$ with inorganic phosphate (represented by P_i) to form degraded starch $(C_6H_{10}O_5)_{x-1}$ and glucose-1-phosphate (G-1-P), at pH 7.0, the equilibrium constant is 0.32 for

$$(C_6H_{10}O_5)_x + P_i \rightleftharpoons (C_6H_{10}O_5)_{x-1} + \text{G-1-P}$$

$$K_{eq} = 0.32$$

$$K_{eq} = \frac{[(C_6H_{10}O_5]_{x-1}][\text{G-1-P}]}{[(C_6H_{10}O_5)_x][P_i]}$$

K_{eq} is independent of the concentrations of starch or degraded starch. If the $[P_i]$ was initially 10 mM, what is the maximum concentration of [G-1-P]?

Solution: The maximum amount of [G-1-P] will be present at equilibrium. If the concentration of P_i is reduced by x M at equilibrium, then [G-1-P] $= x$ M, since

$$K_{eq} = \frac{[\text{G-1-P}]}{[P_i]} = 0.32$$

Substituting, we have

$$\frac{x}{1 \times 10^{-2} - x} = 0.32$$

Rearranging, we obtain

$$x = (3.2 \times 10^{-3}) - 0.32x$$

$$1.32x = 3.2 \times 10^{-3}$$

or

$$x = 2.42 \times 10^{-3}\ M$$

Maximum [G-1-P] = 2.42 mM. (Checking: Residual $P_i = (10.0 - 2.42) = 7.58$ mM and therefore $K_{eq} = 2.42/7.58 = 0.32$).

COUPLED REACTIONS

An important extension of the role of reversible reactions in biochemistry relates to the common situation in which two or more reversible reactions are occurring in the same system with at least one component common to each. Such reactions are said to be coupled. It is perhaps true to say that all the reactions in a living cell are related to each other through coupled reactions, and for this reason a change in any component may be reflected in all the others. However, since the cell is not a closed system (see Chapter 10), compensating mechanisms react to the change so that a steady-state equilibrium is maintained. The point to remember is that when a coupled system is at equilibrium the concentrations of the components in each reversible reaction satisfy the same ratio as they would have when occurring alone; the equilibrium constants do not change. To take a simple general case for illustration,

$$\text{A} \rightleftharpoons \text{B} \rightleftharpoons \text{C}$$

This reaction is composed of two reversible reactions,

$$\text{A} \rightleftharpoons \text{B}$$

for which

$$K_{A \rightleftharpoons B} = \frac{[B]}{[A]}$$

and

$$B \rightleftharpoons C$$

for which $K_{B \rightleftharpoons C} = [C]/[B]$. Hence $[B] = [A]K_{A \rightleftharpoons B}$ and $[B] = [C]/K_{B \rightleftharpoons C}$. Therefore

$$\frac{[C]}{[A]} = K_{A \rightleftharpoons B} \times K_{B \rightleftharpoons C} \tag{6.21}$$

In general, the ratio of concentrations of the final product and initial reactant is the product of all the intermediate equilibrium constants.

The Law of Mass Action enables one to appreciate a system of coupled reactions. A reaction at equilibrium, $A \rightleftharpoons B$, can be forced to convert A into B by either adding more A from the system $X \rightleftharpoons A$ (Equation (6.22)), or drawing off some of the B by $B \rightleftharpoons C$. In either event the system will shift to attempt to re-establish equilibrium. The increased conversion of A to B can be brought about by coupling,

$$X \rightleftharpoons A \rightleftharpoons B \rightleftharpoons C \tag{6.22}$$

if the equilibrium $X \rightleftharpoons A$ favors A or $B \rightleftharpoons C$ favors C. Here, if the conversion of $A \rightleftharpoons B$ strongly favors A, the couples with X or C will add to the formation of B. By visualizing the coupled systems in this manner it is easier to figure whether they have overall equilibrium constants of the right order.

Problems involving coupled reactions usually involve the following steps.

1. Unless the question makes a specific statement, choose a direction for the overall reaction.

In the reactions written as

$$X \rightleftharpoons Y \qquad K_{eq} = 0.5 \tag{i}$$

$$X \rightleftharpoons Z \qquad K_{eq} = 1.5 \tag{ii}$$

since the couple is going to be through the common intermediate X, one or other of the reactions must be written in the reverse manner. To end up with an overall coupled reaction with K_{eq} greater than one, it is suggested that the reaction to be reversed be the one with the smallest K_{eq}:

$$Y \rightleftharpoons X \tag{i'}$$

2. When a reaction is written in a reverse manner, correct to the new equilibrium constant:

$$\mathrm{Y} \rightleftharpoons \mathrm{X} \qquad K_{eq} = \frac{1}{0.5} = 2 \qquad \text{(i}'\text{)}$$

3. Add the properly written reactions together and multiply their equilibrium constants. Check that the units of each K_{eq} are proper:

$$\mathrm{Y} \rightleftharpoons \mathrm{X} \qquad K_{eq} = 2 \qquad \text{(i}'\text{)}$$

$$\mathrm{X} \rightleftharpoons \mathrm{Z} \qquad K_{eq} = 1.5 \qquad \text{(ii)}$$

$$\mathrm{X} + \mathrm{Y} \rightleftharpoons \mathrm{Z} + \mathrm{X} \qquad K_{eq} = 2 \times 1.5 \qquad \text{(i}'\text{)} + \text{(ii)}$$

That is, $\mathrm{Y} \rightleftharpoons \mathrm{Z}$ with $K_{eq} = 3.0$

Problem 6.7: The phosphorolysis of sucrose with inorganic phosphate to form free fructose and glucose-1-phosphate has an equilibrium constant of 0.05 at pH 6.6 and 35°C:

$$\frac{[\text{sucrose}][\text{Pi}]}{[\text{fructose}][\text{G-1-P}]} = 0.05 \qquad \text{(i)}$$

The reaction favors phosphorolysis. As was calculated in an earlier problem, phosphoglucomutase enzyme catalyzes the equilibrium

$$\text{G-1-P} \rightleftharpoons \text{G-6-P}, \qquad K_{eq} = 19$$

Commencing with sucrose, inorganic phosphate (P_i), fructose and G-1-P, each at 1-millimolar concentration, what is the equilibrium concentration of G-6-P in the coupled system?

Solution: If a is the fraction of sucrose split at equilibrium (this value must be less than one), then from a materials balance

$$[\text{G-1-P}] + [\text{G-6-P}] = 1 + a \qquad \text{(ii)}$$

However,

$$\frac{[\text{G-1-P}]}{[\text{G-6-P}]} = \frac{1}{19} \qquad \text{(iii)}$$

and therefore

$$[\text{G-6-P}] = 19[\text{G-1-P}] \qquad \text{(iv)}$$

Substituting for [G-6-P] from (iv) in (ii),

$$[\text{G-1-P}] + 19[\text{G-1-P}] = 1 + a$$

$$[\text{G-1-P}] = \frac{1 + a}{20}$$

$$[\text{G-6-P}] = \frac{19(1 + a)}{20}$$

From the original equilibrium constant expression (i),

$$\frac{[1-a][1-a]}{\underline{[1+a][1+a]}} = 0.05$$
$$20$$

and therefore

$$\frac{[1-a]^2}{[1+a]^2} = \frac{1}{400}$$

Cross-multiplying, we have

$$400 - 800a + 400a^2 = 1 + 2a + a^2$$

Collecting terms, we obtain

$$399a^2 - 802a + 399 = 0$$

Therefore

$$a = \frac{802 \pm \sqrt{802^2 - 4 \times 399^2}}{2 \times 399} = \frac{802 \pm \sqrt{643204 - 636804}}{798}$$

$$= \frac{802 \pm 80.0}{798} = \frac{722}{798} = 0.905$$

(The second root of the equation is impossible, giving $a > 1$.) Therefore

$$\text{G-6-P} = \frac{19}{20} \times 1.905$$

If we had started the reaction with one mmole of each reactant, then at equilibrium

G-6-P = 1.810 mmole
G-1-P = 0.095 mmole
sucrose = 0.095 mmole
P_i = 0.095 mmole
fructose = 1.905 mmole

sum = 4.000 mmole (material balance checks)

Problem 6.8: What is the overall equilibrium constant at 38°C and pH 7.0 for the glucokinase reaction?

Given:

$$\text{G-6-P} + H_2O \rightleftharpoons \text{glucose} + P_i, \qquad K_{eq} = 1.30 \times 10^2 \tag{i}$$

$$H_2O + \text{ATP} \rightleftharpoons \text{ADP} + P_i, \qquad K_{eq} = 4.30 \times 10^5 \tag{ii}$$

Solution:

(a) The glucokinase reaction is

$$\text{glucose} + \text{ATP} \rightleftharpoons \text{G-6-P} + \text{ADP} \tag{iii}$$

The required direction is thus chosen.

(b) Reverse reaction (i)

$$\text{glucose} + P_i \rightleftharpoons \text{G-6-P} + H_2O, \qquad K_{eq} = \frac{1}{1.30 \times 10^2} = 7.71 \times 10^{-3} \qquad \text{(i}'\text{)}$$

(c) Add (i′) and (ii)

$$\text{glucose} + P_i + \text{ATP} + H_2O \rightleftharpoons \text{G-6-P} + \text{ADP} + P_i + H_2O$$

or reaction (iii).

(d) Multiply the constants for (i′) and (ii):

$$4.30 \times 10^5 \times 7.71 \times 10^{-3} = 3.31 \times 10^3$$

Problem 6.9: Given the coupled systems

$$\text{G-1-P} \rightleftharpoons \text{G-6-P} \rightleftharpoons \text{Fru-6-P}$$

$$K_{eq} = 19 \qquad K_{eq} = 0.5$$

and starting with 1 mmole of G-6-P, what are the amounts of G-1-P and Fru-6-P at equilibrium?

Solution: From the coupled reactions,

$$\frac{[\text{Fru-6-P}]}{[\text{G-1-P}]} = 19 \times 0.5 = 9.5 \qquad \text{(i)}$$

If y mmole of G-6-P react to give G-1-P and x mmole of G-6-P give Fru-6-P, then $1 - x - y$ of G-6-P is remaining. From (i),

$$\frac{x}{y} = 9.5$$

$$x = 9.5y$$

$(1 - 10.5y)$ of G-6-P is remaining and substituting the values

$$\frac{[\text{G-6-P}]}{[\text{G-1-P}]} = \frac{1 - 10.5y}{y} = 19$$

Therefore

$$1 - 10.5y = 19y$$

Collecting terms, we have

$$1 = 29.5y$$

$$y = \frac{1}{29.5}$$

That is, G-1-P = 0.034 mmole. Therefore

$$x = \frac{9.5}{29.5},$$

That is, Fru-6-P = 0.322 mmole, and

$$\text{G-6-P} = 1 - \frac{9.5}{29.5} - \frac{1}{29.5} = 1 - \frac{10.5}{29.5}$$

$$= \frac{19}{29.5} = 0.644 \text{ mmole}$$

Does the material balance check?

CONVENTIONS FOR WRITING EQUILIBRIUM CONSTANTS

Equilibrium constants expressions can be formulated in several ways each of which is valid and convenient for a particular situation. To illustrate this point we use the hydrolysis reaction of ethyl acetate as described by the equation

$$CH_3COOCH_2CH_3 + H_2O \rightleftharpoons CH_3COOH + CH_3CH_2OH \quad (6.23)$$

$$K_1 = 0.33 = \frac{[CH_3COOH][CH_3CH_2OH]}{[CH_3COOCH_2CH_3][H_2O]} \quad (6.24)$$

For dilute reactants in aqueous solution the water concentration is essentially constant. One can then formulate a new equilibrium constant.*

$$K_2 = 55 \times 0.33 = 18 = \frac{[CH_3COOH][CH_3CH_2OH]}{[CH_3COOCH_2CH_3]} \quad (6.25)$$

When this constant is used, the water concentration is excluded even though it is a reactant. It is valid when all reactants are at low concentration in aqueous solution. This formulation is often used in biochemical systems as this solvent condition is generally satisfied.

The acetic acid formed as a result of hydrolysis—see equation (6.23)—in the highly aqueous solution will ionize for which the following equilibrium constant can be written

$$K_3 = \frac{[CH_3COO^-][H^+][CH_3CH_2OH^-]}{[CH_3COOCH_2CH_3]} = 18 \times 1.8 \times 10^{-5}$$

$$= 3.2 \times 10^{-4} \quad (6.26)$$

---

*This is a change of standard state from 1 *M* water to pure liquid. See the discussion in Chapter 10.

This is a coupling of ionization with the hydrolysis reaction (K_a for acetic acid is 1.8×10^{-5}).

These two equilibrium constants, given by (6.25) and (6.26), are pH independent and hold when the acid-base species indicated are taken into account.

For biochemical systems it is convenient to specify a particular reference pH, usually 7, and formulate a pH dependent equilibrium constant as given by Equation (6.27).*

$$K_4 = \frac{[CH_3COO^-][CH_3CH_2OH]}{[CH_3COOCH_2CH_3]} = K_3 \times 10^7 \tag{6.27}$$

The equilibrium constant can also be expressed in terms of total concentrations of reactants in Equation (6.28).

$$K_5 = \frac{[CH_3COO^- + CH_3COOH][CH_3CH_2OH]}{[CH_3CH_2COOCH_2CH_3]} \cong K_3 \times 10^7 \tag{6.28}$$

Since the acid in this example is completely ionized at pH 7, K_4 and K_5 are equal. This formulation is restricted to a single pH, in this case 7, and dilute solute in aqueous solution. It refers back to Equation (6.25).

These different formulations of equilibrium constants correspond to different values of the standard free energy (see Chapter 10 for more discussion). This is really an introduction to standard states in terms of equilibrium constants.

pK_{eq}

As in the case of any other value or constant, such as $[H^+]$, the equilibrium constant can be expressed as its negative logarithm to the base 10:

$$pK_{eq} = -\log K_{eq} \tag{6.29}$$

where K_{eq} is the equilibrium constant calculated from the concentrations of the reacting species expressed as molarities. In such a form it is valuable for later calculations and has the advantage of abbreviating the scientific notation in which K_{eq} is usually expressed.

Problem 6.10: Express the equilibrium constant 2.0×10^3 as pK_{eq}.

*This procedure is analogous to changing the standard state for hydrogen ion from 1 *M* to 10^{-7} *M*. (See Chapter 10 for more discussion).

Solution: $pK_{eq} = -\log K_{eq}$

$= -\log 2.0 \times 10^3 = -3.30$

Problem 6.11: Express the equilibrium constant 3.09×10^{-5} as pK_{eq}.

Solution: $pK_{eq} = -\log K_{eq}$

$= -\log 3.09 \times 10^{-5} = -(0.4900 - 5.0000)$

$= 4.51$

REFERENCES

A. J. Bard, *Chemical Equilibrium.* Harper & Row, New York, 1966.

P. H. Carnell and R. N. Reusch, *Molecular Equilibrium.* Saunders, Philadelphia, 1963. (A programmed course.)

T. R. Hogness and W. C. Johnson, *Qualitative Analysis and Chemical Equilibrium.* Holt, New York, 1964.

K. B. Morris, *Principles of Chemical Equilibrium.* Reinhold, New York, 1965.

---

6. PROBLEMS A* Answers on page 340

1. From your knowledge of biochemistry, what reaction would be required to couple reaction (i) with reaction (ii)?

$$\text{glutamate} + \text{pyruvate} \rightleftharpoons \alpha\text{-ketoglutarate} + \text{alanine} \quad \text{(i)}$$

$$\text{oxalacetate} + \text{GTP} \rightleftharpoons \text{phosphoenolpyruvate} + CO_2 + \text{GDP} \quad \text{(ii)}$$

2. At 38°C the early reactions of glucose metabolism involve the equilibria

$$\underset{18 \text{ mmole}}{\text{D-Fructose 6-P}} \underset{}{\overset{K_{eq_1}}{\rightleftharpoons}} \underset{38 \text{ mmole}}{\text{D-Glucose 6-P}} \overset{K_{eq_2}}{\rightleftharpoons} \underset{2 \text{ mmole}}{\text{D-Glucose 1-P}}$$

In one experiment the amounts of each sugar phosphate in the equilibrium mixture is shown. Calculate K_{eq_1}, K_{eq_2} and K_{eq_3} for the overall reaction written below.

$$\text{D-Glucose 1-P} \overset{K_{eq_3}}{\rightleftharpoons} \text{D-Fructose 6-P}$$

3. The reaction

$$\text{citrate} \rightleftharpoons \textit{iso}\text{-citrate}$$

reaches equilibrium when the proportions of the two substances are 92.7 mole % citrate and 7.3 mole % *iso*-citrate. Calculate the pK_{eq}.

---

*Unless otherwise stated, the pH should be assumed to be 7.0.

4. D-Fructose 6-P + ATP $\rightleftharpoons$ D-Fructose 1,6-diphosphate + ADP

The equilibrium constant is 7.8×10^3.

(a) What is the pK_{eq}?

(b) If the initial concentration of fructose 6-P was 6 mM and at equilibrium the molar ratio of ADP/ATP was 3.9×10^3 what would be the molarities of the two sugar phosphates at equilibrium?

5. The enzyme, mutarotase, catalyzes the interconversion of α-D-glucose and β-D-glucose. If the β-D-anomer is present at equilibrium to an extent of 64% of the mixture, what is the equilibrium constant and pK_{eq}? Define the direction of the equilibrium for your constants.

6. The enzyme enolase catalyzes the reaction

$$\text{2-Phospho-D-glycerate} \rightleftharpoons \text{2-phospho}\textit{enol}\text{pyruvate} + H_2O$$

$$K_{eq} = 0.2$$

How much 2-phospho*enol*pyruvate would you need to produce 32 mg of 2-phospho-D-glycerate?

7. uracil + D-ribose-1-phosphate $\rightleftharpoons$ uridine + P_i

$$K_{eq} = 11 \text{ at pH 7.0 and 37°C}$$

Starting with 1.5 mmole of uridine and 2.5 mmole of P_i dissolved in a total volume of 5 ml, what will be the amounts of the other two components at equilibrium?

8. D-glucose + ATP $\rightleftharpoons$ D-glucose 6-phosphate + ADP

$$K_{eq} = 4 \times 10^3 \text{ at 38°C}$$

$$\text{ADP + creatine P} \rightleftharpoons \text{ATP + creatine}$$

$$K_{eq} = 10 \text{ at 38°C.}$$

(a) Write the coupled reaction and give the K_{eq} value.

(b) If the coupled reactions at equilibrium had a mole ratio of [creatine]/[creatine P] equal to 3/1, what would be the mole ratios of [G-6-P]/[G] and [ATP]/[ADP]?

9. The conversion of carbohydrates to triglycerides involves the following reactions:

D-fructose 1,6-diphosphate $\rightleftharpoons$

D-glyceraldehyde 3-phosphate + dihydroxyacetone phosphate (i)

$$K_{eq} = 1.0 \times 10^{-4}$$

dihydroxyacetone phosphate + $NADH_2^+$ $\rightleftharpoons$

L-glycerol phosphate + NAD^+ (ii)

$$K_{eq} = 2.0 \times 10^6$$

$$\text{D-glyceraldehyde 3-phosphate} + NAD^+ + P_i \rightleftharpoons \text{1,3-diphospho-D-glycerate} + NADH_2^+ \quad \text{(iii)}$$

$$K_{eq} = 1.0 \times 10^{-3}$$

(a) Does the coupling of these reactions greatly favor the formation of L-glycerol phosphate?
(b) What is the overall K_{eq}?
(c) What reaction can be coupled to reaction (iii) to increase the concentration of L-glycerol phosphate in the coupled system?
(d) If the equilibrium constant for the proposed reaction in (c) is 3.2×10^3 what would be the new overall K_{eq}?

10. The equilibrium constant for the decarboxylation of sodium oxalacetate to CO_2 and pyruvate is 4.6×10^{-4} liter mole^{-1}. What is the concentration of pyruvate at equilibrium, commencing with a 5 mg% solution of sodium oxalacetate?

11. (a) Couple the reactions and give the overall K_{eq} value of

$$\text{glutamate} + \text{pyruvate} \rightleftharpoons \alpha\text{-ketoglutarate} + \text{alanine} \qquad K_{eq} = 1.5$$

$$\text{glutamate} + \text{oxalacetate} \rightleftharpoons \alpha\text{-ketoglutarate} + \text{aspartate} \qquad K_{eq} = 6.7$$

(b) If at equilibrium the mole ratio of pyruvate/oxalacetate is 2, what will be the mole ratio of the amino acids alanine and aspartate?

12. Couple the following reactions and give the pK_{eq} value for the overall reaction:

$$\text{citrate} \rightleftharpoons \text{oxalacetate} + \text{acetate} \qquad K_{eq} = 2.0$$

$$\text{citrate} \rightleftharpoons cis\text{-aconitate} + H_2O \qquad K_{eq} = 0.07$$

$$iso\text{-citrate} \rightleftharpoons cis\text{-aconitate} + H_2O \qquad K_{eq} = 0.50$$

13. The following two reactions are occurring together in solution:

$$\text{ATP} + \text{3-phosphoglycerate} \rightleftharpoons \text{ADP} + \text{1,3-diphosphoglycerate}$$

$$K_{eq} = 3.1 \times 10^{-4}$$

$$\text{2-phosphoglycerate} \rightleftharpoons \text{3-phosphoglycerate}$$

$$K_{eq} = 4.5$$

(a) Write the coupled reaction and its K_{eq}.
(b) If in the equilibrated coupled system there are 3 mmole of 3-phosphoglycerate, calculate how many millimoles of 2-phosphoglycerate would be present.

6. PROBLEMS B* No answers given

1. $$\text{phosphoenolpyruvate} + \text{ADP} \rightleftharpoons \text{pyruvate} + \text{ATP} \quad \text{(i)}$$

$$K_{eq} = 2 \times 10^3 \text{ at } 30°\text{C}$$

$$\text{ATP} + \text{3-phosphoglycerate} \rightleftharpoons \text{ADP} + \text{1,3-diphosphoglycerate} \quad \text{(ii)}$$

$$K_{eq} = 3.1 \times 10^{-4} \text{ at } 30°\text{C}$$

(a) Write the coupled reaction and give the K_{eq} value.
(b) If the coupled reactions at equilibrium had a mole ratio of pyruvate/phosphoenolpyruvate equal to 2000/1, what would be the mole ratio of the two phosphoglycerates in Equation (ii)?
(c) What ratio of ATP/ADP must be maintained in the system in order to drive reaction (i) from right to left, assuming that by associated reactions the other reactants in (i) are maintained equimolar? What effect would this ratio of ATP/ADP have on reaction (ii)?

2. Given the reversible reaction

$$\text{uridine diphosphate galacturonic acid} \rightleftharpoons \text{uridine diphosphate glucuronic acid}$$

$$K_{eq} = 1.1$$

How much uridine diphosphate glucuronic acid would you need to start with in order to produce 50 mg of uridine diphosphate galacturonic acid? What is the pK_{eq} of the equilibrium?

3. Consider the reactions in the pentose pathway:

$$\text{ribose-5-phosphate} \rightleftharpoons \text{ribulose-5-phosphate} \qquad K_{eq} = 0.44$$

$$\text{xylulose-5-phosphate} \rightleftharpoons \text{ribulose-5-phosphate} \qquad K_{eq} = 1.49$$

(a) Write the coupled reaction and give the overall K_{eq} value.
(b) Starting with 5 mmole of ribose-5-phosphate, what would be the concentrations of each component at equilibrium if the reaction volume was 3 ml?

4. In the carboxylation reaction

$$\text{A} + CO_2 \rightleftharpoons \text{A—COO}^-$$

the pCO_2 at equilibrium was 0.3 atm over the aqueous reaction solution (5 ml), containing 0.5 mmole of A and 0.03 mmole of A—COO$^-$. What is K_{eq}? Take the Bunsen coefficient as 0.5 and consider the CO_2 concentration as the total amount in all forms in the water.

5. The following represents an important biochemical reaction without the couple with coenzyme A:

*Unless otherwise stated, the pH should be assumed to be 7.0.

$$\text{oxalacetate} + \text{acetate} \rightleftharpoons \text{citrate}$$

$$K_{eq} = 0.5 \text{ at pH } 7.0$$

(a) What would be the units of K_{eq}?
(b) What is the value of pK_{eq}?
(c) What would be the molarity of the acetate at equilibrium if the initial reaction solution were 4.2 mM in citrate and 5.6 mM in oxalacetate?

6. An alcohol and a carboxylic acid react in a nonaqueous medium to produce the corresponding ester and water. All products and reactants are soluble. At equilibrium the amounts of the components are shown in the equation

$$\underset{0.58\ \text{mmole}}{\text{ROH}} + \underset{0.58\ \text{mmole}}{\text{R'COOH}} \rightleftharpoons \underset{1.0\ \text{mmole}}{\text{R'CO}\cdot\text{OR}} + \underset{1.0\ \text{mmole}}{H_2O}$$

The total volume is 18 ml (neglect volume changes).
(a) What is the equilibrium constant?
(b) What would be the corresponding constant in which the alcohol and acid were the products?
(c) How much ester would remain after 6 ml of water was added to the above equilibrium mixture and a new equilibrium established?
(d) If the water was totally insoluble in the medium, what would be the percent hydrolysis of the ester?

7. As seen in Problem 6.4, the value of K_{eq} in the common equilibrium expression (i) or (ii),

$$\frac{x^2}{a - x} = K_{eq} \qquad \text{(i)}$$

$$x = \frac{-K_{eq} \pm K_{eq}\sqrt{1 + \dfrac{4a}{K_{eq}}}}{2} \qquad \text{(ii)}$$

where x is the variable and a is a constant, can be approximated by ignoring x in the denominator in (i). If a is set equal to one, plot the error of the above approximation against $(1 - x)$ and find a value of a/K above which x produces an error of less than 1%. This is discussed further by Meeks.*

8. $$\text{polynucleotide}_n + H_2O \xrightleftharpoons{\text{specific enzyme}} \text{polynucleotide}_{n-1} + \text{nucleotide}$$

A specific enzyme catalyzes the hydrolysis of the terminal 3′-nucleotide in a polynucleotide. For dilute aqueous solution what would be the apparent equilibrium constant?

9. An enzyme catalyzes the reduction of fumarate to succinate with molecular hydrogen at pH 7.0. Under an atmosphere of hydrogen, 760 mm, and at 37°C (Bunsen coefficient 0.01666), the equilibrium concentrations

*F. R. Meeks, *J. Chem. Ed.* **42**:609 (1965).

were 0.53 M in fumarate and 4.01 mM in succinate. Calculate K_{eq} in liter mole^{-1}.

10. In the reaction

$$\text{L-malate} + \text{NAD}^+ \rightleftharpoons \text{oxalacetate} + \text{NADH} + \text{H}^+ \qquad \text{(i)}$$

the apparent equilibrium constant $K_{eq} = 2.33 \times 10^{-5}$ at pH 7.0 and 22°C.
(a) In the presence of 10 mM L-malate and a molar ratio of [NAD/NADH] equal to 3, how concentrated must the solution be in terms of oxalacetate in order to have no reaction?
(b) Couple reaction (i) with the following reaction (ii), giving the overall reaction and its K_{eq} value, using proper units:

$$\text{oxalacetate} + \text{acetate} \rightleftharpoons \text{citrate} \qquad \text{(ii)}$$

$$K_{eq} = 0.5 \text{ at pH } 7.0$$

(c) What is the pK_{eq} of the overall reaction?

11. The following reactions center upon the metabolism of pyruvate.

$$\text{phosphoenolpyruvate} + \text{ADP} \rightleftharpoons \text{pyruvate} + \text{ATP}$$

$$K_{eq} = 2.0 \times 10^3$$

$$\text{pyruvate} + \text{CO}_2 + \text{ATP} \rightleftharpoons \text{oxalacetate} + \text{ADP} + P_i$$

$$K_{eq} = 4.1 \times 10^3$$

$$\text{oxalacetate} + \text{GTP} \rightleftharpoons \text{phosphoenolpyruvate} + \text{CO}_2 + \text{GDP}$$

$$K_{eq} = 0.37$$

Comment on the coupling of these reactions.

12. The following reactions are involved in glycogenolysis.

$$\text{D-G-6-P} + \text{H}_2\text{O} \rightleftharpoons \text{Glucose} + P_i \qquad K_{eq} = 2.1 \times 10^2 \text{ at pH } 7.0$$

$$\text{D-G-6-P} \rightleftharpoons \text{G-1-P} \qquad K_{eq} = 19 \text{ at pH } 7.0$$

$$(\text{glycogen}) + P_i \rightleftharpoons \text{G-1-P} + (\text{glycogen}) \qquad K_{eq} = 0.32 \text{ at pH } 7.0$$

What are the relative concentrations of the glucose phosphates and glucose at equilibrium?

7

Weak Acid and Bases; Buffers

DISSOCIATION OF WEAK ACIDS AND WEAK BASES

Of particular interest in the class of reversible reactions are those involving the hydrogen ion. These include both weak acids and weak bases which have been defined in general by Brønsted as follows:

$$\underset{\text{acid}}{HA} \rightleftharpoons H^+ + \underset{\text{conjugate base}}{A^-} \tag{7.1}$$

$$\underset{\text{base}}{B} + H^+ \rightleftharpoons \underset{\text{conjugate acid}}{BH^+} \tag{7.2}$$

An acid (Brønsted) can ionize to form a proton and its conjugate base, whereas a base can react with a proton to form its conjugate acid. Thus acetate ion is the conjugate base of acetic acid and acetic acid is the conjugate acid of acetate; the couple HAc and Ac^- are called a conjugate acid-base pair:

$$CH_3COOH \rightleftharpoons H^+ + CH_3COO^- \tag{7.3}$$

The incomplete dissociations of these acids and bases distinguish them from the strong acids such as HCl and HNO_3, and from the strong bases such as sodium hydroxide. Clearly there will be intermediate examples and the demarcation may be placed at those acids and bases with equilibrium constants at 10^{-2} and 10^{-12}, respectively. In this context the equilibrium constants are called dissociation constants and are represented as K_a or K_b.

Problem 7.1: The pK_a values of acetic acid and trichloroacetic acid are 4.74 and 0.7, respectively. Which is the stronger acid and what are the corresponding dissociation constants?

Solution: From Equation (6.23), p$K_a = -\log K_a$, or

$$\log K_a - \mathrm{p}K_a$$

For acetic acid,

$$\log K_a = -4.74$$

Since we cannot have a negative mantissa, this must be rearranged:

$$\begin{aligned}\log K_a &= \bar{5}.00 + 0.26\\ K_a &= \text{antilog } \bar{5}.00 + \text{antilog } 0.26\\ &= 10^{-5} \times 1.82\end{aligned}$$

Similarly, for trichloroacetic acid:

$$\begin{aligned}\mathrm{p}K_a &= -\log 0.7\\ K_a &= \text{antilog } (\bar{1}.00 + 0.300)\\ &= 10^{-1} \times 2.00\end{aligned}$$

Thus trichloroacetic acid, $K_a = 2.00 \times 10^{-1}$, is more completely dissociated and thus a stronger acid than acetic acid, $K_a = 1.82 \times 10^{-5}$.

Since most biochemical reactions take place in aqueous solutions, it is always necessary to consider the ionization of water:

$$H_2O \rightleftharpoons H^+ + OH^- \tag{7.4}$$

At ordinary temperatures and pressures the concentration of undissociated water is constant and very large (1000/18 moles liter^{-1}). We can consider this constant value to be incorporated into the equilibrium constant, which is identified as K_w:

$$\begin{aligned}K_w &= [H^+] \cdot [OH^-] \qquad (7.5)\\ &= 1.0 \times 10^{-14} \text{ at } 25°\text{C}\end{aligned}$$

This is the basis of the Sørenson pH scale of 0 – 14 (see Chapter 5). No matter what the concentration of H^+ and OH^-, which may be contributed by ionizations of acids or bases, the following relationship holds under normal conditions:

$$[H^+] \times [OH^-] = 1.0 \times 10^{-14}$$

which in logarithmic form is

$$\log [H^+] + \log [OH^-] = -14 \tag{7.6}$$

and since

$$pH = -\log [H^+] \tag{7.7}$$

$$pOH = -\log [OH^-] \tag{7.8}$$

then

$$pH + pOH = 14 \tag{7.9}$$

Problem 7.2: What is the $[OH^-]$ of 0.1 N HCl solution?

Solution: The $[H^+]$ of the strong acid is

$$0.1\ N\ \text{HCl} = 10^{-1}\ M$$

or

$$pH = -\log [H^+] = 1$$

Therefore

$$pOH = 14 - 1 = 13$$

$$[OH^-] = 10^{-13}\ M$$

As stated earlier, the dissociation constant of a weak acid, K_a, may be expressed as pK_a. Similarly, for a weak base there is a corresponding value K_b and its derivative, pK_b. Consider the weak organic base RNH_2 which reacts with the water, acting as a weak acid, to generate hydroxyl ions and the conjugate acid of the weak base RNH_3^+; thus

$$RNH_2 + H_2O \rightleftharpoons RNH_3^+ + OH^- \tag{7.10}$$

The equilibrium constant of this reaction is designated K_b and is a measure of the extent to which OH^- is formed in aqueous solution:

$$K_b = \frac{[RNH_3^+][OH^-]}{[RNH_2]} \tag{7.11}$$

Substituting $\frac{K_w}{[H^+]}$ for $[OH^-]$ from Equation (7.5),

$$K_b = \frac{[RNH_3^+]K_w}{[RNH_2][H^+]} \tag{7.12}$$

Since the conjugate acid RNH_3^+ can ionize as a weak acid as follows,

$$RNH_3^+ \rightleftharpoons H^+ + RNH_2 \tag{7.13}$$

we can write

$$K_a = \frac{[H^+][RNH_2]}{[RNH_3^+]} \tag{7.14}$$

Then from Equations (7.12) and (7.14),

$$K_b = \frac{K_w}{K_a} \tag{7.15}$$

$$K_a K_b = K_w \tag{7.16}$$

$$pK_a + pK_b = pK_w = 14 \tag{7.17}$$

Problem 7.3: The K_b for ammonia is 1.8×10^{-5} mole liter^{-1}. What is the K_a value?

Solution: The reaction is written

$$\underset{\text{free base}}{NH_3} + H_2O \rightleftharpoons \underset{\substack{\text{protonated}\\\text{base}}}{NH_4^+} + OH^-$$

$$K_b = \frac{[NH_4^+][OH^-]}{[NH_3]}$$

The ammonium ion, dissociating as an acid, is written

$$NH_4^+ \rightleftharpoons H^+ + NH_3$$

$$K_a = \frac{[H^+][NH_3]}{[NH_4^+]}$$

Since, from (7.16),

$$K_a K_b = K_w = 1.0 \times 10^{-14}$$

$$K_a = \frac{1.0 \times 10^{-14}}{K_b}$$

then substituting the values given

$$K_a = \frac{1.0 \times 10^{-14}}{1.8 \times 10^{-5}} = 0.55 \times 10^{-9}$$

$$= 5.5 \times 10^{-10}$$

By Equations (7.15) or (7.17) it is possible to put the relative ionizations of acids and protonated bases onto the same scale, expressed as K_a or

pK_a. This is particularly valuable for molecules such as amino acids, which have acidic and basic groups.

Problem 7.4: The ionization constants for glycine, CH_2NH_2COOH, are pK_a 2.35 for the —COOH group and pK_b 4.22 for the —NH_2 group. What are the ionization constants for each expressed as reactions involving ionization of the proton?

Solution: Converting to the same type of ionization

$$pK_b = 4.22 = 14.00 - pK_a$$

or

$$pK_a = 14.00 - 4.22 = 9.78$$

Consider glycine in its completely protonated form as a dibasic acid:

$$\begin{matrix} CH_2NH_3^+ \\ | \\ COOH \end{matrix} \underset{pK_a\ 2.35}{\rightleftharpoons} H^+ + \begin{matrix} CH_2NH_3^+ \\ | \\ COO^- \end{matrix} \underset{pK_a\ 9.78}{\rightleftharpoons} H^+ + \begin{matrix} CH_2NH_2 \\ | \\ COO^- \end{matrix}$$

From Equation (6.23),

$$K_a \text{ (COOH group)} = \text{antilog } (-pK_a) = \text{antilog } (-2.35)$$
$$= \text{antilog } (\bar{3}.00 + 0.65) = 10^{-3} \times 4.47$$

Similarly,

$$K_a \text{ (NH}_3^+\text{ group)} = \text{antilog } pK_a = \text{antilog } (-9.78)$$
$$= \text{antilog } (\overline{10}.00 + 0.22) = 10^{-10} \times 1.70$$

The dissociation constants of the simple weak acids and bases in water may be determined experimentally by preparing solutions of known concentration and measuring the resulting pH. The equilibrium reaction may again be represented:

$$HA \rightleftharpoons H^+ + A^-$$

and

$$K_a = \frac{[H^+][A^-]}{[HA]} \tag{7.18}$$

Assume a total of c moles of pure acid added to 1 liter of water. Since a material balance must be maintained, which in this case requires that the total number of dissociated A^- and undissociated HA remain constant, then

$$c = [HA] + [A^-] \tag{7.19}$$

The ion product constant, K_w, is assumed to be that of pure water, 1×10^{-14}. It is also necessary to maintain electrical neutrality by balancing the positively and negatively charged species, or

$$[H^+] = [A^-] + [OH^-] \tag{7.20}$$

The contribution of $[OH^-]$ in the acidic solution will be very small. For example, in 0.1 M acetic acid the pH is 2.8, which means that $[H^+]$ is 1.6×10^{-3} mole per liter, and the pOH is 11.2, or $[OH^-]$ is 6.3×10^{-12} mole per liter. It is reasonable, therefore, to consider as a first approximation

$$[H^+] = [A^-] \tag{7.21}$$

and by substituting for $[A^-]$ from (7.19) in (7.21)

$$c = [HA] + [H^+]$$

or

$$[HA] = c - [H^+] \tag{7.22}$$

Substituting Equations (7.21) and (7.22) in (7.18),

$$K_a = \frac{[H^+]^2}{(c - [H^+])} \tag{7.23}$$

Since $[H^+]$ can be determined on a carefully standardized pH meter and c is known, then K_a can be calculated. When K_a is known, the pH can be predicted for any concentration. By rearranging (7.23),

$$[H^+]^2 + K_a[H^+] - K_a c = 0$$

and therefore

$$[H^+] = \frac{-K_a \pm \sqrt{K_a^2 + 4K_a c}}{2} \tag{7.24}$$

In many cases the concentration of the pure acid in the aqueous solution is so much greater than the $[H^+]$ that, by further approximation,

$$c - [H^+] \approx c \tag{7.25}$$

Equation (7.23) thus simplifies to

$$K_a = \frac{[H^+]^2}{c} \tag{7.26}$$

or

$$[H^+] = \sqrt{K_a c} \tag{7.27}$$

The derivations of Equations (7.23), (7.24), and (7.27), although given in most biochemistry textbooks, are repeated here to stress their approximations and limitations, which are the following.

1. The ion product constant, K_w, is assumed to be that of pure water. This may not be valid in concentrated solutions, in particular where there are high concentrations of other solutes, such as urea. These situations are difficult for other reasons, too, but for these the reader is referred to more advanced texts.

2. The contribution of $[OH^-]$ was ignored in the charge balance. As the pH approaches neutrality, this approximation becomes more significant.

3. The most serious approximation for the everyday situation is the derivation of Equation (7.27) from (7.23). In very dilute solutions, particularly of acids in which the pK_a approaches 2, this approximation is not valid.

Problem 7.5: Calculate the pH of a 0.001 M solution of dichloroacetic acid, K_a 5.0×10^{-2}, by Equations (7.24) and (7.27).

Solution: By Equation (7.24),

$$[H^+] = \frac{-5.0 \times 10^{-2} \pm \sqrt{(5.0 \times 10^{-2})^2 + 4 \times 5.0 \times 10^{-2} \times 10^{-3}}}{2}$$

$$= \frac{-5.0 \times 10^{-2} \pm \sqrt{25 \times 10^{-4} + 20 \times 10^{-2} \times 10^{-3}}}{2}$$

$$= \frac{-5.0 \times 10^{-2} \pm \sqrt{25 \times 10^{-4} + 2.0 \times 10^{-4}}}{2}$$

$$= \frac{-5.0 \times 10^{-2} \pm \sqrt{27 \times 10^{-4}}}{2}$$

$$= \frac{-5.0 \times 10^{-2} \pm 5.2 \times 10^{-2}}{2}$$

$$= \frac{-10.2 \times 10^{-2}}{2} \quad \text{or} \quad \frac{+0.2 \times 10^{-2}}{2}$$

$[H^+]$ can only be positive and so equals 0.1×10^{-2} mole liter^{-1}:

$$\text{pH} = -\log [H^+]$$

$$= -\log 10^{-3} = 3.0$$

By Equation (7.27),

$$[H^+] = \sqrt{5.0 \times 10^{-2} \times 10^{-3}}$$

$$= \sqrt{5.0 \times 10^{-5}} = \sqrt{50.0 \times 10^{-6}}$$

$$= 7.1 \times 10^{-3} \text{ mole liter}^{-1}$$

$$\text{pH} = -\log 7.1 \times 10^{-3} = -(0.85 - 3.00)$$

$$= 2.15$$

Note the difference of 0.85 pH units by the use of two methods in this example.

Common Ion Effect and Buffers. By Le Châtelier's Principle, the equilibrium of the dissociation of a weak acid or weak base will be shifted by the addition of a common ion or alternatively the removal of one of the components. As given by Equation (7.1),

$$\text{HA} \rightleftharpoons \text{H}^+ + \text{A}^-$$

Thus if H^+ is removed by titration with alkali, then more HA will dissociate until no undissociated molecules remain. It follows that the amount of alkali required to titrate the same volume of all acid solutions of equal normality is the same. Such a titration measures the *total acidity* of a solution.

Problem 7.6: 0.1 *M* HCl has a pH 1.0. 0.1 *M* acetic acid has a pH 2.8. What volume of 0.1 *N* NaOH would be required to titrate 10 ml aliquots of each acid solution to their respective end points?

Solution: Since both HCl and CH_3COOH are monovalent acids, equimolar solutions would have the same total acidity and would, therefore, require equal volumes of 0.1 *N* NaOH; in fact, 10 ml of 0.1 *N* NaOH.

If to the system in Equation (7.1) some anion, A^-, were to be added in the form of a salt, such as the completely ionized sodium salt, then the reaction would be driven to the left and the $[H^+]$ would decrease. In other words, the addition of sodium acetate to a solution of acetic acid would be similar to titrating partially the acid with sodium hydroxide, since both procedures would reduce $[H^+]$. Consider therefore the mixture of a weak acid (HA) and its salt (XA) of a strong base in aqueous solution. The weak acid partially dissociates and the ions mix with those from the salt: $\boxed{X^+} + \boxed{A^-}$ so that $HA \rightleftharpoons H^+ + \boxed{X^+} + A^- + \boxed{A^-}$. It is of prime importance to relate the total anion concentration $[A^-]$ to $[H^+]$ and $[HA]$. For the pure acid, as given by Equation (7.18),

$$K_a = \frac{[\text{H}^+][\text{A}^-]}{[\text{HA}]}$$

or

$$K_a = [\mathrm{H}^+] \times \frac{[\mathrm{A}^-]}{[\mathrm{HA}]} \tag{7.28}$$

As has been seen above, the amount of a pure weak acid that dissociates in solution is quite small. For 0.1 *M* acetic acid, pK_a 4.74, only 1.6% of the molecules dissociate. Thus the concentration [HA] is given approximately by the total weight of pure acid that was initially measured into the water. By the same token, the contribution of [A^-] by the weak acid will be small compared with that from the salt, so that the [A^-] may be taken as that of the salt initially added to the water. Equation (7.28) may therefore be expressed

$$K_a = [\mathrm{H}^+] \times \frac{[\text{salt}]}{[\text{acid}]} \tag{7.29}$$

or

$$\log K_a = \log\, [\mathrm{H}^+] + \log \frac{[\text{salt}]}{[\text{acid}]} \tag{7.30}$$

but since pK_a = −log K_a, from Equation (6.23), and pH = −log [H^+], from Equation (7.7), the rearrangement of (7.30) gives

$$-\log\, [\mathrm{H}^+] = -\log\, K_a + \log \frac{[\text{salt}]}{[\text{acid}]}$$

or

$$\mathrm{pH} = \mathrm{p}K_a + \log \frac{[\text{salt}]}{[\text{acid}]} \tag{7.31}$$

This relationship, known as the *Henderson-Hasselbalch* equation, has been derived here to emphasize the assumptions and limitations.

Because the amount of acid that dissociates was assumed to be small, it follows that Equation (7.31) applies best to the weaker acids. It cannot be used with any confidence for acids with pK_a < 2.0. The condition must also be met that little anion is contributed by the acid added, and this is not true when trying to calculate the initial course of the titration of a weak acid with a strong base, or a weak base with a strong acid (see below).

In a similar way, an equivalent expression to (7.31) can be derived for a base from the equilibrium constant for the ionization of the salt of the base:

$$K_a = \frac{[\mathrm{B}][\mathrm{H}^+]}{[\mathrm{HB}^+]} \tag{7.32}$$

where [B] is the concentration of free base and $[HB^+]$ is the concentration of the protonated base. Then

$$pH = pK_a + \log \frac{[B]}{[HB^+]} \tag{7.33}$$

In other words, we are considering the system as the ionization of an acid, which is the protonated base HB^+. The Henderson-Hasselbalch equation (7.31), or (7.33), thus fits all weak acids and weak bases.

The conjugate acid-base pairs of weak acids or weak bases act, by virtue of Le Châtelier's Principle, as buffers to a change in hydrogen ion concentration. The pH of the buffer is related to the ratio of salt to acid by Equation (7.31) and it will be seen that, since the relative amounts of the salt and acid must be expressed in the same units, their ratio is a dimensionless pure number. The pH of the buffer solution is not determined by its concentration at our level of approximation. We consider that a buffer will have the same pH if it is diluted with water. Some buffers do indeed show little concentration dependence, while others are quite sensitive. However, as will be seen later (p. 121) the buffer capacity is proportional to concentration.

Problem 7.7: A solution of 0.05 *M* acid, pK_a 6.10, is mixed with an equal volume of a 0.1 *M* solution of its sodium salt. What is the pH of the final mixture?

Solution: Since equal volumes of the two solutions were mixed, the final concentration of the acid is 0.025 *M* and the salt is 0.05 *M*.
The molar ratio of the concentrations of salt to acid is

$$\frac{[\text{salt}]}{[\text{acid}]} = \frac{0.05}{0.025} = 2.0$$

From a knowledge of this ratio and the pK_a value we take equation (7.31):

$$\begin{aligned} pH &= pK_a + \log \frac{[\text{salt}]}{[\text{acid}]} \\ &= 6.10 + \log 2.0 = 6.10 + 0.30 \\ &= 6.40 \end{aligned}$$

Problem 7.8: If 10 ml of 0.09 *M* acid, when mixed with 20 ml of 0.15 *M* potassium salt of that acid, gave a solution with a pH of 5.85, what is the pK_a for the acid?

Solution: Total volume of the final mixture is 30 ml. Therefore, the final concentration of acid = $10/30 \times 0.09\ M = 0.03\ M$, and the final concentration of salt = $20/30 \times 0.15\ M = 0.10\ M$. Thus

$$\frac{[\text{salt}]}{[\text{acid}]} = \frac{0.10\ M}{0.03\ M} = 3.33$$

From Equation (7.31)

$$5.85 = \text{p}K_a + \log 3.33$$
$$= \text{p}K_a + 0.52$$

When we rearrange,

$$\text{p}K_a = 5.33$$

Problem 7.9: How many moles of sodium acetate and acetic acid are required to prepare 1 liter of a buffer, pH 5.0, which is 0.1 M in total available acetate (dissociated and undissociated). Acetic acid has a pK_a of 4.74.

Solution: From Equation (7.31)

$$5.0 = 4.74 + \log\frac{[\text{salt}]}{[\text{acid}]}$$

or

$$\log\frac{[\text{salt}]}{[\text{acid}]} = 0.26$$

The ratio of the concentration of sodium acetate to acetic acid is the antilog 0.26, or 1.82; there are 1.82 molecules of NaAc for each molecule of HAc. The mole fraction of salt is 1.82/2.82 and the mole fraction of acid is 1.00/2.82. We are told that the total molar concentration of acetate is 0.1 M (0.1 mole in 1 liter, which is the volume of buffer required). Therefore, of the 0.1 mole, the salt must provide its fraction,

$$\frac{1.82}{2.82} \times 0.1 \text{ mole} = 0.065 \text{ mole}$$

and the acid must provide its fraction,

$$\frac{1.00}{2.82} \times 0.1 \text{ mole} = 0.035 \text{ mole}$$

It will be noted that many buffer problems involve a determination of the ratio of the [salt] to [acid], each of which must be expressed in the same units, and the total of the [salt] and [acid] together. The general situation calls, therefore, for two simultaneous equations

$$\frac{[\text{salt}]}{[\text{acid}]} = a, \qquad [\text{salt}] + [\text{acid}] = b$$

It is essential in any buffer problem to identify from the given description the values of pK_a, pH, a, and b, which can be expressed in many ways. pK_a can be given as K_a or the equilibrium concentrations of the weak acid. pH can be given as $[H^+]$, pOH, or $[OH^-]$. [Salt] or [acid] may be derived by titration, but in all cases the four unknowns must be obtained from the given information or by calculation. The most comprehensive buffer problem asks for the preparation of a volume of buffer of a certain concentration and pH from a weak acid of given pK_a and a strong base. The general approach may be by the following steps.

1. Since pH and pK_a are given, determine the [salt]/[acid] ratio, which is a pure number.

2. The concentration of the buffer is usually expressed as the sum of the salt and acid forms.

If the salt (or the acid) concentration is given, then the corresponding acid (or salt) can be calculated directly from the ratio of [salt] to [acid] in step 1. If the total concentration, [salt] + [acid], is given, then the solution of the two simultaneous equations is required. Note that the unit of concentration of the salt and acid so calculated will be the one used for the [salt] + [acid] concentration. For example, if

$$\frac{[\text{salt}]}{[\text{acid}]} = 2 \qquad \text{(i)}$$

$$[\text{salt}] + [\text{acid}] = 0.3\ M \qquad \text{(ii)}$$

Then, since [salt] = 2[acid]

and by substitution in (i),

$$2[\text{acid}] + [\text{acid}] = 0.3\ M$$

$$[\text{acid}] = \frac{0.3}{3} M = 0.1\ M$$

Now, if we substitute this value in (i),

$$[\text{salt}] = 2 \times 0.1\ M = 0.2\ M$$

Check these values in (i) and (ii):

$$\frac{[\text{salt}]}{[\text{acid}]} = \frac{0.2\ M}{0.1\ M} = 2$$

$$[\text{salt}] + [\text{acid}] = 0.2\ M + 0.1\ M = 0.3\ M$$

3. From the concentrations of salt and acid the actual amounts of salt and acid can be calculated for the volume required from their molecular weights. In the most general case, where the given materials are pure acid

and pure base, the salt will be generated from equivalent amounts of acid and base; it is necessary to measure an amount of pure acid equivalent to salt and acid and add to it an amount of base equivalent to the salt.

In the example given in step 2 above, if 1 liter of buffer were required and if the acid were monobasic, then to 0.3 mole of acid would be added 0.2 mole of base and the volume brought to 1 liter. This would give 0.2 *M* salt and leave 0.1 *M* free acid.

An alternative approach to that developed above for the solution of problems involving equilibria is to use the *fraction of dissociation*, α, which is defined for any system as

$$\alpha = \frac{\text{moles of HA dissociated}}{\text{total moles of species containing A}}$$

$$= \frac{\text{moles HA dissociated}}{\mathrm{A_T}}$$

For the dissociation of a weak acid

$$\mathrm{HA} \rightleftharpoons \mathrm{H^+} + \mathrm{A^-}$$

Since

$$[\mathrm{A_T}] = [\mathrm{HA}] + [\mathrm{A^-}]$$

then

$$\alpha = \frac{[\mathrm{A^-}]}{[\mathrm{HA}] + [\mathrm{A^-}]}$$

This applies equally to the salt of a weak acid and, in such cases as buffer systems,

$$\frac{[\text{salt}]}{[\text{acid}]} = \frac{[\mathrm{A^-}]}{[\mathrm{HA}]} = \frac{\alpha}{1-\alpha}$$

Equation 7.31 therefore becomes

$$\mathrm{pH} = \mathrm{p}K_a + \log\frac{\alpha}{1-\alpha}$$

Problem 7.10: To 100 ml of 0.2 *M* formic acid, pK_a 3.77, is added 10 mmole of solid NaOH. What will be the final pH?

Solution: The given data are

$$pK_a = 3.77 \quad \text{(i)}$$

$$\text{The total millimoles of formic acid} = 0.2 \times \frac{100}{0.2 \times 100} \quad \text{(ii)}$$

$$= 20 \text{ mmole}$$

Because the salt is formed from the original acid present,

$$\text{salt} + \text{acid} = 20 \text{ mmole}$$

The ratio of [salt]/[acid] must be calculated:

10 mmole of NaOH will neutralize 10 mmole of formic acid

Therefore (20 − 10) mmole of acid will be converted to salt:

$$\text{salt} = 10 \text{ mmole in } 100 \text{ ml}$$

$$\text{acid} = 10 \text{ mmole in } 100 \text{ ml}$$

Therefore

$$\frac{[\text{salt}]}{[\text{acid}]} = \frac{100 \text{ m}M}{100 \text{ m}M} = 1$$

Substituting in Equation (7.31),

$$\text{pH} = 3.77 + \log 1$$

Since log 1 = 0,

$$\text{pH} = 3.77$$

Alternate solution:

Total concentration of formic acid and formate

$$= 20 \text{ mmole}$$

Total concentration of formate is that equivalent to the base added

$$= 10 \text{ mmole}$$

$$\therefore \alpha = \frac{\text{moles of formate}}{\text{total moles of formic acid and formate}}$$

$$= \frac{10}{20}$$

$$= 0.5$$

$$\therefore \text{pH} = 3.77 + \log \frac{0.5}{1 - 0.5}$$

$$= 3.77$$

Problem 7.11: The dissociation constant for lactic acid is 1.38×10^{-4}. How much lactic acid must be added to 2 g of NaOH to make 500 ml of buffer at pH 4.00?

Solution: We are given

$$\text{pH} = 4.00 \tag{i}$$

$$\begin{aligned} \text{p}K_a &= -\log 1.38 \times 10^{-4} \\ &= -(0.1399 - 4.00) \\ &= -(-3.86)\,(\text{to 2 decimals}) \\ &= 3.86 \end{aligned} \tag{ii}$$

$$\begin{aligned} \text{salt} &= \frac{2}{40} \times \frac{1000}{500} M \\ &= 0.10\ M\ \ (0.10\ N) \end{aligned} \tag{iii}$$

We do not know the acid concentration. When we substitute the known values in Equation (7.31),

$$4.00 = 3.86 + \log \frac{0.10}{[\text{acid}]}$$

or

$$0.14 = \log \frac{0.10}{[\text{acid}]}$$

Therefore

$$\frac{0.10}{[\text{acid}]} = 1.38$$

or

$$[\text{acid}] = \frac{0.10}{1.38} M = 0.0725\ M$$

Thus 500 ml requires 0.0363 mole of the free acid and to this must be added the amount to react with the 2 g of NaOH, or 0.05 mole. Therefore

$$[\text{salt}] + [\text{acid}] = (0.0363 + 0.05)\ \text{mole/500 ml.}$$

and the total acid needed is 0.0863 mole or 0.09 mole (2 significant figures).

Alternate solution:
As noted in the first solution

$$\text{p}K_a = 3.86$$

$$\begin{aligned} \text{Concentration of lactate} &= \text{concentration of NaOH} \\ &= 0.10\ M \end{aligned}$$

When we substitute the appropriate values

$$4.0 = 3.86 + \log \frac{0.10/[\mathrm{A_T}]}{1 - 0.10/[\mathrm{A_T}]}$$

$$0.14 = \log \frac{0.10}{([\mathrm{A_T}] - 0.10)}$$

$$\therefore\ 1.38 = \frac{0.10}{[\mathrm{A_T}] - 0.10}$$

$$\therefore\ 1.38[\mathrm{A_T}] - 0.138 = 0.10$$

$$[\mathrm{A_T}] = \frac{0.238}{1.38}\,M$$

$$= 0.1725\,M$$

But $[\mathrm{A_T}] = [\text{salt}] + \text{HA}$

$$\therefore\ [\text{lactic acid}] = (0.1725 - 0.10)\ M$$

$$= 0.0725\ M$$

The solution continues as in the first calculation and

$$\text{total acid needed} = 0.09 \text{ mole}$$

Problem 7.12: 0.10 *N* ammonium hydroxide, K_b 1.8×10^{-5}, is added to an equal volume of 0.03 *N* HCl. What is the resulting pH?

Solution: For the $NH_4^+ \rightleftharpoons NH_3 + H^+$ reaction, (i)

$$K_b = 1.8 \times 10^{-5}$$

Therefore

$$K_a = \frac{1.0 \times 10^{-14}}{1.8 \times 10^{-5}}$$

$$= 5.5 \times 10^{-10}$$

From Equation (6.23), $pK_a = -\log 5.5 \times 10^{-10} = 9.26$: $[\text{acid}] = 0.015\ M$, since the acid is the $[NH_4^+]$ ion produced by reacting a volume of 0.03 *N* HCl with an *equal* volume of ammonia and so diluting the acid to

one-half its normality, or 0.015 *M* (ii)

In this situation the [salt] is the ammonia concentration, as can be seen from (i):

$$[NH_3] = \frac{0.10 - 0.03}{2}\,M$$

by the same reasoning as used in (ii), or

$$[NH_3] = 0.035\ M$$

Therefore

$$pH = 9.26 + \log\frac{0.035}{0.015} = 9.26 + \log 2.333$$

$$= 9.26 + 0.37 \text{ (to 2 decimal places)}$$

$$= 9.63$$

TITRATION CURVES OF WEAK ACIDS AND WEAK BASES

To appreciate fully the problems of polyvalent acids and bases, titration of peptides and proteins, charges of polyions, buffer capacity and respiration chemistry, it is well to consider in detail the titration curves of acids and bases, in particular the weak types.

The pH of a 0.1 *N* strong acid (Figure 7.1), which would be completely ionized, pH 1.0, does not change dramatically upon adding strong base, except for the effect of dilution, until nearly all the hydrogen ions are neutralized. At this point there is an abrupt change to that approximating the pH of the diluted titrating base.

Figure 7.1

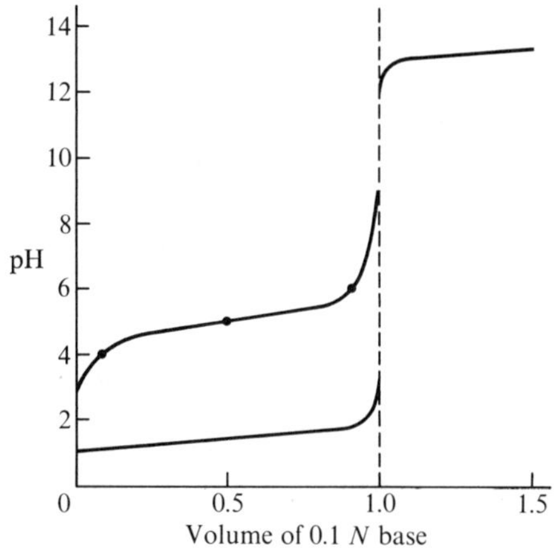

As could be calculated from the Henderson-Hasselbalch equation, the change in pH upon adding alkali to a weak acid is much less abrupt. The initial pH for a 0.1 *M* weak acid, pK_a 5.0, is approximately

$$pH = -\tfrac{1}{2}[\log K_a + \log c] \tag{7.34}$$

$$= \tfrac{1}{2}[-\log K_a - \log c]$$

If we substitute for $c = 0.1\ M$, pK_a 5.0,

$$pH = \tfrac{1}{2}[5.0 + 1.0]$$
$$= 3.0$$

When exactly equivalent amounts of base have been added, the pH is that due to the completely ionized salt, BX:

$$BX \rightleftharpoons B^+ + X^- \tag{7.35}$$

The anion acting as a base reacts with H_2O to give undissociated HX and an equivalent amount of OH^-:

$$X^- + H_2O \rightleftharpoons HX + OH^- \tag{7.36}$$

For the acid formed from reaction (7.36), rearrange Equation (7.18),

$$K_a = \frac{[H^+][X^-]}{[HX]}$$

to give

$$[H^+] = \frac{K_a[HX]}{[X^-]} \tag{7.37}$$

The amount of $[HX] + [X^-] \approx [X^-]$ and $[HX] = [OH^-]$. When we substitute in (7.19)

$$[H^+] = \frac{K_a[OH^-]}{[salt]} \tag{7.38}$$

When we substitute for $[OH^-]$ in (7.38) and rearrange,

$$[H^+]^2 = \frac{K_a K_w}{[salt]} \tag{7.39}$$

or logarithmically,

$$2 \log [H^+] = \log K_a + \log K_w - \log [salt] \tag{7.40}$$

Multiply (7.40) by -1

$$pH = \tfrac{1}{2}[pK_a + pK_w + \log [salt]] \tag{7.41}$$

As indicated above, for the condition of neutralization of the weak acid, pK_a 5.0, the salt concentration $[X^-]$ is 0.05 M. From (7.41),

$$pH = \tfrac{1}{2}[5.0 + 14.0 + \log 0.05]$$
$$= \tfrac{1}{2}[19.0 + \log (5.0 \times 10^{-2})]$$

$$= \tfrac{1}{2}[19.0 + (0.699 - 2.0)] = \tfrac{1}{2}[17.699]$$

$$= 8.85$$

It will be noted that a calculation of the pH of an aqueous solution of a weak acid or its salt could not be made from Equation (7.31).

As some of the acid is neutralized by the base to form the salt, so the pH changes with the ratio of [salt]/[acid], as given by Equation (7.31). For an acid, pK_a 5.0, some values of pH are calculated in Table 7.1 (see also Figure 7.1).

Table 7.1
Titration of weak acid, pK_a 5.0

Percent acid neutralized	$\frac{[salt]}{[acid]}$	*pH (calculated)*
~9%	1/10	(pK_a − 1) = 4.0
50%	1/1	(pK_a − 0) = 5.0
~91%	10/1	(pK_a + 1) = 6.0

The Henderson-Hasselbalch equation fits the experimental observations best in the range of one pH unit either side of the midpoint of the titration. Outside of these limits the assumptions made in its derivation are not valid.

The pH at the midpoint of the titration of a weak acid, at which the salt concentration is equal to the acid concentration, is also important because it is a direct experimental measure of the pK_a of the acid. By the same token, if the pK_a and concentration of an acid are given, then the titration curve can be quickly constructed from the calculation of the five points outlined in the example for Figure 7.1. This approach would also apply to weak bases titrated with strong acids, since we have seen earlier that the weak bases, expressed in the protonated form, can be considered as weak acids by the dissociation of that proton, as demonstrated in Problem 7.13:

$$BH^+ \rightleftharpoons B + H^+ \tag{7.42}$$

Problem 7.13: Construct the titration curve of 10 ml of 0.1 *M* methylamine, pK_a 10.72, with 0.1 *N* HCl.

Solution: If we remember that the $CH_3NH_3^+$ form is the "acid" and CH_3NH_2 is the "salt" in Equation (7.33), and use the pK_a scale, the five easily calculated points are the following.

(a) Free base:

$$pH = \tfrac{1}{2}[pK_a + pK_w + \log 0.1]$$

$$= \tfrac{1}{2}[10.72 + 14.00 + (-1.00)]$$

$$= 11.86$$

(b) Completely titrated (protonated) base:

$$\begin{aligned} pH &= \tfrac{1}{2}[pK_a - \log 5.0 \times 10^{-2}] \\ &= \tfrac{1}{2}[10.72 - (0.699 - 2.00)] \\ &= 6.01 \end{aligned}$$

(c) ~0.9 ml acid added: $pH = pK_a + 1 = 11.72$
(d) 5.0 ml acid added: $pH = pK_a = 10.72$
(e) ~9.1 ml acid added: $pH = pK_a - 1 = 9.72$

These values are shown in Figure 7.2. The graph indicates that the pH of the solution of the free base is greater than that calculated in (a) because it is moderately strong.

Figure 7.2

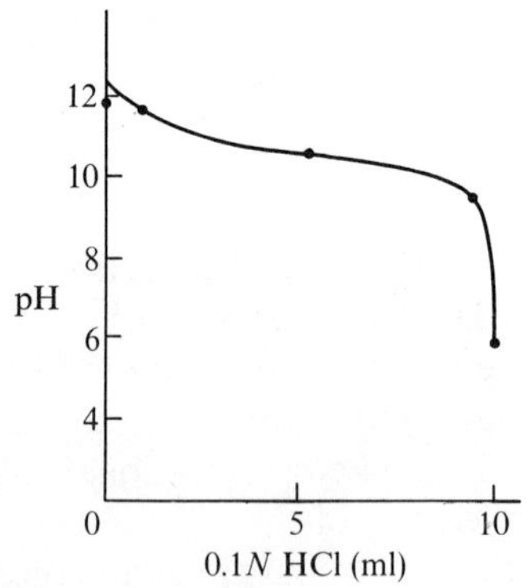

It will be immediately obvious that the calculations in (a) and (b) present anomalies due again to the limitations of the Henderson-Hasselbalch equation. At this extreme of pH the contribution of OH^- from the water is significant. The reader is referred to advanced physical chemistry texts for further consideration of this point, which will arise again at the other end of the scale when considering moderately strong acids.

BUFFERS AND BUFFER CAPACITY

The degree to which a solution resists a change in pH following the addition of acids or of bases varies greatly as can be seen in Figures 7.1 and 7.2. If the solution is buffered then the extent of this change is reduced because the buffer salts that are present resist the effect of the added acid or alkali. A buffer contains a weak acid and its salt.

A solution is acting best as a buffer at the point at which the rate of change of pH with the addition of $[H^+]$ or $[OH^-]$ to the solution is least, that is at the pK_a. When pH equals pK_a, there is an equal concentration of the salt of the weak acid and the acid; it follows that solutions of weak acids, or weak bases, or salts alone, will not have good buffering action.

The *buffering capacity,* β, has been defined by Van Slyke as the ratio of the gram equivalents per liter of strong base (ΔB) to the corresponding increment produced in the pH (ΔpH).

$$\beta = \frac{\Delta \text{B}}{\Delta \text{pH}} \tag{7.43}$$

The buffering capacities of solutions may be determined from titration curves by measuring ΔB and ΔpH at any segment. In effect this measures the average slope of the curve over the increments of ΔB and ΔpH taken. If these measurements are infinitesimally small then the slope at the point on the curve is its tangent and

$$\beta = \frac{d\text{B}}{d\text{pH}} \tag{7.44}$$

It is of interest to note the following points:

1. A solution has a buffer capacity of one if the addition of one mole equivalent of $[H^+]$ or $[OH^-]$ to one liter of buffer produces a change of 1.0 pH unit.

2. It can be shown* that the buffer capacity at constant pH is directly proportional to the buffer concentration.

3. As a consequence of (1) and (2) the buffer capacity is sometimes defined as the mole equivalents of $[H^+]$ or $[OH^-]$ that are required to change 1 liter of 1 *M* buffer by 1.0 pH unit.

Problem 7.14: The pH of 10 ml of 1 *M* buffer, pH 3.9, was decreased to pH 2.9 when 1.3 ml *N* HCl was added. What is the buffer capacity of the buffer at pH 3.9?

$$2.1 \text{ ml of } 1.3\ N \text{ HCl contains } (2.1/1000) \times 1.3 \text{ equiv. of } H^+$$

$$= 2.73 \text{ meq of } H^+$$

$$1000 \text{ ml of buffer would require } (1000/10) \times 2.73 \times 10^{-3} \text{ equiv. of } H^+$$

$$= 2.73 \times 10^{-1} \text{ equiv. of } H^+$$

Therefore, the buffer capacity is 0.273.

The same buffer, pH 3.9, at 0.5 *M* concentration would have a buffer capacity of 0.137 equivalent of H^+ per liter.

---

*See H. B. Bull, *Introduction to Physical Biochemistry.* Davis, Philadelphia, 1964, page 106.

4. When the pK_a of the acid and the ratio [salt]/[acid] in the buffer is known, the Henderson-Hasselbalch equation can be used to calculate the buffer capacity. As demonstrated in Problem 7.13, the addition of 4.1 ml 0.1 N HCl between points (c) and (d) produced a change of 1.0 pH unit

$$4.1 \text{ ml } 0.1\ N \text{ HCl} = \frac{4.1}{1000} \times 0.1 \text{ mole equivalent of } H^+$$

$$= 4.1 \times 10^{-4} \text{ mole equivalent of } H^+$$

This was added to 10.9 ml of buffer. 1000 ml of buffer would require $1000/10.9 \times 4.1 \times 10^{-4}$ equivalent of $H^+ = 3.76 \times 10^{-2}$ equivalent of H^+. Therefore the buffer capacity is 3.76×10^{-2}.

5. The buffer capacity is different in the acid direction from the alkaline direction except at the pK_a.

Table 7.2
The pK_a values of some weak acids in conjugate acid-base pairs

Acid	pK_{a_1}	pK_{a_2}	pK_{a_3}
Acetic	4.73		
[Ammonium]$^+$	9.26		
Barbituric	4.98		
Carbonic (in plasma, *see text*)	6.10	10.22	
Citric	3.13	4.76	6.40
[Diethylammonium]$^+$	11.10		
[Ethylenediammonium]$^{++}$	7.52	10.65	
Formic	3.77		
Fumaric	3.03	4.47	
Glutamic	2.30	4.30	9.70
Glycerophosphoric	1.47	6.19	
Glycine	2.35	9.78	
Histidine	1.82	6.10	9.20
Lactic	3.86		
Malic	3.40	5.05	
Malonic	2.85	6.10	
Nicotinic	4.85		
Oxalic	1.19	4.21	
Phenol	9.89		
Phosphoric	2.10	6.70	12.30
[Pyridinium]$^+$	8.64		
Succinic	4.19	5.57	
[TRIS]$^{+*}$	8.40		
[Xanthine]$^{++}$	7.53	11.63	
Uric	3.89		

* [Tris-(hydroxymethyl)-aminomethane · H]$^+$

REFERENCES

R. G. Bates, *Determination of pH*. Wiley, New York, 1964.

H. N. Christensen, *pH and Dissociation* (2nd ed.). Saunders, Philadelphia, 1964. (A programmed course)

J. T. Edsall and J. Wyman. *Biophysical Chemistry*. Academic Press, New York, 1958. Chapters 8 and 9.

7. PROBLEMS A

Answers on page 341

1. A 2% (w/v) solution of a monoprotonic acid, pK_a 5.2, had a pH of 3.5. What was the molecular weight of the acid?

2. A 0.01 M solution of a phenol is 0.05% ionized at 25°C. What is the pK_a of the acid?

3. What strength of solution of bartituric acid (K_a 1.05×10^{-5}) would have pH 5.20?

4. What strength of formic acid (K_a 1.76×10^{-4}) solution should be prepared to give a pH of 3.80?

5. What is the pH of a solution of 10 g of pyridine (K_a 2.3×10^{-9}) in 500 ml of solution?

6. Construct a titration curve of 10 ml of 2% aqueous phenol solution, using 0.2 N NaOH. What is the buffer capacity at pH 9.5?

7. Which solution would have a lower pH, 0.02% acetic acid or 0.005% lactic acid?

8. Which is the stronger acid—a one-tenth molar acid having a pH of 3 or one having a pH of 5? How much stronger is the stronger acid? What is the M concentration of hydrogen ion in each?

9. Which is stronger—a twentieth normal H_2SO_4 solution or a tenth normal acetic acid solution? Explain.

10. Which solution would have the lower pH—0.05% aqueous pyridine (K_a 2.3×10^{-9}) or 5 mM ethanediamine (K_b 2.8×10^{-5})?

11. (a) A 0.1 M solution of formic acid is 4.6% ionized at 25°C. Calculate the ionization constant of formic acid.
    (b) What would be the pH of a solution prepared by mixing 100 ml 0.1 M formic acid and 5.0 ml 1 N NaOH?

12. Make up one liter of 0.01 M sodium acetate buffer at pH 4.53. Calculate the grams of anhydrous sodium acetate (mol. wt. 82.04) and number of milliliters of glacial acetic acid (mol. wt. 60.05, density 1.049).

13. Calculate the pH of the following solutions:
    (a) 0.1 *M* tertiary amine R_3N, pK_a 8.0
    (b) 5.19×10^{-6} *M* lactic acid, K_a 1.38×10^{-4}
    (c) 0.137 *M* diethylamine in 1000 ml of water; K_b 1.26×10^{-3}
    (d) 2% aqueous phenol; K_a 1.3×10^{-10}
    (e) 17.6 mg/dl pyruvic acid; pK_a 2.50

14. What is the pH of a 0.1 *M* sodium acetate solution?

15. What is the pH and buffer capacity of a 0.03 *M* lactic acid solution to which has been added two volumes of 0.01 *M* sodium lactate?

16. Calculate the H^+ concentration and pH of a solution formed by adding 200 ml of 0.1 *M* acetic acid to 200 ml of 0.05 *N* sodium hydroxide.

17. A weak organic monobasic acid serves as an indicator. A color change is observed when 10% of the indicator has been converted to ions and at this point the pH of the solution is 10.4. What is the ionization constant of the indicator?

18. A weak acid has a dissociation constant of K_a 1.6×10^{-5}.
    (a) What is the pH of a 0.01 *M* solution of this acid?
    (b) To 200 ml of a 0.1 *M* solution of this acid are added 50 ml of 0.1 *M* NaOH and the solution made up to 1 liter. What is the hydrogen ion concentration and the pH of the final solution?

19. Prepare a titration curve of 10 ml 0.1 *N* NH_4OH (K_b 1.8×10^{-5}) with 0.1 *N* HCl. What is the buffer capacity at pH 8.0?

20. Calculate the weights of 80% (w/v) lactic acid solution (density 1.2 g ml^{-1}) and NaOH needed to prepare 1 liter of Na lactate buffer, pH 5.5, 0.05 *M* in Na^+.

21. If you added 27 ml of a 0.1 *N* acetic acid (K_a 1.85×10^{-5}) solution to 75 ml of 0.033 *N* NH_4OH solution, what would be the pH of the resultant solution?

22. Calculate the amounts of acetic acid and sodium hydroxide required to prepare 1 liter of buffer, pH 3.8, 0.05 *M* in acetate. K_a 1.82×10^{-5}.

23. If 2.3 g of sodium metal were added in small portions to one liter of a buffer of ammonium chloride, pH 10.5, and 1 *M* in NH_4^+, what would be the resulting pH?

24. If one mixed 100 ml of 0.1 *N* sodium acetate with 100 ml of 0.1 *N* acetic acid, the pH of the solution would be 4.7. If 80 ml of 0.1 *N* NaOH were added to this solution, what would be the pH?

25. How many grams of solid tris-(hydroxymethyl)-aminomethane (mol. wt. 121) and how many milliliters of 1 *N* HCl must one use to prepare 500 ml of Tris buffer 0.05 *M* of pH 7.6? pK_a 8.4.

26. Calculate the buffer capacity of 0.1 M barbiturate buffer at pH 5.3 using both 0.1 N HCl and 0.1 N NaOH.

---

7. PROBLEMS B No answers given

1. The pK_{a_2} values of D-fructose-6-P and D-glucose-6-P are 6.59 and 6.69, *respectively*.

$$\text{D-Fru-6-PO}_3^-\text{H} \rightleftharpoons \text{D-Fru-6-PO}_3^{--} + \text{H}^+ \qquad pK_a\ 6.59$$

$$\text{D-G-6-PO}_3^-\text{H} \rightleftharpoons \text{D-G-6-PO}_3^{--} + \text{H}^+ \qquad pK_a\ 6.69$$

The isomerase enzyme reacts with the completely ionized species and at pH 7.0 the apparent equilibrium constant is 0.5 for the following

$$\text{G-6-P} \rightleftharpoons \text{Fru-6-P}$$

$$K_{eq}\ 0.5$$

How does the equilibrium constant vary with pH (see page 93)?

2. Isotonic sodium lactate, pH 7.4, is commonly administered intravenously to combat metabolic acidosis.
   (a) How many milliliters of concentrated lactic acid (85% by weight, density 1.2) and how many grams of sodium hydroxide would be used to prepare 2500 ml of this solution?
   (b) If the sodium lactate in one liter of this solution were completely burned in the body, how many liters of oxygen at SC would be consumed?

3. You have a buffer, pH 3.5, of potassium formate which is 0.05 M with respect to formate (acid and salt).
   (a) What would be the pH of a mixture of equal volumes of this buffer and 0.02 M formic acid?
   (b) A salt gradient is required for ion exchange column chromatography. A constant-flow pump was delivering the potassium formate buffer, pH 3.5, 0.05 M, at 25 ml/hr to a flask that initially held 100 ml of 0.002 M potassium formate, pH 3.5, while at the same rate (25 ml per hour) the mixed solution was being pumped to the column. Assuming efficient mixing, what would be the rate of change of buffer concentration with time of the solution being delivered to the column, starting at zero time with the 0.002 M buffer?

4. In a study of lactic acid dehydrogenase it was necessary to analyze the kinetics by considering the effective concentration of undissociated lactic acid and lactate ion. Without making any approximations, such as were made in the derivation of the Henderson-Hasselbalch equation, calculate the concentrations of all components in a sodium lactate solution, pH 7.0, 0.1 M in sodium ion.

5. Calculate the ionic strength of ammonium acetate, pH 6.5, 0.1 *M* in acetate ion.

6. Construct a titration curve of 100 ml of 0.1 *N* lactic acid with 0.05 *N* NaOH; estimate the relative amounts of the two solutions present at pH 4.5 and then calculate what they should be, assuming the buffer to be 0.02 *M* in lactate (acid and salt).

7. A series of buffers of pH 5.2 is required for stepwise chromatography on DEAE-cellulose. The ionic strength should increase from 0.01 to 0.1 in two even steps. Ignoring the effect of ionic strength on the effective concentrations (activities) of the buffer components, calculate the composition of the three buffer solutions, in moles per liter, and give the grams of glacial acetic acid and anhydrous sodium acetate required.

8. Given two solutions, *A*, which is 1 *M*-HCl, and *B*, which is 1 *M*-tris-(hydroxymethyl)-aminomethane, how many milliliters of *B* should be mixed with 50 ml of *A* and brought to 1 liter in order to prepare buffers of pH 7.4, 8.0, and 8.6 all at 0.05 ionic strength? You are also given a solution of 1 *M* NaCl.

9. The apparent equilibrium constant for the reduction of acetaldehyde to alcohol by NADH is 1.71×10^5 at pH 7. What is the true equilibrium constant? (See page 184.)

10. A phenylated fatty acid was fed to an Irish setter and from her urine a crystalline acid metabolite was isolated. A sample weighing 142 mg was dissolved in water and titrated with 0.05 *N* NaOH. After 5.28 ml of alkali was added, the pH was 3.50; 15.84 ml was required to completely titrate the acid to a phenolphthalein end point (pH 8.3). What were the equivalent weight, the dissociation constant, and the probable formula of this acid metabolite?

11. It is necessary to conduct an enzymatic hydrolysis of one liter of 0.01 *M* butyryl monoglyceride at pH 7.3–7.4 in the complete absence of free Cu^{++} ion. The reaction will go to completion; K_a butyric acid 1.5×10^{-5}. What buffer might be chosen?

8

Polyprotonic Acids and Bases; Buffers; Complex Ions

DISSOCIATION OF POLYPROTONIC ACIDS AND BASES

Many of the acids in biochemical systems have more than one ionizable proton and are said to be polyprotonic. The equilibrium constants for each of these protons is different and, to the first order of approximation, each ionizing species may be considered as a separate acid that will dissociate completely before the next proton is removed. When the pK_a values differ by two or more units the error of this assumption is less than 1%. For example, phosphoric acid has three ionizable protons, the first ionization being represented as

$$H_3PO_4 \rightleftharpoons H_2PO_4^- + H^+ \qquad (8.1)$$

$$pK_{a_1}\ 2.1$$

We will assume that all the H_3PO_4 goes to $H_2PO_4^-$ before any of this monovalent anion ionizes. The complete ionization is, therefore, in three discrete steps, the last two being

$$H_2PO_4^- \rightleftharpoons HPO_4^{--} + H^+ \qquad (8.2)$$

$$pK_{a_2}\ 6.7$$

$$HPO_4^{--} \rightleftharpoons PO_4^{---} + H^+ \tag{8.3}$$

$$pK_{a_3}\ 12.3$$

The various points may be illustrated by calculating the titration curve of phosphoric acid, considering the particular case of 10 ml of 0.1 *M* H_3PO_4 being titrated with 0.1 *N* NaOH. The first ionization constant, pK_a 2.1, represents a moderately strong acid and the starting pH would be nearly 1.0. A meaningful value cannot be found from the approximate Equation (7.27),

$$[H^+] = \sqrt{K_{a_1} c}$$

and we should calculate the pH more accurately from Equation (7.24). When 5 ml of 0.1 *N* NaOH has been added, the pH is approximately 2.1 and half the original phosphoric acid is present as the monosodium salt. The conversion is complete when one equivalent of alkali, 10 ml of 0.1 *N* NaOH, has been added. The pH, calculated as before for salts of weak acids, would be, from (7.40),

$$\begin{aligned} pH &= \tfrac{1}{2}[pK_a + pK_w + \log [\text{salt}]] \\ &= \tfrac{1}{2}[2.1 + 14.0 + \log [0.05]] \\ &= 7.4 \end{aligned}$$

Clearly this is not possible because the second pK_a is 6.7, the pH where half of the $H_2PO_4^-$ is neutralized to HPO_4^{--}. The discrepancy in the calculated pH of the solution of NaH_2PO_4 results because the parent acid is moderately strong and is thus appreciably ionized in solution.

In the construction of the titration curve of polyprotonic acids, such as H_3PO_4, it is better to consider only those points within one pH unit of the pK_a values and to join such segments by a series of S-shaped curves, as in Figure 8.1. From this curve one would approximate the pH of the NaH_2PO_4 solution as 4.5 rather than the 7.4 calculated.

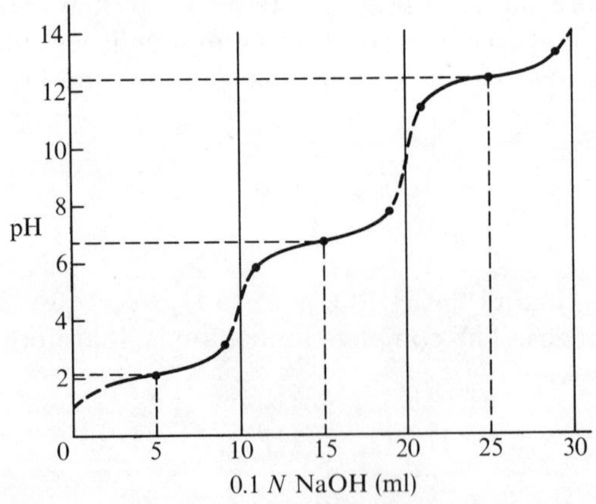

Figure 8.1

The approximation is equally poor for calculating the pH of Na_3PO_4 solutions which involve the pK_a of HPO_4^{--}, a very weak acid; that is, its conjugate base PO_4^{---} is a strong base. In contrast the calculated pH of Na_2HPO_4 using Equation (7.40) closely approximates that indicated in Figure 8.1. The pK_a used in the calculations is that of $H_2PO_4^-$, a weak acid. It is clear, therefore, that Equation (7.40) applies only to the salts of weak acids and weak conjugate bases.

Problem 8.1: What is the titration curve of 10 ml 0.05 M glycine hydrochloride at pH 1.0 with 0.05 N NaOH?

Solution: As seen in an earlier problem, the two pK_a values for glycine are 2.35 and 9.78. From an inspection of Equation (7.31), at pH 1.0 nearly all the glycine would be completely protonated (student should check this point). The ionic species would be

$$\underset{A}{H_3^+NCH_2COOH} \underset{pK_{a_1}\ 2.35}{\rightleftharpoons} \underset{B}{H_3^+NCH_2COO^-} \underset{pK_{a_2}\ 9.78}{\rightleftharpoons} \underset{C}{H_2NCH_2COO^-}$$

At pH 2.35 half the glycine is in the form of A and half as B. This would require the addition of 5.0 ml of 0.05 N NaOH, since the complete titration of the first proton requires 10.0 ml of 0.05 N NaOH. Similarly, when [B] = [C] a further 5.0 ml of alkali must be added and the pH will be 9.78. From the Henderson-Hasselbalch equation the other points can be calculated (see Chapter 7 for methylamine):

$$\begin{aligned} &0.9 \text{ ml alkali added,} && \text{pH} = 2.35 - 1 \\ &9.1 \text{ ml alkali added,} && \text{pH} = 2.35 + 1 \\ &10.9 \text{ ml alkali added,} && \text{pH} = 9.78 - 1 \\ &19.1 \text{ ml alkali added,} && \text{pH} = 9.78 + 1 \end{aligned}$$

The S-segments symmetrically join the above curves (see Figure 8.2).

Figure 8.2

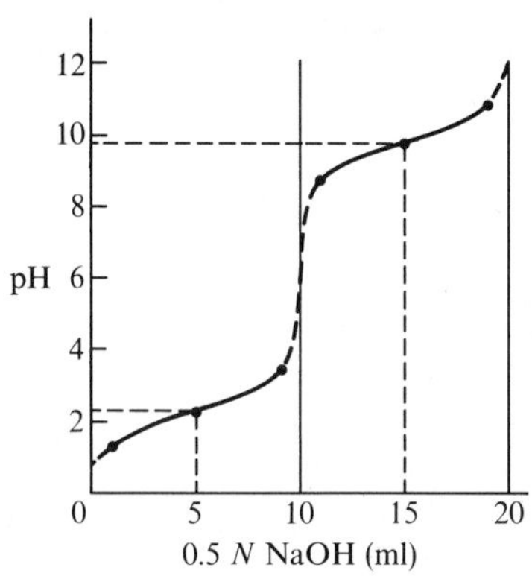

TITRATION CURVE OF PROTEINS AND PEPTIDES

The amino acid residues in polypeptide chains that present ionizable groups are listed in Table 8.1, together with the average pK_a values. It is appreciated that each of the individual groups will have pK_a values that are influenced by the environment. In the proteins of larger molecular weight these differences are difficult to measure or calculate and we can consider at this point the ionization values indicated, from which one can obtain an approximate titration curve from the chemical composition. To simplify the problem further, as in the case of the other polyprotonic acids, the ionizations are considered independent of each other, except for those groups that have close pK_a values. These are end and side chain carboxyl groups and ϵ-NH_2, phenolic —OH, and sulphydryl —SH groups.

Table 8.1
Average pK_a values for functional groups on proteins

Ionizable groups	*pK_a values*	
	Range	Average
α-COOH (end group)	3.0 to 3.6	3.3
δ or γ-COOH (Asp, Glu)	3.0 to 4.7	3.8
Imidazole (His)	5.6 to 7.0	6.3
α-NH_2 (end group)	7.5 to 8.4	8.0
ϵ-NH_2 (Lys)	9.4 to 10.6	10.0
Guanidino (Arg)	12.0 to 13.0	12.5
Phenolic-OH (Tyr)	9.8 to 10.4	10.1
—SH (Cys)	9.1 to 10.8	10.0

Problem 8.2: What is the titration curve for one mole of the peptide glycyl-aspartyl-lysine?

Solution: Consider the molecule in its completely protonated form:

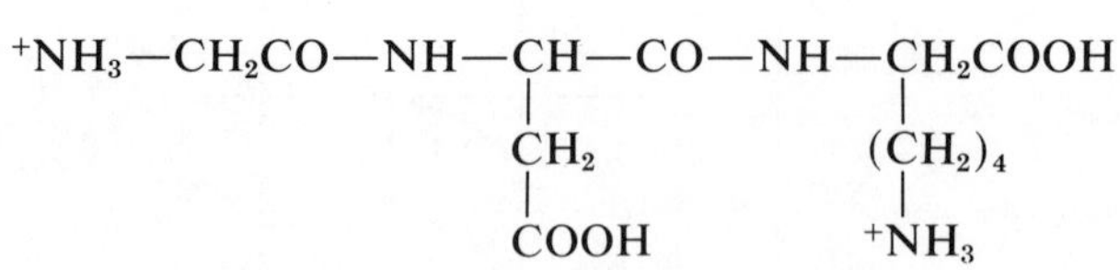

There are four ionizable groups

	Average pK_a (from Table 8.1)
—COOH terminal	3.3
—COOH Asp	3.8
—NH_3^+ terminal	8.0
—NH_3^+ ϵ-Lys	10.0

Each of the four protons in one mole of the protonated peptide will require one equivalent of base, and when each is half titrated the pH will be equal to the pK_a of the acid concerned. The segments of the titration curve around each of the pK_a values are joined by S-shaped curves as in the case of H_3PO_4 (Figure 8.1), with the exception of the first two —COOH groups, which have similar pK_a values, 3.3 and 3.8. As mentioned earlier, such groups will not ionize one after the other and may be considered to be two equivalents of the same acid with pK_a 3.5. The titration curve would thus show only the slightest inflection after one equivalent of base was added. The two NH_3^+ groups are far enough separated in their pK_a values to be treated independently, the pH of the titration solution being 8.0 after 2.5 equivalents of base are added and 10.0 after the addition of 3.5 equivalents of base. The titration curve is shown in Figure 8.3.

Figure 8.3

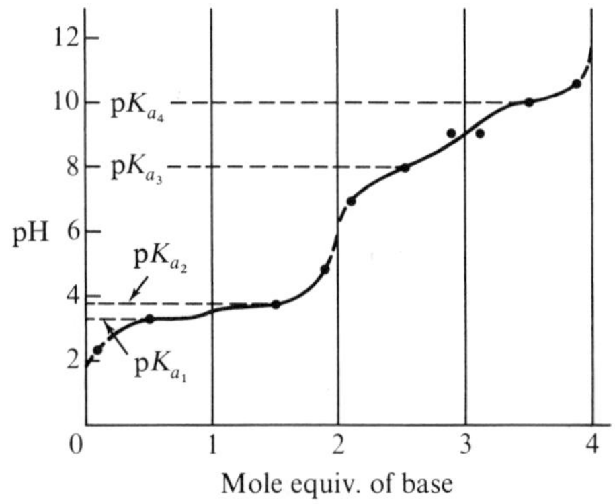

In proteins and peptides some of the —COOH groups of aspartyl and glutamyl residues are in the form of amides and as such may be considered nontitratable. Similarly, cysteine—SH groups may be in the disulfide —S—S— form and some tyrosine residues may be "buried" in the large molecule and so not accessible to titration. The titration curve of a protein is thus more approximate than that of the smaller peptides.

Problem 8.3: Ovalbumin, mol. wt. 45,000, has the following polyfunctional amino acid residues:

Tyrosine 9	Histidine 7
Cystine 2	Lysine 20
Cysteine 5	Aspartic 32
Arginine 15	Glutamic 52
N-acetylated end group 1	Amide 31

What would be the general shape of the titration curve?

Solution: By inspection of the composition,

(a) cystine (2 residues) will have no ionizable H^+;

(b) of the total Asp + Glu, 84 residues, 31 are as the amide so that the total side chain —COOH groups is 53;

(c) the $—NH_2$ group at the end of the peptide chain is blocked by an acetyl group and is therefore unavailable for protonation; however, the —COOH end is free;

(d) the total ionizable groups in the completely protonated molecule is, therefore,

$$9 + 5 + 15 + 7 + 20 + 53 + 1 = 110$$

(e) in order of pK_a values we would have

	Average pK_a
End and side —COOH 54	3.8
imidazole (His) 7	6.3
ϵ-NH_3^+ (Lys) 20	10.0
—OH and —SH (Tyr + Cys) 14	10.1
guanidino (Arg) 15	12.5

The approximate titration curve is shown in Figure 8.4.

Figure 8.4

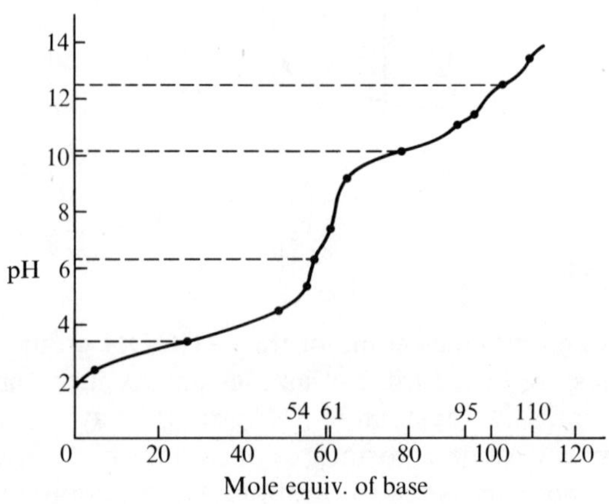

ISOELECTRIC POINT

At some pH in the titration of amphoteric molecules, such as peptides and proteins, there will be an equal number of negatively and positively charged groups. The pH at which the net charge is zero is known as the isoelectric point, pI. In simple molecules, this can be calculated from the average of the pK_a values of the appropriate pair of ionizing groups:

$$pI = \frac{pK_{a_x} + pK_{a_y}}{2}$$

The only problem is to select the proper pK_a values, and the procedure we will follow is illustrated by the amino acid L-aspartic acid. The steps are the following.

1. Write the molecule in its completely protonated form (A):

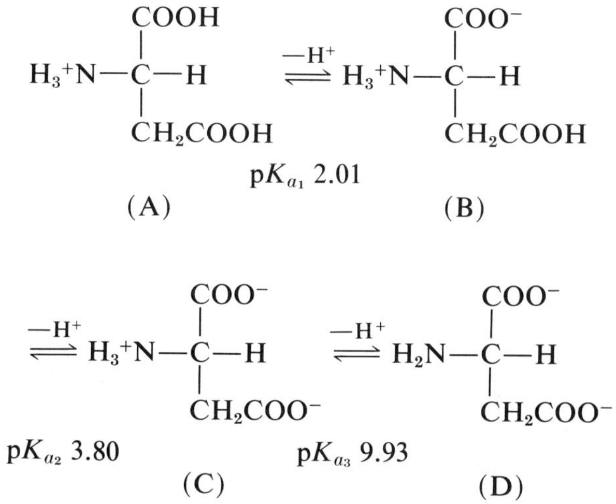

2. Ionize each H^+ in turn, in the order of increasing pK_a values, which will be known from physical tables, to give (B), (C), and (D) for L-aspartic acid.

3. Calculate the net charge on each ionized species:

(A) net charge +1 ($-NH_3^+$)
(B) net charge 0 ($-NH_3^+$, COO^-)
(C) net charge −1 ($-NH_3^+$, 2(COO^-))
(D) net charge −2 (2(COO^-))

4. Take the average of the two pK_a values associated with the species of zero charge, which in our base is (B). Species (B) is formed by an ionization characterized by pK_a, 2.01, and it also loses a proton by an ionization with a pK_a of 3.80. The average of these two pK_a values is the isoelectric point:

$$pI = \frac{2.01 + 3.80}{2} = 2.91$$

With increasing experience one can carry out some of the above steps by inspection, but the critical step is 4, in which the pair of pK_a values to be averaged is recognized.

Problem 8.4: Calculate the isoelectric point of L-alanine.

Solution:

Ionic species

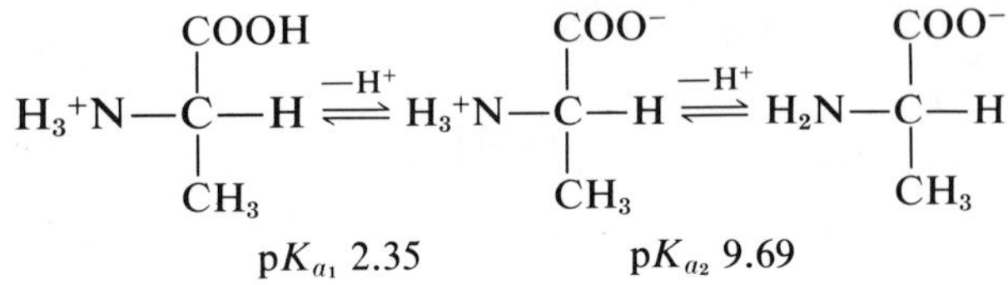

Net charge +1 0 −1

$$pI = \frac{2.35 + 9.69}{2} = \frac{12.04}{2} = 6.02$$

Problem 8.5: Calculate the isoelectric point of glycyl-aspartyl-lysine (see earlier problem in titration).

Solution: In an abbreviated form we may represent the tripeptide:

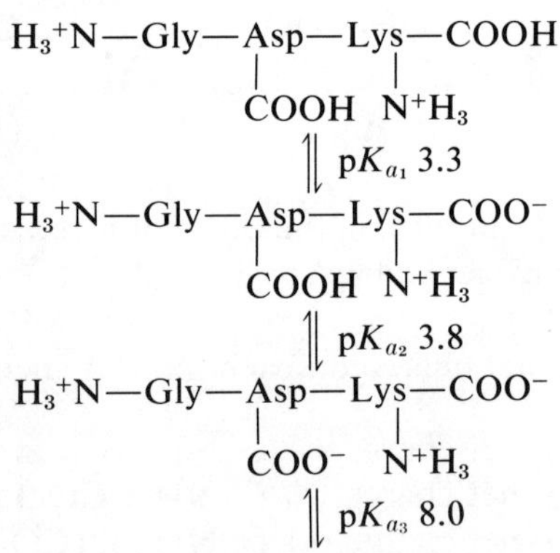

Net charge = +2

Net charge = +1

Net charge = 0

Since we have identified the species of zero net charge and know the values of the proper pK_a's, 3.8 and 8.0, there is little need to progress further in the ionized species:

$$pI = \frac{3.8 + 8.0}{2} = \frac{11.8}{2} = 5.9$$

It may be noted from Figure 8.3 that this pH is reached after adding two equivalents of base to a mole of completely protonated peptide.

In the case of complex molecules, such a microscopic identification of the charged species is not possible. An estimate can be made, however, from the titration curve, knowing the number of negatively charged COO^- groups that must be generated to neutralize the $—NH_3^+$ groups in the molecule. For example, in the case of ovalbumin (Figure 8.4) the *completely protonated* molecule has 110 ionizable groups, composed of

$$\left.\begin{array}{l} 54\ \text{—COOH} \\ 20\ \text{—NH}_3^+ \\ 7\ \text{His}^+ \\ 15\ \text{Arg}^+ \\ 5\ \text{—SH} \\ 9\ \text{Phenolic —OH} \end{array}\right\} \text{Net charge} = 42^+$$

This would require the ionization (neutralization) of 42 COOH groups, forming a countering 42 negative charges. The pH in the calculated titration curve where 42 equivalents of base have been added is 4.1. A more direct approach to obtaining the pI would be to calculate the pH where 42 of the 54 available —COOH groups have been titrated using the Henderson-Hasselbalch equation.

$$\text{pH} = 3.7 + \log \frac{42}{12}$$

$$\text{pH} = 4.2$$

The actual pI is 4.6. If one appreciates the fact that individual pK_a values should be used and not averages, as we have done, this agreement is good. For proteins with isoelectric points greater than about pI 6 the carboxyl content may not be sufficient to change the charge to zero. An example is given in Problem 8.6.

This approach can be made more precise by calculating, as a function of pH, the fractional charges associated with each ionizable group that is within about one pH unit of the estimated pI. The charge contributions of all the groups are then summed to give the net charge which is plotted versus pH. The pH where the net charge is zero is interpolated from the graph.

Problem 8.6: Estimate the pI of a thymus nucleohistone fraction with the composition of polyfunctional amino acids as follows:

Glutamic acid	3	Tyrosine	1
Aspartic acid	2	Terminal COOH	1
Lysine	28	Terminal NH_3^+	1
Arginine	2		

	Average pK	Number of groups
—COOH (end and side)	3.7	6
Terminal NH_3^+	8.0	1
ϵ-NH_3^+ and phenolic —OH	10.0	29
Guanidino$^+$	12.5	2
Total ionizable groups		38

The total charge on the completely protonated protein (1 + 28 + 2) is +31. When the 6 —COOH groups have ionized to give 6 negatively charged COO^- groups, the net positive charge will be 31 − 6, or +25. If, in addition, 19 of the ϵ-NH_3^+ groups ionize and thus lose their charge, there will remain on the protein 6 positive and 6 negative groups, or a net charge of zero. Thus a total of 25 protons have been removed.

From the titration curve (Figure 8.5), constructed as described earlier for ovalbumin (Figure 8.4), the pH after 25 protons have been titrated would be approximately 10.2. The true value is 10.8, which is in good agreement since this protein has an unusual charge density.

Figure 8.5

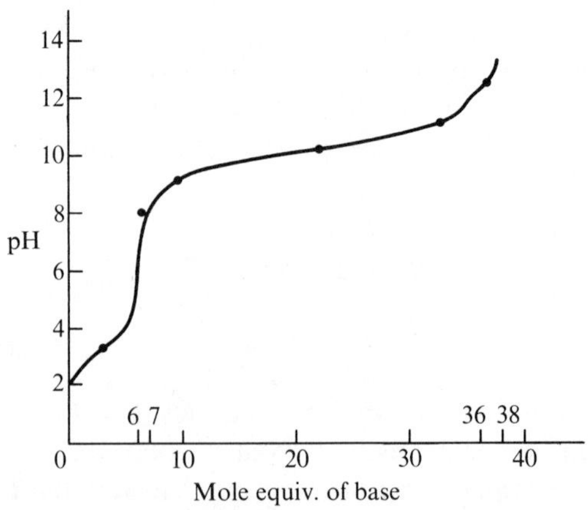

BUFFERS OF POLYPROTONIC ACIDS AND BASES

The questions concerning the ionization of compounds with several dissociating hydrogen ions, the titration curves of such compounds, and buffers, are brought to focus in a consideration of the preparation of buffer solutions. As will have been noted from the titration curves of H_3PO_4, glycine, and the like, the areas of greatest buffer capacity are those in which the ratio of salt to acid is near unity—that is, near to the pK_a value of the dissociating species in question. The polyprotonic acids and bases obviously have as many good buffer regions as they have ionizable protons or pK_a values. It is only necessary, therefore, to decide the nature of the conjugate acid-base pair for the buffer pH in question and thus the appropriate pK_a value to apply to the Henderson-Hasselbalch equation.

In this simple treatment we still assume that each ionization step goes to completion before the next ionization commences. We therefore look upon these buffer problems exactly as those of the monoprotonic acids and bases, but in this case the pK_a (and therefore the associated salt/acid pair) is that which is closest to the pH of the buffer desired. For example, if we wish to prepare a phosphate buffer of pH 7.4, then the closest pK_a is 6.7 (see earlier section) and this pK_a value is for the $H_2PO_4^- \rightleftharpoons HPO_4^{--}$ pair.

As was also noted earlier for simple buffers, if the required pH is outside the range p$K_a \pm 1$, then the buffer capacity will be low and the Henderson-Hasselbalch calculation will be quite approximate.

The suggested steps in the buffer calculations involving polyprotonic acids or bases may be summarized.

1. From the known pK_a value for each ionization, select the one closest to the required pH for the buffer. If this pH is more than 1 pH unit away, then a better buffer system should be sought.
2. Write the ionized forms of the polyprotonic acid, base, or salt for the ionization that relates to the chosen pK_a value in step 1.
3. Calculate the amount of acid or alkali required to change the given starting material to one of the selected conjugate acid-base pairs.
4. Calculate the [salt]/[acid] ratio for this selected conjugate acid-base pair from the Henderson-Hasselbalch equation.
5. Calculate the [salt + acid] concentration relationship.
6. From 4 and 5 calculate the [salt] and [acid] for the buffer system required.
7. From 6 calculate the acid or alkali to be added over and above that in step 3 to give the ratio in step 4.
8. Calculate the total components of the buffer, including the water, required to make the proper concentration.

These steps are illustrated in the following problem.

Problem 8.7: Calculate the components of 1 liter of a sodium glycinate buffer, pH 8.80, that is 0.10 *M* in total glycine. You are given solid glycine hydrochloride and solid NaOH.

Solution: Glycine, pK_{a_1} 2.35, pK_{a_2} 9.78.

1. Select the proper pK_a; pK_{a_2} 9.78 is closest to the desired pH.
2. The nature of the salt/acid pair in relation to the given materials is

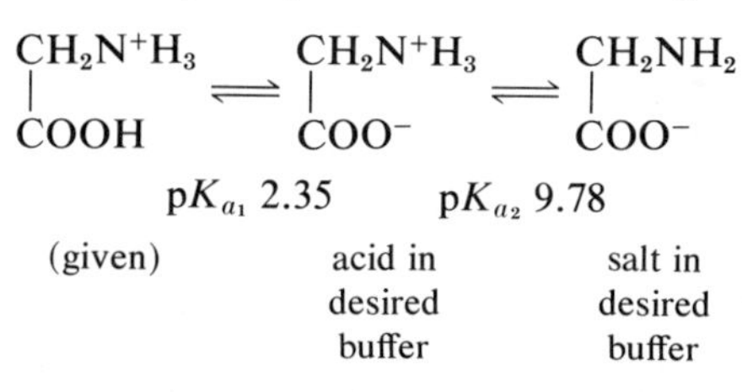

3. Calculate the extent of the titration necessary to get from the given material to the acid form in the buffer. Since we were given the completely

protonated amino acid and 0.1 mole of acid are needed (this is the strength of the buffer solution with reference to the glycine), then the first ionization will require 0.1 mole of NaOH.

4. Calculate the [salt]/[acid] ratio in the buffer. From Equation (7.31),

$$8.80 = 9.78 + \log \frac{[\text{salt}]}{[\text{acid}]}$$

therefore

$$\log \frac{[\text{salt}]}{[\text{acid}]} = -1.02 = +0.98 - 2.00$$

and

$$\frac{[\text{salt}]}{[\text{acid}]} = \text{antilog}\ (+0.98 - 2.00) = 9.55 \times 10^{-2}$$

But we know that the total amount of glycine is 0.1 mole liter^{-1}; that is,

$$[\text{salt}] + [\text{acid}] = 0.10$$

Since from the ratio we also know that

$$[\text{salt}] = 9.55 \times 10^{-2}\ [\text{acid}]$$

substituting for [salt] we get

$$9.55 \times 10^{-2}\ [\text{acid}] + [\text{acid}] = 0.10$$

When we collect like terms,

$$(1.0 + 9.55 \times 10^{-2})\ [\text{acid}] = 0.10$$

$$1.0955\ [\text{acid}] = 0.10$$

$$[\text{acid}] = \frac{0.10}{1.0955} \text{ mole liter}^{-1}$$

$$= 0.0913 \text{ mole liter}^{-1}$$

Therefore

$$\text{salt} = 0.10 - 0.0913 = 0.0087 \text{ mole liter}^{-1}$$

5. From step 4 calculate the extent of the titration needed to give the correct [salt]/[acid] ratio. Since the [salt] is 0.0087 mole liter^{-1}, it would require this amount of NaOH to produce it from the desired acid form. The total NaOH is therefore (0.10 + 0.0087) mole.

6. Total components of buffer:

$$\text{glycine HCl} = 0.10 \text{ mole}$$
$$(\text{mol. wt. } 101.6)$$
$$= (0.10 \times 101.6) \text{ g}$$
$$= 10.2 \text{ g}$$

$$\underset{\text{(mol. wt. 40.0)}}{\text{NaOH}} = 0.1087 \text{ mole}$$
$$= (0.1087 \times 40) \text{ g}$$
$$= 4.4 \text{ g}$$

Water is added to the mixture of the above compounds to make 1000 ml of solution.

CO_2, H_2CO_3, and HCO_3^- IN RESPIRATION

The transport of the CO_2 in blood and the control of blood pH involves principally hemoglobin and the HCO_3^-/H_2CO_3 buffer systems. As noted in Chapter 6, CO_2 hydrates in solution to form carbonic acid, which is a medium strong acid:

$$CO_2 + H_2O \rightleftharpoons H_2CO_3 \tag{8.4}$$

$$H_2CO_3 \rightleftharpoons H^+ + HCO_3^-, \qquad K_{a_1} = 1.3 \times 10^{-4};\ pK_{a_1}\ 3.88 \tag{8.5}$$

$$HCO_3^- \rightleftharpoons H^+ + CO_3^{--}, \qquad K_{a_2} = 6 \times 10^{-11};\ pK_{a_2}\ 10.22 \tag{8.6}$$

The K_{a_1} refers to the small fraction (0.3%) of CO_2 dissolved in water that is hydrated. If Equations (8.4) and (8.5) are combined and an overall equilibrium constant is used, then the concentration of dissolved CO_2, not just the hydrated form, can be used for calculations:

$$CO_2 + H_2O \rightleftharpoons H^+ + HCO_3^-, \qquad K_a = 4.4 \times 10^{-7};\ pK_a\ 6.36 \tag{8.7}$$

The total volume of gas in solution is directly proportional to the partial pressure of gas in the liquid and to its solubility coefficient. The solubility coefficient of CO_2 in blood plasma is 0.51 at 38°C and 760 mm Hg CO_2 pressure, so that 1 ml of plasma dissolves 0.51 ml CO_2 (SC) under these conditions. This relationship can be expressed in various ways (Chapter 5).

Since the solubility of CO_2 in plasma is greater than in water, the pK_a for the coupled reaction (8.7) is 6.1 and the pH of the plasma solution is therefore

$$pH = 6.1 + \log \frac{[HCO^-]}{[CO_2 + H_2CO_3]} \tag{8.8}$$

For each millimeter of Hg partial pressure of CO_2 in plasma, the dissolved CO_2 is

$$\frac{0.51 \text{ ml } CO_2(SC)}{\text{ml plasma } (38°)} \times \frac{1}{760 \text{ mm Hg}}$$

$$= 6.71 \times 10^{-4} \text{ ml } CO_2(SC)/\text{ml plasma } (38°C)/\text{mm}$$

$$= \frac{6.71 \times 10^{-1} \text{ ml } CO_2(SC) \text{ m}M}{22.26 \text{ ml } CO_2(SC)} = 0.030 \text{ m}M/\text{mm}$$

since 1 mmole of CO_2, a nonideal gas, at 760 mm Hg pressure occupies 22.26 ml at 0°C(SC). At 38°C the total dissolved CO_2 in plasma is therefore at a concentration (millimolar) equal to 0.030 × partial pressure of CO_2 in millimeters of mercury. When we rewrite Equation (8.8),

$$\text{pH} = 6.1 + \log \frac{[HCO_3^-]}{0.030\ (P_{CO_2})} \tag{8.9}$$

[*Note:* The $[HCO_3^-]$ concentration must be expressed as millimolar.]

Problem 8.8: Calculate $[HCO_3^-]$ concentration in blood plasma of pH 7.4, 40 mm Hg P_{CO_2}.

Solution: From Equation (8.9),

$$7.4 = 6.1 + \log \frac{[HCO_3^-]}{40 \times 0.030}$$

When we rearrange,

$$\log\ [HCO_3^-] - \log 1.20 = 7.4 - 6.1$$

$$\log\ [HCO_3^-] = 1.30 + \log 1.20 = 1.30 + 0.08 = 1.38$$

$$[HCO_3^-] = 24 \text{ m}M$$

Problem 8.9: By direct measurement, the dissolved CO_2 in a plasma was 2.8 ml (SC) per 100 ml. What is the P_{CO_2}? The solubility coefficient is 0.51.

Solution:

$$CO_2 = 2.8 \text{ ml}/100 \text{ ml plasma} = 0.028 \text{ ml}/1 \text{ ml plasma}$$

But at 1 mm P_{CO_2}, 0.000671 ml CO_2 dissolves per 1 ml plasma. Therefore

$$P_{CO_2} = \frac{0.028}{0.000671} \text{ mm Hg} = 41.8 \text{ mm}$$

Under normal conditions the pH of blood is 7.4 (with P_{CO_2} being 40 mm Hg) and the ratio of $[HCO_3^-]$ to [dissolved CO_2] is 20. Changes in this ratio, as well as the absolute values of $[HCO_3^-]$ and [dissolved CO_2], are indicative of the nature of various physiological imbalances. Equation (8.9) and its variations (see Problem 8.10) permit the calculation of these critical values.

Problem 8.10: By direct measurement, the total volume of CO_2 liberated upon acidification of 2 ml of plasma at 38°C was 1.114 ml (SC). The pH of the plasma was 7.35. What is the $[HCO_3^-]/[\text{dissolved } CO_2]$ ratio?

Solution: 1.114 ml CO_2 = 1.114/22.26 mmole.
Since this amount is in 2 ml plasma, the millimolarity of the total CO_2 is

$$\frac{1.114}{22.26} \times \frac{1000}{2}\ \text{m}M = 25.0\ \text{m}M \tag{i}$$

$$[\text{total } CO_2] = [HCO_3^-] + [\text{dissolved } CO_2]$$

$$[HCO_3^-] = [\text{total } CO_2] - [\text{dissolved } CO_2] \tag{ii}$$

Equation (8.8) may be written

$$\text{pH} = 6.1 + \log \frac{[\text{total } CO_2] - [\text{dissolved } CO_2]}{[\text{dissolved } CO_2]} \tag{iii}$$

Substituting numbers in (iii), we have

$$7.35 = 6.1 + \log \left(\frac{[\text{total } CO_2]}{[\text{dissolved } CO_2]} - 1 \right)$$

$$1.25 = \log \left(\frac{25.0}{[\text{dissolved } CO_2]} - 1 \right)$$

Therefore

$$17.8 = \frac{25.0}{[\text{dissolved } CO_2]} - 1$$

$$[\text{dissolved } CO_2] = \frac{25.0}{(17.8 + 1)} = 1.3\ \text{m}M \tag{iv}$$

From (i) and (iv),

$$[HCO_3^-] = (25.0 - 1.3) = 23.7\ \text{m}M$$

and

$$\frac{[HCO_3^-]}{[\text{dissolved } CO_2]} = \frac{23.7}{1.3} = 18$$

COMPLEX METAL IONS

A particular kind of equilibrium reaction is one in which metal ions are associated with groups (called ligands) in a molecule to form a complex ion. Water, a common ligand, forms numerous aquocomplexes, for example,

$Cu^{++}(4H_2O)$. Many metals are actually in this form in aqueous solution even though we write them in equations as the uncomplexed ion:

$$M^{++} + 4X \rightleftharpoons M(X)_4^{++} \tag{8.10}$$

The equilibrium constant, called here the stability constant K_{st}, is defined as usual:

$$K_{st} = \frac{[M(X)_4^{++}]}{[M^{++}][X]^4} \tag{8.11}$$

Equally accurate and common is the inverse equilibrium constant, called the dissociation or instability constant K_{inst}, which considers the same reaction as

$$M(X)_4^{++} \rightleftharpoons M^{++} + 4X \tag{8.12}$$

$$K_{inst} = \frac{[M^{++}][X]^4}{[M(X)_4^{++}]} = \frac{1}{K_{st}} \tag{8.13}$$

In critical tables it is obviously important to know which constant is being used. A stable complex is one with a large stability constant and from your everyday experience of stable complexes, such as the metal cyanides, you can evaluate the type of equilibrium constant being tabulated.

Problem 8.11: Calculate $[Cu^{++}]$ in a 0.1 M solution of $CuSO_4$ that is 0.5 M with respect to uncomplexed NH_3; $K_{st} = 3.9 \times 10^{12}$.

Solution: $[NH_3] = 0.5\ M$ and $[Cu(NH_3)_4^{++}] \approx 0.1\ M$, since nearly all the Cu^{++} will be complexed. Then, when we substitute in (8.11),

$$\frac{0.1}{[Cu^{++}][0.5]^4} = 3.9 \times 10^{12} \tag{i}$$

Rearranging (i), we have

$$[Cu^{++}] = \frac{0.1}{3.9 \times 10^{12} \times 0.0625} = \frac{0.1}{0.23 \times 10^{-12}} = 4.2 \times 10^{-13}\ M$$

Complex metal ions play a most important role in biochemistry, examples being ligands in simple molecules, such as in Mg^{++}—ATP, and in biopolymers like the metalloproteins. The only unequivocal way of finding the composition and properties of a complex ion is by experimental determination, but some hints may be gathered from the following.

1. The number of ligands (coordination number) that can associate with a metal ion is often twice its ionic charge. Where two or more oxidation

states exist, the coordination number of the higher oxidation state applies to all; however, a coordination number of more than 6 is rare.

Usual maximum coordination number	*Metal ion*
4	Cu^+ or Cu^{++}
4	Mg^{++}
6	Fe^{++} or Fe^{+++}
6	Sn^{++} or Sn^{++++}

2. Metal ion-ligand systems involve reactions of the successive addition of each ligand or, perhaps more precisely, the successive replacement of a water ligand by the new ligand:

$$M^+(H_2O)_2 + 2X \underset{K_{st_1}}{\rightleftharpoons} M(H_2O)^+X + X + H_2O \underset{K_{st_2}}{\rightleftharpoons} MX_2^+ + 2H_2O \tag{8.14}$$

The relative concentrations of each species is related to the concentration of M^+ and X as follows:

$$K_{st_1} = \frac{[M(H_2O)X^+]}{[M^+(H_2O)_2][X]} \tag{8.15}$$

$$K_{st_2} = \frac{[MX_2^+]}{[M(H_2O)X^+][X]} \tag{8.16}$$

These successive constants tend to decrease as the number of ligands increase, particularly when the ligand carries a charge, and electrostatic repulsion of the entering ligand occurs. We will usually only deal with the maximally coordinated species, all the intermediate reactions and constants being coupled. The problem is related, however, to a consideration of the relative stabilities of complex ions.

Problem 8.12: An aqueous solution of a copper-protein complex, with a K_{st} 1.5×10^{10}, is treated with an equimolar amount of a simple organic copper-complexing agent with a K_{st} of 3.0×10^8. What is the distribution of copper between the two complexes?

Solution:

$$Cu^{++} + \text{protein} \rightleftharpoons (\text{Cu-protein}) \qquad K_{st}\ 1.5 \times 10^{10}$$

$$Cu^{++} + \text{organic agent} \rightleftharpoons (\text{Cu-organic}) \qquad K_{st}\ 3.0 \times 10^{8}$$

The ratio of the copper-complexes would be in proportion to the stability constants:

$$\frac{[\text{Cu-protein}]}{[\text{Cu-organic}]} = \frac{1.5 \times 10^{10}}{3.0 \times 10^8} = \frac{100}{2} = \frac{50}{1}$$

---

3. The complexing agent with more than one ligand forms more stable complex ions (chelates) than would be gained by the summation of the same ligands in separate molecules. For example,

$$Zn^{++} + EDTA^{-4} \rightleftharpoons ZnEDTA^{-2} \qquad pK_{st} = -16.5$$

Ethylenediamine tetraacetate, EDTA, complexes through two nitrogen and two oxygen atoms

$$Zn^{++} + 4NH_3 \rightleftharpoons Zn(NH_3)_4^{+2} \qquad pK_{st} = -9.06$$

$$Zn^{++} + 4CH_3COO^- \rightleftharpoons Zn(CH_3COO)_4^{-2} \qquad pK_{st} = -1.03$$

4. Some complexes are formed with the liberation of a proton. In these cases the equilibria will be pH-dependent. To look at this another way, only the base of a conjugate pair may chelate or complex so that the concentration of base and therefore the equilibrium of the complex will be pH-dependent.

Problem 8.13: What is the concentration of uncomplexed Zn^{++} in a solution of 0.01 M $ZnSO_4$ and 0.1 M cyanide at pH 8.0 as compared to pH 6.0?

Solution:

$$Zn^{++} + 4CN^- \rightleftharpoons Zn(CN)_4^{--} \qquad \text{(i)}$$

$$K_{st} = \frac{[Zn(CN)_4^{--}]}{[Zn^{++}][CN^-]^4} = 8.3 \times 10^{17} \qquad \text{(ii)}$$

$$K_a = \frac{[H^+][CN^-]}{[HCN]} = 7.2 \times 10^{-10} \qquad \text{(iii)}$$

The material balances are

for Zn^{++}: $\quad [Zn^{++}] + [Zn(CN)_4^{--}] = 0.01\ M$ (iv)

for CN^-: $\quad [HCN] + [CN^-] + 4[Zn(CN)_4^{--}] = 0.1\ M$ (v)

Because of the high stability constant, $[Zn^{++}]$ is going to be small and to a first approximation we could consider $[Zn(CN)_4^{--}]$ as 0.01 M. (vi)

The total cyanide (uncomplexed) is, from (v) and (vi),

$$[HCN] + [CN^-] = (0.1\ M - 4 \times 0.01\ M)$$

Substituting for [HCN],

$$[CN^-] + \frac{[H^+][CN^-]}{7.2 \times 10^{-10}} = 0.06\ M, \qquad [CN^-]\left(1 + \frac{[H^+]}{7.2 \times 10^{-10}}\right) = 0.06\ M$$

at pH 8 or $[H^+] = 10^{-8}$,

$$[CN^-] = \frac{0.06}{1 + \dfrac{1.0 \times 10^{-8}}{7.2 \times 10^{-10}}} = \frac{6.0 \times 10^{-2}}{1 + 14}$$

$$= 4.0 \times 10^{-3}\ M$$

When we substitute in (ii),

$$\frac{0.1}{[Zn^{++}][4.0 \times 10^{-3}]^4} = 8.3 \times 10^{17}$$

$$[Zn^{++}] = \frac{1.0 \times 10^{-1}}{8.3 \times 10^{17} \times 256 \times 10^{-12}} = \frac{1.0 \times 10^{-6}}{0.21 \times 10^{-4}}$$

$$= 4.8 \times 10^{-10}\ M$$

at pH 6 or $[H^+] = 10^{-6}$,

$$[CN^-] = \frac{0.06}{1 + \dfrac{10 \times 10^{-7}}{7.2 \times 10^{-10}}} = \frac{6.0 \times 10^{-2}}{1 + 1.4 \times 10^3}$$

$$= 4.3 \times 10^{-5}\ M$$

When we substitute in (ii),

$$[Zn^{++}] = \frac{1.0 \times 10^{-1}}{8.3 \times 10^{17} \times 4.3 \times 10^{-20}} = \frac{1.0 \times 10^{-1}}{8.3 \times 342.3 \times 10^{-3}}$$

$$= \frac{1.0 \times 10^{-1}}{0.28 \times 10^{+4} \times 10^{-3}} = 3.5 \times 10^{-2}\ M$$

In acidic solutions, pH < 6, the $[CN^-]$ is nearly zero and virtually no complex ion forms.

BINDING EQUILIBRIA OF MACROMOLECULES

Analogous to the equilibria that occur in the complexation of metal ions by ligands are those equilibria involving the binding of one molecule to other molecules. In biochemistry this type of interaction is found in the interaction of molecules, such as hormones, with receptor molecules on cell surfaces, antigen-antibody interactions and the effects of small inhibitor molecules, the interaction of oxygen with hemoglobin and the binding of substrate and inhibitor molecules with enzymes, which are large molecular weight proteins.

Consider a protein, P, that has several binding sites for ligand molecules, L.

$$P + L \rightleftharpoons PL \qquad K_1 = \frac{[PL]}{[P][L]}$$

$$PL + L \rightleftharpoons PL_2 \qquad K_2 = \frac{[PL_2]}{[PL][L]}$$

$$PL_{n-1} + L \rightleftharpoons PL_n \qquad K_n = \frac{[PL_n]}{[PL_{n-1}][L]}$$

Determination of the individual equilibrium constants in such a complex mixture is not possible in most cases. It is possible, however, to determine experimentally the average number of moles of L bound per mole of protein.

$$\bar{\nu} = \frac{[L]_{\text{bound}}}{[P]_{\text{total}}} \tag{8.17}$$

$$= \frac{[PL] + 2[PL_2] \cdot \cdot \cdot n[PL_n]}{[P] + [PL] + + [PL_2] \cdot \cdot \cdot [PL_n]}$$

The various equilibria can be described in terms of $\bar{\nu}$, n, the number of binding sites, and a single constant K, which is an intrinsic binding constant. The n sites are noninteracting and equivalent to n single sites. The equation for this particular binding is

$$\bar{\nu} = \frac{nK[L]}{1 + K[L]} \tag{8.18}$$

Since there are n sites on P then the fraction of sites occupied, θ, is given by

$$\theta = \frac{[L]_{\text{bound}}}{n[P]_{\text{total}}} \tag{8.19}$$

$$= \frac{\bar{\nu}}{n}$$

In some cases θ is easier to obtain experimentally than $\bar{\nu}$.

The binding of one ligand may activate or inhibit the binding of the next ligand molecule. The first ligand is then called an *effector* molecule. When the effector molecule increases the affinity of P for the next ligand there is said to be *cooperative binding*. Such models may be expressed

$$\bar{\nu} = \frac{nK[L]^m}{1 + K[L]^m} \tag{8.20}$$

or

$$\frac{\bar{\nu}}{n - \bar{\nu}} = \frac{\theta}{1 - \theta} = K[L]^m \tag{8.21}$$

where n is the number of binding sites and m is a measure of interaction between sites, often referred to as the Hill coefficient (see Equation 8.20). The values of m and n are greater than one and, although m theoretically could equal n, it is rare that the value is attained.

Where there is no cooperativity between the sites the Hill coefficient is unity and

$$\frac{\theta}{1-\theta} = K[\mathrm{L}] \tag{8.22}$$

Evaluation of *K*, *n* and *m*. An evaluation of the constants in the equilibrium binding Equations 8.18 and 8.21 can be made graphically. The experimental data can be obtained in various ways that result in the determination of the concentration of the bound ligand and the total concentration of the protein, the ratio of these two values being $\bar{\nu}$. The total protein in the system is determined by analysis before adding any ligand. After a known amount of the ligand has been added and equilibrium has been established, the free ligand can be measured by virtue of some spectroscopic property, which changes upon binding, or by equilibrium dialysis if the ligand will diffuse through a membrane, or by molecular sieve chromatography. The experimental details for such studies are described in various texts.*

From the experimental data

$$\bar{\nu} = \frac{[\mathrm{L}]_{\text{total}} - [\mathrm{L}]_{\text{free}}}{[\mathrm{P}]_{\text{total}}} \tag{8.23}$$

Since Equations 8.18 and 8.20 are either hyperbolic or sigmoidal functions, the direct plots are used less frequently than some rearranged forms that give a linear relationship.

For Equation 8.18 the simple rearrangement (8.24) gives a hyperbolic

$$\bar{\nu} = \frac{n}{1 + \dfrac{1}{K[\mathrm{L}]}} \tag{8.24}$$

curve by plotting $\bar{\nu}$ versus [L] (Figure 8.6), but the double reciprocal plot of $1/\bar{\nu}$ and $1/[\mathrm{L}]$ (Figure 8.7) from the linearized form 8.25

$$\frac{1}{\bar{\nu}} = \frac{1}{n} + \frac{1}{nK} \cdot \frac{1}{[\mathrm{L}]} \tag{8.25}$$

$$(y = b + m.x)$$

*See, for example, K. E. Van Holde, *Physical Biochemistry*. Prentice-Hall, Englewood Cliffs, New Jersey, 1971.

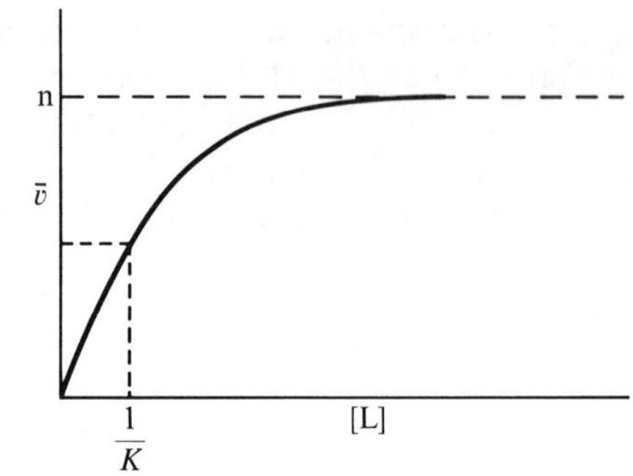

Figure 8.6

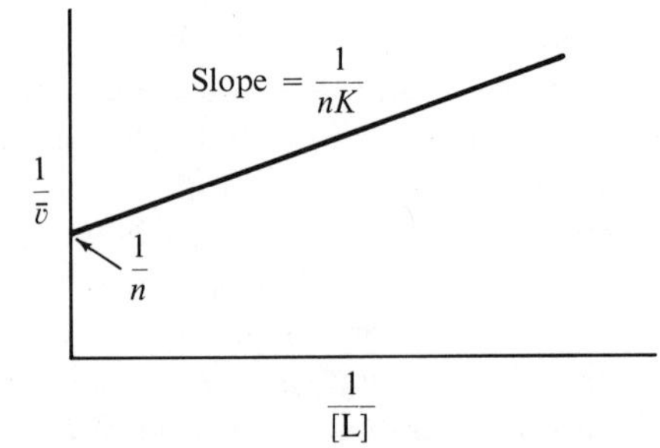

Figure 8.7

gives a more accurate estimate of n from the intercept $1/n$ and K from the slope $1/nK$.

An alternate form of Equation 8.18 gives 8.26,

$$\frac{\bar{\nu}}{[\mathrm{L}]} = nK - K\bar{\nu} \tag{8.26}$$

$$(y = b + m.x)$$

which is called a Scatchard plot* (Figure 8.8), with slope $-K$ and intercept n on the $\bar{\nu}$ axis (abscissa).

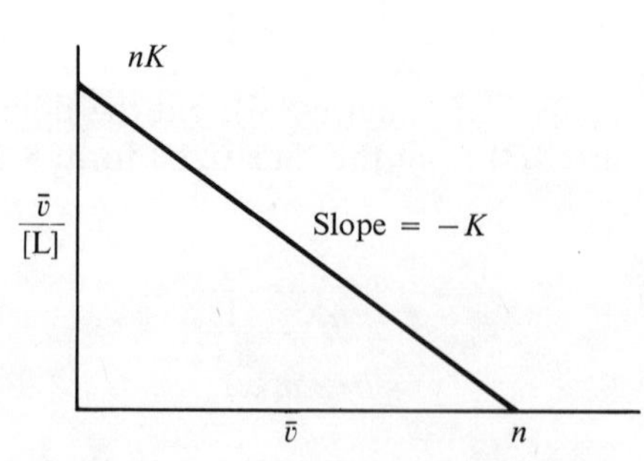

Figure 8.8

*G. Scatchard, J. S. Coleman, and A. L. Shen, *J. Amer. Chem. Soc.* **79**:12 (1957).

Problem 8.14: D-Mannose in concentrations 1.25 to 20 mM and a purified mitogenic lectin protein (molecular weight 49000) from garden peas (*Pisum sativum*) in concentrations of 20 to 30 mg/ml were placed in opposite compartments of equilibrium dialysis cells. The cells were rotated for 24 hours to establish equilibrium after which time the free mannose concentration in the protein-free cell was determined. The amount of bound mannose was determined by difference and the ratio, moles bound mannose/mole lectin, was calculated. The results for two series of experiments are given in Table 8.2. Plots of $\bar{\nu}$ versus [free mannose], double reciprocal plots, and $\bar{\nu}/[L]$ versus $\bar{\nu}$ were made to calculate K and n; K is the intrinsic association constant for the binding of sugar to each site.

Solution: The necessary reciprocal values for the graphical solutions are calculated and summarized in Table 8.2.

Table 8.2
Data for graphical solution of problem

$[Mannose]_{free}$ M $\times 10^{-3}$	$\frac{1}{[Mannose]_{free}}$ M^{-1}	$\bar{\nu}$	$\frac{1}{\bar{\nu}}$	$\frac{\bar{\nu}}{[Mannose]_{free}}$
		Series 1		
0.47	2110	0.735	1.361	1550
0.96	1040	1.072	0.933	1120
1.98	525	1.518	0.659	797
4.48	223	1.880	0.532	420
		Series 2		
0.45	2229	0.735	1.361	1640
1.00	997	1.133	0.883	1130
1.88	530	1.530	0.654	812
4.08	245	1.892	0.529	464

It is seen from Figure 8.9 that the number of binding sites is approaching two. The value of $1/K$ is found at 50% binding to be equivalent to $K = 1200$ liters mole^{-1}. By a double reciprocal plot (Figure 8.10), the value of n is 2.27 binding sites and $K = 1000$ liters mole^{-1} as compared to the Scatchard plot, where $n = 2.35$ and $K = 970$ liters mole^{-1}. If in the hyperbolic plot (Figure 8.11), the value of n had been set at 2.30, which is the average of the values from the two other plots, then K would be 909 liters mole^{-1}.

[Problem from Ian S. Trowbridge; compare with the original paper for further discussion.*]

*Ian S. Trowbridge, *J. Biol. Chem.* **249** 6004–6012 (1974).

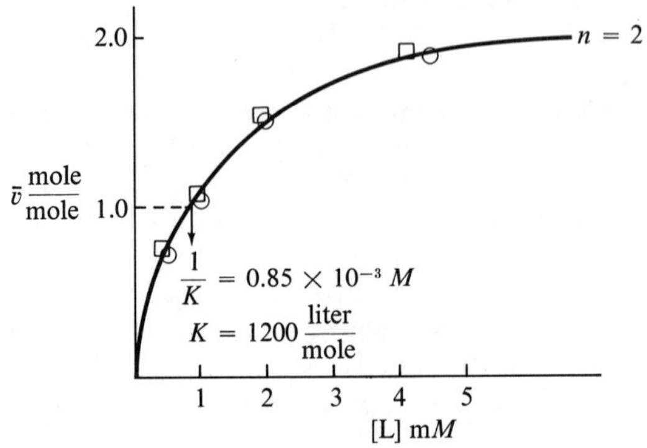

Figure 8.9

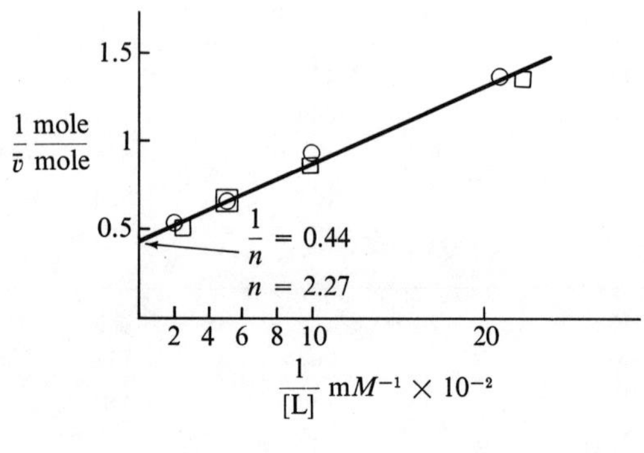

Figure 8.10

$$\text{Slope} = \frac{0.7}{1600}\,\frac{\text{mmole}}{\text{liter}}$$

$$= 4.38 \times 10^{-4}\,\frac{\text{mmole}}{\text{liter}}$$

$$= \frac{1}{nK}\,\frac{\text{mmole}}{\text{liter}}$$

$$\therefore\ \frac{1}{K} = 2.27 \times 4.38 \times 10^{-4}$$

$$= 9.94 \times 10^{-4}\,\frac{\text{mmole}}{\text{liter}}$$

$$K = 1000\,\frac{\text{liter}}{\text{mole}}$$

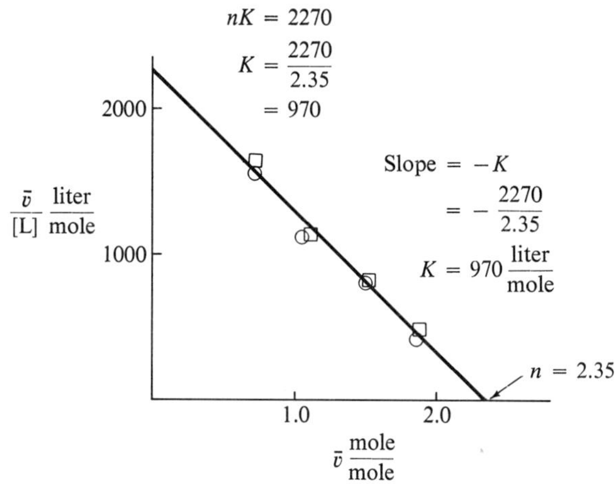

Figure 8.11

The equations for cooperative binding, 8.20 and 8.21, where $m > 1$, do not give linear double reciprocal or Scatchard plots. The Hill equation (8.22)

$$\frac{\bar{\nu}}{n - \bar{\nu}} = K[\mathrm{L}]^m$$

is linear in the logarithmic form

$$\log \frac{\bar{\nu}}{n - \bar{\nu}} = m \log [\mathrm{L}] + \log K \tag{8.27}$$

$$(y = m.x + b)$$

This leads to a Hill plot with the slope equal to the Hill coefficient m and the intercept is log K (Figure 8.12).

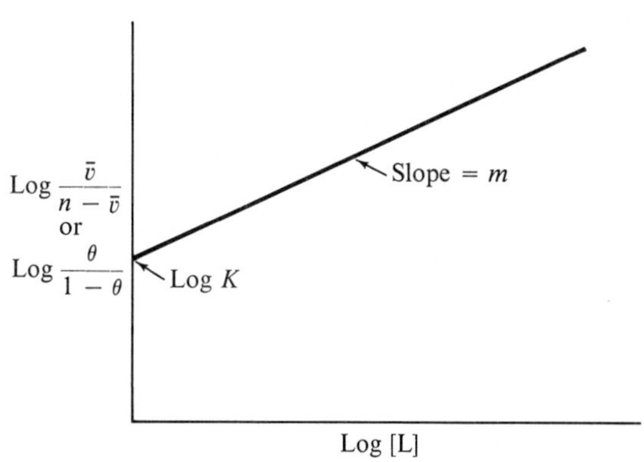

Figure 8.12

Equation 8.27 is also seen in terms of the fraction of sites occupied, θ.

$$\log \frac{\theta}{1-\theta} = m \log [\mathrm{L}] + \log K \tag{8.28}$$

It is of importance to note that the first and last binding sites of a macromolecule with multiple sites are filled with no cooperativity possible, that is, m is unity. In the Hill plot, this is seen at the first and last few percent of binding, the maximum value for m being where $\log \theta/1 - \theta$ is zero.

In terms of the Monod-Wyman-Changeux model where two conformations R and T of protein are in equilibrium, the two values of M at either extreme of the Hill plot reflect first the binding to the T-form and finally to the R-form. This is seen particularly well in the binding of oxygen to hemoglobin (Figure 8.13).

Figure 8.13

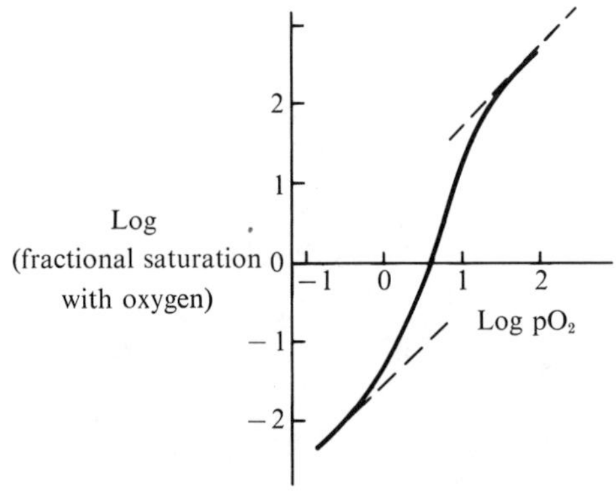

REFERENCES

H. B. Bull, *An Introduction to Physical Biochemistry*. Davis, Philadelphia, 1964. Chapter 5.

H. N. Christensen, *Body Fluids and the Acid-Base Balance*. Saunders, Philadelphia, 1964. (A programmed course.)

H. W. Davenport, *The ABC of Acid-Base Chemistry*. University of Chicago Press, Chicago, 1958.

K. E. Van Holde, *Physical Biochemistry*. Prentice-Hall, Englewood Cliffs, New Jersey, 1971.

E. S. West, W. R. Todd, H. S. Mason, and J. T. VanBruggen, *Textbook of Biochemistry* (4th ed.). Macmillan, New York, 1966. Chapters 15, 16, and 17.

F. Wold, *Macromolecules: Structure and Function*. Prentice-Hall, Englewood Cliffs, New Jersey, 1971.

8. PROBLEMS A

Answers on page 343

Buffers, Complex Ions, and Titration Curves.

1. Compare the buffer capacities of an acetate buffer (pH 6.6, [acetate] + [acetic acid] = 0.4 *M*) and a phosphate buffer (pH 7.5, [HPO_4^{--}] + [$H_2PO_4^-$] = 0.6 *M*).

2. A buffer is prepared by mixing 500 ml of 0.4 *M* Na_2HPO_4 and 500 ml of 0.2 *M* NaH_2PO_4, pK_a 6.7 at this ionic strength.
   (a) Calculate the pH of this buffer.
   (b) 10 ml of 1 *N* HCl are added to this buffer. Calculate the change in pH. (Ignore the change in volume.)
   (c) Calculate the pH change if the same quantity of HCl were added to 1 liter of pure water (pH 7.0).

3. Given 100 ml of 0.10 *M* sodium bicarbonate solution, calculate the volume of 0.10 *N* HCl and the volume of water to be added to give a solution with a pH of 7.4 and which contains 25 m*M* total carbon dioxide. Provided no gas were permitted to escape, what would be the partial pressure of carbon dioxide?

4. Calculate the relative amounts of the components in a 0.05 *M* buffer of ethylene diammonium ($C_2H_4(NH_3^+)_2$), pH 9.5.

5. You need a phosphate buffer of pH 7.2 (0.1 *M*). How would you make 1 liter of such a buffer in the laboratory, given only crystalline Na_3PO_4 and concentrated HCl?

6. Calculate how you would prepare 1 liter of a 0.1 *M* histidine buffer, pH 7.0, starting from solid histidine dihydrochloride, 0.5 *N* NaOH, and water.

7. How many milliliters of 0.1 *N* HCl are needed to shift the pH of 1200 ml of a phosphate buffer from pH 5.6 to 5.0 if the concentration is 57 m*M*?

8. How would you prepare 500 ml of a buffer, pH 7.4, which is 0.2 *M* in total phosphate, from a solution of 1 *M* Na_3PO_4 and one of 1 *M* NaH_2PO_4? You may use as much distilled water as you need.

9. (a) Make up 100 ml of 0.05 *M* phosphate buffer, pH 7.4. Calculate the grams of $Na_2HPO_4 \cdot 12H_2O$ and of KH_2PO_4 required.
   (b) Add 10 ml of 0.10 *M* HCl to the above buffer and calculate the resulting pH and hydrogen ion concentration.

10. The purine, xanthine, has two ionizable protons with pK_{a_1} 7.53 and pK_{a_2} 11.63. Using 0.1 *N* NaOH or 0.1 *N* HCl, distilled water, and xanthine, calculate how to prepare a solution (1000 ml) at pH 7.40 which is 0.001 *M* in xanthine.

11. A solution of 0.1 *M* riboflavin (10 liters) is accidentally mixed with 1 liter of 1 m*M* Fe^{++}. What percentage of the riboflavin would be in the complexed form? pK_{st}, − 7.1.

12. You have 75.1 mg of glycine (mol. wt. 75.1) dissolved in 10 ml of 0.10 *M* HCl. 0.10 *M* NaOH is added in small increments and the pH determined after each addition of base. The titration is continued until both the carboxyl and amino groups of the glycine have been titrated. Draw an approximate titration curve. Clearly indicate the following.
    (a) The isoelectric point.
    (b) The points of maximum buffer capacity.
    (c) The pH values at which glycine exists principally as a cation, a zwitter ion (dipolar ion), and as an anion.

13. What volume of 0.06 *N* KOH should be added to 80 ml of 0.12 *N* KH_2PO_4 to give a solution with pH 5.86?

14. Calculate how much citric acid (anhydrous) and NaOH you need to prepare 1 liter of citrate buffer, pH 5.2, which is 0.2 *M* in Na^+.

15. What is the composition of a 0.01 *M* phosphate buffer, pH 5.50, expressed as grams per liter of the anhydrous sodium salts?

16. What volume of 0.2 *M* phosphate buffer, pH 6.47, should be added to 10 ml of 0.1 *N* NaOH to give a buffer with pH 6.65?

17. A 0.2 *N* solution of a weak acid is mixed with an equal volume of 0.05 *N* NaOH. The final pH is 4.2. What is the acid?

18. Calculate the p*I* for the peptide Tyr-Asn-Leu-Thr.

19. Draw the titration curve and calculate the p*I* for vasopressin

    Cys-Tyr-Phe-Gln-Asn-Cys-Pro-Arg-Gly-amide
    └——S——S——┘

    Use the average pK_a values for the functional groups on proteins.

20. What volume of 0.1 *N* $H_2PO_4^-$ would need to be mixed with 25 ml of 0.1 *N* HPO_4^{--} to give pH 7.70?

21. A sample (155 mg) of an amino acid was titrated with 0.03 *N* NaOH. The titration curve in Figure 8.14 (top of the facing page) was obtained.
    (a) What was the equivalent weight of the acid?
    (b) Write the formula for a possible amino acid.
    (c) Calculate the isoelectric point.

22. How would you prepare 1 liter of 0.1 *M* glycylglycine buffer, pH 7.40, pK_a 3.06, pK_{a_2} 8.13, starting with the hydrochloride salt and solid KOH.

23. How much (in moles) EDTA need be added to 1 liter of a solution containing 0.1 mole Ni^{++}·riboflavin complex to free all the vitamin?

$$Ni^{++} + \text{riboflavin} \rightleftharpoons \text{Ni·riboflavin} \qquad pK_{st} - 4.1$$

$$Ni^{++} + \text{EDTA} \rightleftharpoons \text{Ni·EDTA} \qquad pK_{st} - 18.6$$

Figure 8.14

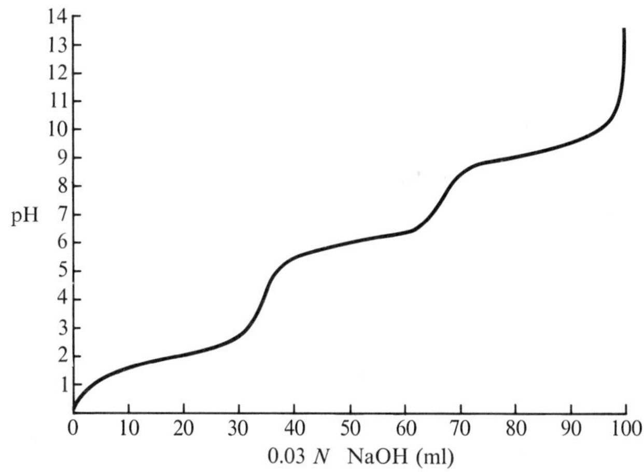

24. Calculate the titration curve of an amino acid, pK_1, 2.5; pK_2, 9.5.

25. Calculate the pI for the peptide Lys-Val-Thr-Asp(NH_2)-Tyr-Glu.

26. (a) Write the structural formula of Lys-His-Tyr.
    (b) What is the approximate pI of this tripeptide?
    (c) In what pH regions would the peptide act as a buffer?

27. A buffer, pH 7.0, prepared from Na_2HPO_4 and KH_2PO_4, is 0.01 M with respect to phosphate. What would be the molar concentration of HPO_4^{--} in the buffer?

28. If you wished to prepare 500 ml of an isotonic sodium phosphate solution of pH 7.40, how much $NaH_2PO_4 \cdot H_2O$ and $Na_2HPO_4 \cdot 7H_2O$ would you have to weigh out?

29. Calculate how much concentrated HCl (11.0 N) should be added to 0.1 M trisodium citrate to give one liter of a buffer, pH 4.25, which is 0.05 M in citrate.

30. Draw the titration curve and calculate the pI for the peptide Lys-Arg-Pro-His-Asp-Glu(NH_2).

31. Glutathione, γ-glutamylcysteinylglycine, has the following pK_a values: 2.12, 2.34, 8.66, and 9.62. Draw the titration curve, indicate which ionic form of glutathione is present at the plateaus of the curve, and calculate the isoelectric point.

32. The pK_a values for arginine are 2.17 (carboxyl), 9.04 (amino), and 12.48 (guanidino). Calculate the relative amounts of the ionic forms of this amino acid at pH 9.04.

33. Draw the titration curve and calculate the p*I* of the α-melanocyte-stimulating hormone (MSH):

    Acetyl-Ser-Tyr-Ser-Met-Glu-His-Phe-Arg-Tyr-Gly-Lys,Pro-Val(NH_2)

34. Calculate how you would prepare 1 liter 0.1 *M* glutamate buffer, pH 10.0, starting with glutamic acid hydrochloride, 0.5 *M* NaOH, and water.

35. If 80 ml of 0.12 *N* KH_2PO_4 is mixed with 20 ml of 0.06 *N* KOH, what is the pH of the final solution?

36. What volume of 0.1 *N* $H_2PO_4^-$ would need to be mixed with 25 ml of 0.1 *N* HPO_4^{--} to give a H^+ concentration of 2×10^{-8} *M*?

37. How many milliliters of 1.0 *N* NaOH is required to titrate 150 ml of 0.5 *M* H_3PO_4 to pH 6.7?

38. Make up 500 ml of a 0.05 *M* malonate buffer of pH 7.1. Calculate the grams of malonic acid, $C_3H_4O_4$ and the milliliters of 0.25 *N* NaOH required.

39. If 10 ml of 0.1 *N* NaOH were added to 50 ml of 0.2 M phosphate buffer, pH 6.47, what would the pH then be?

40. How would you prepare 1 liter of 0.1 *M* glycine buffer, pH 9.4, starting with glycine hydrochloride, sodium hydroxide pellets, and water?

41. (a) Write the structural formula of the tetrapeptide, Glu-Gly-Tyr-Lys.
    (b) What would be the approximate isoelectric point of this peptide?
    (c) In what pH regions would the peptide act as a buffer?

42. From the binding equilibrium data given in Table 8.3, what do you conclude about the nature of the binding? Calculate the appropriate binding constants. Concentration of macromolecule, 15 μM.

Table 8.3

Fraction of binding sites occupied by ligand	$[L]$mM
0.14	0.021
0.27	0.029
0.50	0.043
0.73	0.058
0.86	0.077
0.95	0.120
0.99	0.278

43. A solution of pure hemoglobin (1.55×10^{-5} *M*) was equilibrated with oxygen and the fractional saturation was calculated from the change in

absorbance at 560 nm. The results are summarized in Table 8.4. From the Hill plot calculate the Hill coefficients at the extremes of the saturation and at 50% saturation of the hemoglobin.

Table 8.4

Percent saturation	P_{O_2} *(mm Hg)*
99.7	83.3
99.1	27.8
97.8	15.5
95.3	12.0
86.4	7.7
50.0	4.3
27.3	2.9
13.6	2.1
6.6	1.3
3.0	0.8
1.3	0.4
0.2	0.2

H^2CO^3 and Respiration

1. Arterial blood, pH 7.40, normally carries about 0.0012 M (dissolved $CO_2 + H_2CO_3$). What is the $[HCO_3^-]$?

2. What is the total CO_2 content in volume % (SC) of a plasma sample from a patient whose blood pH is 7.39 and with a blood P_{CO_2} of 40 mm? What would be the total CO_2 if his blood pH were 7.25 but the P_{CO_2} were unchanged?

3. A plasma sample from a patient with diabetic acidosis had P_{CO_2} 26 mm Hg and $[HCO_3^-]$ 14 mM. What is the pH and the $[HCO_3^-]/[CO_2]$ ratio? Is the patient compensated?

4. The average values for $[HCO_3^-]$ and [dissolved CO_2] in arterial blood are 25 mM and 1.23 mM, respectively. Venous blood contains about 1.1 mM more $[HCO_3^-]$ and a P_{CO_2} about 4.65 mm Hg higher. What is the pH of the venous blood?

5. The pK_a of carbonic acid in plasma is 6.1 and the pH of plasma is 7.4. The concentration of HCO_3^- in plasma is 28 meq liter^{-1}.
   (a) How many milliequivalents of acid would be necessary to drop the plasma pH to 7.0 if the $[H_2CO_3]$ is kept constant?
   (b) How many millimoles of H_2CO_3 would be required to change the plasma pH to 7.0 if the $[HCO_3^-]$ is kept constant?

6. What is the pH of a plasma sample which contained 29.3 volumes percent (SC) of [total CO_2] after equilibration with 40 mm Hg pressure of CO_2. The solubility of CO_2 in plasma is 0.03 mmole/liter/mm Hg pressure of CO_2 gas. One mole of CO_2 occupies 22.26 liters at SC.

7. A sample of venous blood plasma showed pH 7.38 and contained a total of 63.0 ml of total CO_2 per 100 ml plasma. Calculate the P_{CO_2} and the bicarbonate concentration.

8. What is the pH of a plasma sample which contains 30 volumes percent of total CO_2 (SC) after equilibration at 38°C with 40 mm P_{CO_2}?

9. Arterial blood plasma normally carries about 0.0240 M HCO_3^- and 0.0012 M H_2CO_3. Calculate the blood pH from these data.

10. Hypoventilation by holding one's breath may increase the [total CO_2] to 30 meq/liter and P_{CO_2} to 60 mm Hg in the blood plasma. What is the pH of the blood?

11. A blood plasma has a pH of 7.4 and a total carbon dioxide of 25 mM. Add sufficient mineral acid to reduce the pH to 7.0. What would be the carbon dioxide tension in millimeters of Hg if no carbon dioxide were allowed to escape?

12. A blood plasma sample has a pH of 7.4 when exposed to a P_{CO_2} of 40 mm Hg. What fraction of the total CO_2 would be released at a P_{CO_2} of 40 mm Hg if sufficient hydrochloric acid were added to produce a pH of 6.1?

13. Hyperventilation may change the blood composition to 27.5 mM [HCO_3^-] and P_{CO_2} 30 mm Hg. What is the pH of the plasma?

14. A sample of human blood plasma has a pH of 7.30 and the total CO_2 was found to be 25 mM. What is the partial pressure of the CO_2? What clinical condition could account for this result?

15. Given a blood plasma whose pH is 7.40 and whose P_{CO_2} is 40 mm Hg, reduce the pH to 7.00 by the addition of acid maintaining the P_{CO_2} at 40 mm Hg. Calculate in milliequivalents the amount of acid that has reacted with the bicarbonate system in 1 liter of plasma.

16. A sample of plasma from arterial blood had pH 7.40 and 1.2 mmole per liter of dissolved CO_2. Calculate the [HCO_3^-] and the [total CO_2], expressing the latter as milliliters of CO_2 per 100 ml plasma at 38°C.

17. A plasma sample had P_{CO_2} 28 mm Hg and [HCO_3^-] 15 mM. What are the pH and the [HCO_3^-]/[CO_2] ratio? Is it more acid than normal?

18. The following conditions, or clinical equivalents, may be experienced daily in a clinical laboratory. Compare the pH, P_{CO_2}, [HCO_3^-]/[CO_2] ratio, and [CO_2] and [HCO_3^-] as mM and [total CO_2] as milliliters of gas at 38°C per 100 ml plasma for each case.
    (a) A patient swallowed 10 g NH_4Cl. After about 1 hour his plasma showed pH 7.36 and [HCO_3^-] 20 mM.
    (b) The pH and total CO_2 of a man who had been breathing deeply and rapidly (hyperventilating) was 7.60 and 20.6 mM respectively.

(c) A patient in hypoventilation showed plasma values of pH 7.30 and $[HCO_3^-]$ 40 m*M*.
(d) A diabetic patient showed plasma pH 7.39 and P_{CO_2} of 30 mm Hg.
(e) A patient swallowed about 2 oz $NaHCO_3$. An hour or so later the plasma showed pH 7.50 and total CO_2 34 m*M*.

---

8. PROBLEMS B No answers given

Buffers, Complex Ions, and Titration Curves

1. How many meq of base are conserved by the kidneys per day in excreting 1600 ml of urine of pH 5.9 which contains phosphate reported as 450 mg % of $H_2PO_4^-$? The blood pH is 7.41.

2. Construct a titration curve for horse heart cytochrome *c* on the acid side of its isoelectric point from the data on its amino acid composition:*

Lysine	19	Tyrosine	4
Histidine	3	Arginine	2
Aspartic acid	8	Amide	8
Glutamic acid	12		

Compare your result with the titration curve given by Bull and Breese.†

3. Hemoglobin and oxyhemoglobin are both weak acids. The pK_a for the former is 7.71 and the latter is 7.16. From the salt/acid ratios at pH 7.4, how much less H^+ is undissociated in oxyhemoglobin than in the same number of moles of hemoglobin?

4. Calculate the concentrations of succinic acid, succinate^{-1}, and succinate^{-2} in sodium succinate, pH 5.0, 0.1 *M* in sodium. Do not approximate. K_{a_1} 6.89×10^{-5}; K_{a_2} 2.47×10^{-6}.

5. (a) Draw the titration curve and calculate the p*I* for the peptide released from trypsinogen in its conversion to the enzyme trypsin,

Val-Asp-Asp-Asp-Asp-Lys

(b) Select and calculate the composition of a buffer system in which the above peptide could be electrophoretically separated from aspartic acid.

6. Given the following complex ion reactions:

$$Cu^{++} + \text{alanine} \rightleftharpoons [\text{Cu·alanine}]^{++} \qquad pK_{st} - 15.10$$

$$Cu^{++} + \beta\text{-alanine} \rightleftharpoons [\text{Cu·}\beta\text{-alanine}]^{++} \qquad pK_{st} - 12.60$$

What is the relative concentration of uncomplexed Cu^{++} in the two systems, assuming identical concentrations of total copper in each case?

---

*E. Margoliash et al., *J. Biol. Chem.* **237**:2148 (1962).
†H. B. Bull and K. Breese, *Biochem. Biophys. Res. Comm.* **24**:74 (1966).

7. A sample (50 mg) of ovalbumin (mol. wt. 45,000) was dissolved in 20 ml of phosphate buffer of pH 7.0. After being subjected to the action of pronase for a time the solution showed a pH drop to 6.8. It required 2.50 ml of 0.01 *N* NaOH to return the pH to 7.0.
   (a) What was the molarity of the phosphate buffer?
   (b) How many peptide bonds per mole of protein were broken?

8. Exactly 100.0 mg of a solid crystalline acid was dissolved in 10 ml of distilled water and then 0.1 *N* NaOH was slowly added. The pH was measured with a pH meter. When 5.0 ml of base had been added, the pH reading was 1.97. The pH rose very rapidly from 4 to 5 when 8.63 ml of base had been added. When 10.0 ml of base had been added the pH was 5.35. Once again the pH rose very rapidly from pH 8 to 10 when 17.26 ml of base had been added.
   (a) What is the equivalent weight of this acid?
   (b) What is its probable molecular weight?
   (c) Assuming that it has two dissociable H^+, what are the pK_{a_1} and pK_{a_2} values of this acid?
   (d) What is the probable chemical structure of this acid?

9. Draw a titration curve for ribonuclease A (for amino acid composition see a biochemistry text) and estimate the p*I*. [See C. Tanford and J. D. Hauenstein, *J. Amer. Chem. Soc.*, **78**:5287 (1956).]

10. The isoelectric points for proteins can be determined from a study of their mobility in an electric field as a function of pH. From the data given below calculate the isoelectric point of β-lactoglobulin at a Γ/2 of 0.01 and 0.03.

Γ/2 = 0.01		Γ/2 = 0.03	
pH	Mobility $\mu sec^{-1}\ volt^{-1}\ cm$	pH	Mobility $\mu sec^{-1}\ volt^{-1}\ cm$
4.13	1.71	4.15	0.93
4.98	0.30	4.62	0.52
4.98	0.01	4.95	0.17
5.58	−1.10	5.52	−0.59
5.84	−1.02	5.79	−0.85
5.94	−1.33	7.09	−1.75

[Data courtesy of L. B. Barnett and H. B. Bull.]

11. Aureomycin became completely complexed with Fe^{++}:

$$Fe^{++} + 2\ \text{aureomycin} \rightleftharpoons [Fe\cdot 2\ \text{aureomycin}]^{++} \qquad pK_{st} - 10.14$$

What mole ratio of EDTA/aureomycin would be required to free 99% of the antibiotic from Fe^{++}?

$$Fe^{++} + \text{EDTA} \rightleftharpoons [Fe\cdot\text{EDTA}]^{++} \qquad pK_{st} = 14.3$$

12. If 100 ml of a 0.05 *M* lactate buffer, pH 4.6, is mixed with 100 ml of a 0.05 *M* phosphate buffer of pH 6.7 what would be the pH of the resulting

solution? Write the equations for the reactions that occurred when the solutions were mixed.

13. The partial composition of human serum albumin, mol. wt. 68,000, in terms of moles of amino acid per 100,000 g protein is reported to be*

Tyrosine	25.7	Aspartic acid	78.2
Tryptophan	0.9	Glutamic acid	11.8
Cystine/2	46.5	Arginine	35.3
Cysteine	5.8	Histidine	22.6
Amide N	62.9	Lysine	84.2

Construct a titration curve on the basis of the pK_a values given in Table 8.1 and compare with that of Tanford et al.†

14. A pronase digestion of a ^{14}C-protein at pH 7.5 in 0.01 *M* phosphate buffer was adjusted to pH 5.0 with glacial acetic acid in order to assay for ^{14}C-labeled L-glutamic acid in the hydrolysate by L-glutamic decarboxylase. Disregarding the buffer effect of the amino acids and peptides, how much acid would need to be added to 10 ml of the digest?

15. 10 ml of an unknown tripeptide is treated with strong HCl to hydrolyze it completely. The excess HCl is removed by vacuum distillation. The solid residue is dissolved in 10 ml of water and titrated with 0.10 *N* NaOH. The titration curve in Figure 8.15 was obtained.

Figure 8.15

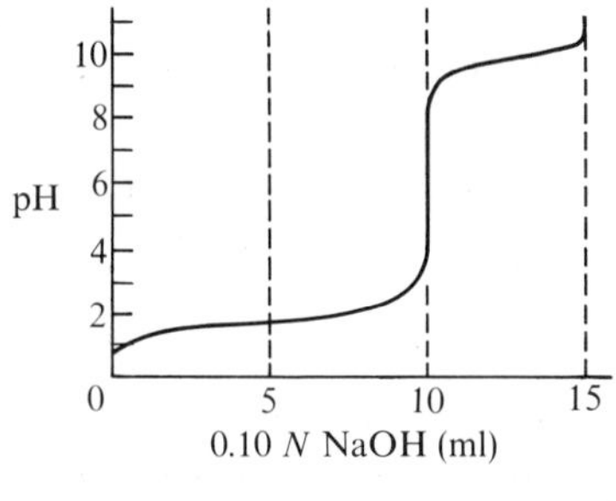

(a) Calculate the molarity of the original tripeptide solution.
(b) Name and write a possible structure of this peptide.

16. Fehling solution, used to test for reducing groups in sugars, contains 34.64 g $CuSO_4 \cdot 5H_2O$ and 173 g sodium potassium tartrate in 1 liter of *N* NaOH:

$$Cu^{++} + 2\ \text{tartrate}^{--} \rightleftharpoons [Cu \cdot 2\ \text{tartrate}]^{--} \qquad pK_{st} - 6.50$$

(a) How much free Cu^{++} is present in the solution?
(b) Why does not the Cu^{++} precipitate as the insoluble hydroxide?

*R. F. Phelps and F. W. Putnam, *The Plasma Proteins*. Academic Press, New York, 1960. Vol. 1, p. 143.
†C. Tanford, S. A. Swanson, and W. S. Shore, *J. Amer. Chem. Soc.*, 77:6414 (1955).

17. A mixture of 8 μmole uracil, 7.2 μmole D-ribose-1-$^{32}PO_4^{--}$ (10,150 cpm per μmole), 40 μmole tris-malate buffer, pH 7.4, and 0.79 mg uridine phosphorylase in a total volume of 1 ml was incubated at 37°C. Periodically, 0.1 ml reaction mixture was withdrawn and added to 0.12 ml magnesia and the suspension centrifuged. The precipitate was dissolved in 0.2 N HCl and the whole sample counted. A parallel blank experiment was performed.

At zero time the counts above background were 50 cpm (blank) and 75 cpm (sample).

At 180 min equilibrium values were obtained at 78 cpm (blank) and 4800 cpm (sample). Assuming the same errors in counting the ribose-1-phosphate as in counting the samples and blanks, what is the equilibrium constant? For further elaboration of this problem consult H. Pontis et al.*

18. In the reaction involving nucleoside diphosphokinase of pea seeds,

$$\text{UDP} + \text{ATP} \rightleftharpoons \text{UTP} + \text{ADP}$$

it was found that the apparent equilibrium constant varied with Mg^{++} concentration as follows.** At pH 8.0 and with $MgCl_2$ concentrations of 1.0, 2.5, and 5.0 mM, the values of K_{eq} were 0.96, 0.91, and 0.98, respectively. If the initial concentrations of ADP and UTP were 2.0 mM and the pK_{st} of the Mg^{++} complexes of ATP, ADP, UTP, and UDP were −4.04, −3.15, −4.02 and −3.17, respectively, what would be the values of K_{eq} assuming that the Mg^{++} complexes were the reactive components? For simplification, take the pK_{st} for both triphosphates to be −4.03 and for both diphosphates to be −3.16.

19. (a) Draw the titration curve and calculate the pI of pressor substance hypertensin I,

Asp-Arg-Val-Tyr-Ileu-His-Pro-Phe-His-Leu

and hypertensin II,

Asp-Arg-Val-Tyr-Ileu-His-Pro-Phe

(b) On the basis of this information, comment on possible methods of separating mixtures of those two peptides.

20. The characterization of phosphoserine (mol. wt. 185.1) in a synthesis† included the titration of 30 mg of material at 27.5°C in 5 ml water with 1.694 N KOH using a recording pH meter and a microburette. The following pK_a values were obtained: pK_{a_1} 1, pK_{a_2} 2.65 ± 0.05, pK_{a_3} 5.91 ± 0.05. Construct a titration curve using 0.2 mmole of phosphoserine in 5 ml water and 2.0 N KOH. What is the equivalent weight of phosphoserine for H‡?

21. The number of sites, n, for binding 5′-AMP to phosphorylase a and the apparent association constant, K_{app}, can be determined from experimental

*H. Pontis, G. Degentedt, and P. Reichard, *Biochim. Biophys. Acta* **51**:138 (1961).
R. J. A. Kirkland and J. F. Turner, *Biochem. J.* **72:716 (1959).
†F. C. Neuhaus and S. Korkes, *Biochem. Preparations*, **6**:78 (1958).
‡E. Helmreich, M. C. Michaelides, and C. F. Cori, *Biochemistry*, **6**:3695 (1967).

measurements of the average number of moles bound to each molecule of enzyme, r, as a function of the concentration, c, of 5′-AMP.

$$\frac{r}{c} = K_{\text{app}}(n - r)$$

If r/c is plotted *versus* r the intercept on the abscissa is n and the slope is K_{app}. The presence of glycogen affects this binding. Calculate the number of binding sites and K_{app} by least-squares fit of the following data:

0.1% glycogen		0.5% glycogen	
r	$c(M \times 10^{+5})$	r	$c(M \times 10^{+5})$
0.7	0.052	0.8	0.036
1.1	0.089	1.5	0.083
2.1	0.225	2.3	0.190
3.1	0.462	3.6	0.450
3.8	1.230	4.3	1.340

The problem is extensively discussed by Helmreich et al.*

22. *p*-Nitrophenyl α-D-mannoside binds to the protein concanavalinA, molecular weight of monomer unit 26,500 daltons, with a resulting change in spectrum. The free ligand absorbs maximally at 305 nm but this absorption maximum is shifted when the ligand is bound, such that the maximum difference between the two forms occurs at 317 nm. In one set of data using 0.038 m*M* concanavalinA the absorbance difference at 317 nm upon binding *p*-nitrophenyl α-D-mannoside was as follows:

p-nitrophenyl α-D-mannoside m*M*	ΔAbs at 317 nm
0.025	0.025
0.050	0.038
0.075	0.047
0.100	0.053
0.125	0.056
0.150	0.060
saturating	0.085

What is the binding constant (equilibrium constant)? How many moles of ligand bind per mole of concanavalinA?

23. A solution of hemoglobin (1 mg/ml) stripped of all effector molecules in buffer at pH 7.4 and 25°C was equilibrated with oxygen. The amount of oxygen bound was calculated from the change in absorbance at 560 nm and the molar absorbance coefficients of deoxyhemoglobin (1.34×10^4/heme) and oxyhemoglobin (0.906×10^4/heme). Compare the oxygen affinities for the two conformers and their Hill coefficients of binding. How does the maximum value of the Hill coefficient compare with the number of binding sites, which is 4. (See Table 8.5 on the next page.)

---

*E. Helmreich, M. C. Michaelides, and C. F. Cori, *Biochemistry,* **6**:3695 (1967).

Table 8.5

Absorbance change at 560 nm/cm light path	P_{O_2} *(mm Hg)*
0.2684	83.30
0.2676	44.75
0.2666	27.83
0.2655	21.54
0.2633	15.51
0.2565	12.00
0.2325	7.74
0.1956	5.78
0.1346	4.31
0.0735	2.99
0.0366	2.08
0.0178	1.29
0.0080	0.81
0.0058	0.62
0.0036	0.40
0.0005	0.21

[Based on data by J. V. Kilmartin, K. Imai, and R. T. Jones; to be published.]

H_2CO_3 and Respiration

1. A patient with suspected encephalitis presented the following blood gas data: P_{CO_2} 18 mm Hg, [total CO_2] 12 meq/liter. Additional lactic acid is being produced by muscles because of the hyperventilation. If the P_{CO_2} remains constant, what would be the effect of adding 2 meq/liter lactic acid to the blood?

2. You are given a solution of 0.10 N NaOH, a cylinder of carbon dioxide gas, as much water as you need, and a manometer. Describe very briefly how you would prepare a liter of bicarbonate buffer with the same concentration and pH as the bicarbonate buffer of blood plasma. Take the pK_a of carbonic acid as 6.10.

3. The concentration of HCO_3^- in a plasma, pH 7.4, is 28 meq per liter.
   (a) How many meq of acid would be necessary to drop the plasma pH to 7.2 if the [H_2CO_3] is kept constant?
   (b) How many millimoles of H_2CO_3 would be required to change the plasma pH to 7.2 if the [HCO_3^-] is kept constant?
   (c) How many meq of acid would be necessary to change the plasma from pH 7.4 to 7.2 if at the same time the H_2CO_3 concentration is reduced by one-half?
   (d) Under which conditions—(a), (b), or (c)—is the system bicarbonate-carbonic acid a better buffer for the pH range 7.4 to 7.2?

4. 1.2 ml of CO_2 at standard temperature and pressure are removed from a 2-ml sample of plasma after addition of acid. What is the CO_2 content of the plasma in volumes percent and in millimoles?

5. Carbon dioxide gas was liberated from 1 ml of plasma by adding acid. The gas was measured and collected over water. The volume of gas liberated was 0.5 ml at 25°C and 753.8 mm Hg pressure. The vapor pressure of water at 25°C is 23.8 mm Hg.
   (a) What is the pressure of dry CO_2 gas under the above conditions?
   (b) What is the volume of dry CO_2 at standard conditions?
   (c) What is the volume percent of CO_2 in the plasma at standard conditions?

6. The following clinical conditions may be experienced with samples brought daily to a clinical laboratory. Compare the pH, P_{CO_2} $[HCO_3^-]$ and $[HCO_3^-]/[CO_2]$ ratio in each sample.
   (a) Moderate respiratory alkalosis from mild hyperventilation, P_{CO_2} 31 mm Hg and $[HCO_3^-]$ 24.5 m*M*.
   (b) Uncompensated respiratory acidosis due to a narcotic overdose, pH 7.22 $[CO_2]$ 1.81 m*M*.
   (c) Partially compensated respiratory acidosis from a long-standing chronic obstructive lung disease, pH 7.37 $[HCO_3^-]$ 35.0 m*M*.
   (d) Diabetic with severely depressed ventilation by heavy sedation, P_{CO_2} 50 mm Hg, $[HCO_3^-]$ 15.0 m*M*.

7. The two most important inorganic buffer systems in the body are the phosphoric and carbonic acid systems. At the pH of blood (7.4), what is the ratio of the following?

$$\frac{[PO_4^{---}]}{[HPO_4^{--}]}, \quad \frac{[HPO_4^{--}]}{[H_2PO_4^-]}, \quad \frac{[H_2PO_4^-]}{[H_3PO_4]}, \quad \text{and} \quad \frac{[CO_3^{--}]}{[HCO_3^-]}, \quad \frac{[HCO_3^-]}{[H_2CO_3]}$$

9

Oxidation-Reduction in Biological Systems

OXIDATION AND REDUCTION

An electrical dry battery is composed essentially of two half cells, between which there is a difference in electrical potential, expressed as a voltage. One of the simplest batteries can be constructed, as shown in Figure 9.1, from an iron electrode immersed in a solution of a ferrous salt connected through a salt bridge to a cupric salt solution into which is placed a copper electrode. The salt bridge is an electrical connection provided by an aqueous solution of a salt (for example, potassium chloride) which plays no role in the chemical reactions at each electrode. Without the completion of this circuit either by the salt bridge or between the metal electrodes, no change occurs at either electrode although each possesses the intensive property* of a potential for change. This potential may be considered as an electron pressure at each electrode, which may be represented

$$Fe^{++} + 2e^- \rightleftharpoons Fe \tag{9.1}$$

$$Cu^{++} + 2e^- \rightleftharpoons Cu \tag{9.2}$$

---

*The properties of a system may be classified as extensive or intensive. Extensive properties are additive—for example, volume and mass. Intensive properties are not additive so that their sum is not equal to the value for the whole system. Examples of intensive properties are temperature, density, and electrical potential, where the amount of material present is not important (see Chapter 10).

This potential for change can be measured experimentally by connecting the two metal electrodes through a voltmeter, which theoretically should have an infinite resistance and, therefore, permits no electron flow or change in each of the half-reactions (9.1) and (9.2). In this example the experimentally measured potential is 0.777 volt under normal conditions of temperature (25°C).

Figure 9.1

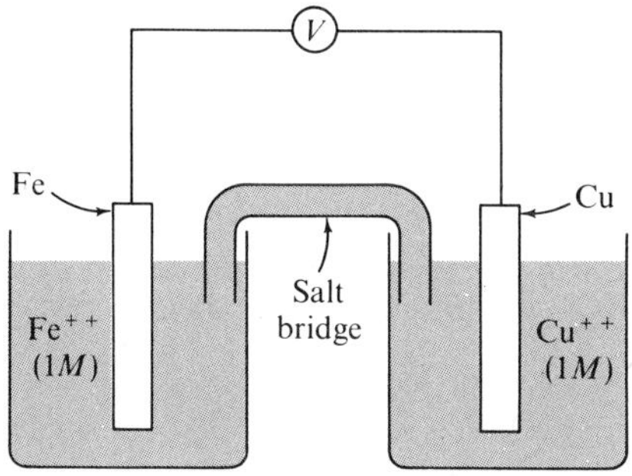

Electrical energy can be derived from chemical oxidation-reduction (redox) systems by permitting an electron flow to occur. In the above example, if the voltmeter in Figure 9.1 were bypassed through a conductor, electrons would flow from the iron electrode to the copper electrode, these electrons being derived from a conversion (oxidation) of the iron to ferrous ions. At the copper electrode the electrons are picked up by the cupric ions, which are converted (reduced) to metallic copper. Since the electrons are negatively charged and are originating from the iron half-reaction, the iron electrode is the negative electrode.

As the electron flow continues between the two half-reactions, the composition of each cell changes. The overall reaction is given for our example by a summation of reaction (9.2) with reaction (9.1) written in the reverse, which we can do because it is a reversible reaction. The summation is

$$Cu^{++} + Fe \rightleftharpoons Cu + Fe^{++} \tag{9.3}$$

Oxidation is defined most generally as the loss of electrons by a chemical species. There may be an actual loss of electrons, as in the oxidation of an atom,

$$Cu \rightleftharpoons Cu^{++} + 2e^- \tag{9.4}$$

or an ion,

$$Fe^{++} \rightleftharpoons Fe^{+++} + e^- \tag{9.5}$$

but the electron may be removed with a hydrogen atom, as in the oxidation of an alcohol to an aldehyde,

$$CH_3CH_2OH \rightleftharpoons CH_3CHO + 2H^+ + 2e^- \tag{9.6}$$

or the addition of an oxygen atom, as in the case of the oxidation of hydrogen,

$$2H^+ + \tfrac{1}{2}O_2 + 2e^- \rightleftharpoons H_2O \tag{9.7}$$

In the latter case we may consider that the valency electrons of the hydrogen are shared by the oxygen atom and in that sense are lost.

From our knowledge of chemical systems it is clear that the electrons and protons derived from hydrogen will not remain free in solution and that they will react with another chemical species present in the solution, which is then said to be reduced. For example,

$$Zn^{++} + 2e^- \rightleftharpoons Zn \tag{9.8}$$

In fact *all oxidation reactions must occur concomitantly with a reduction.* By the same token, an oxidation-reduction reaction can always be divided into two parts—the oxidation and the reduction. These two associated reactions are called half-reactions; a similar reaction to (9.3) is the oxidation-reduction reaction

$$Cu^+ + Fe^{+++} \rightleftharpoons Cu^{++} + Fe^{++} \tag{9.9}$$

We have the copper half-reaction

$$Cu^+ \rightleftharpoons Cu^{++} + e^- \tag{9.10}$$

which is written in the form of an oxidation, and the iron half-reaction

$$Fe^{+++} + e^- \rightleftharpoons Fe^{++} \tag{9.11}$$

which is written in the direction of a reduction. Obviously, the copper half-reaction (9.10) could also be written in the form of a reduction,

$$Cu^{++} + e^- \rightleftharpoons Cu^+ \tag{9.12}$$

As was discussed under reversible reactions, it is conventional to read these equations as reactants on the left side going to products on the right side. Reactions (9.1), (9.2), (9.7), (9.8), (9.11), and (9.12) are written in the form of a reduction, the potentials for which express the relative tendencies for these reactants to be reduced and therefore to oxidize some other material. This potential is equal but opposite to the tendencies for these same reactions to go in the opposite direction, in the form of an oxidation, as shown for the reduction (9.12), going in the direction of oxidation (9.10).

The potential for reduction is obviously one of two equally suitable ways to discuss redox systems and is the most common in biochemistry. The half-reactions are tabulated with least confusion by giving the complete half-reaction which, as a reduction, will have the electrons on the left side. When the reactions involve reduction by adding hydrogen, the hydrogen is represented as a proton plus an electron, as in (9.7). The redox half-reactions used in this text are collected in Table 9.1. In general,

$$\text{oxidized form} + ne^- \rightleftharpoons \text{reduced form}$$

From a knowledge of half-reactions it is possible to combine many oxidation-reduction pairs to give representations of known chemical events. In the area of biochemistry these oxidation-reduction reactions are mediated by a series of common intermediate molecules (called coenzymes), usually with an eventual reaction of the last reduced specie with oxygen. A common sequence of oxidation-reduction steps might be as represented in Figure 9.2. The oxidation of the substrate usually commences with a series of transfers of hydrogen atoms, followed by oxidations that involve electron transfers, and finally the reduction of oxygen to water.

Figure 9.2

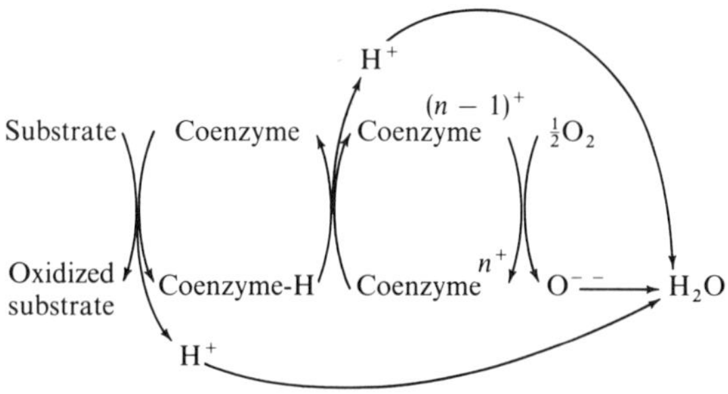

Equivalent Weight. The addition of the two component half-reactions demands that each be balanced to involve the same number of electrons. This is the same as saying that the oxidation-reduction equivalent of each half-cell is reacted. Such an equivalence is referred to in terms of a normal redox solution, which is the weight (in grams) of the chemical species that will reduce one mole of H^+, or is involved in the loss or gain of one mole of electrons. For example,

$$\underset{\text{pyruvic acid}}{\begin{array}{c} CH_3 \\ | \\ C{=}O \\ | \\ COOH \end{array}} + 2H^+ + 2e^- \rightleftharpoons \underset{\text{lactic acid}}{\begin{array}{c} CH_3 \\ | \\ CHOH \\ | \\ COOH \end{array}} \qquad (9.13)$$

$$\text{redox equiv. wt.} = \frac{\text{g-mol. wt.}}{2}$$

$$\underset{\text{ferric}}{Fe^{+++}} + e^- \rightleftharpoons \underset{\text{ferrous}}{Fe^{++}} \tag{9.14}$$

Table 9.1
Standard reduction potentials

System	E_0(*pH* 0 *at* 30°*C*) *volt*	E_0'(*pH* 7.0 *at* 30°*C*) *volt*
$\frac{1}{2}O_2 + 2H^+ + 2e^- \rightarrow H_2O$	+1.229	+0.816
$Fe^{+++} + e^- \rightarrow Fe^{++}$	+0.771	+0.771
$Br_2 + 2e^- \rightarrow 2Br^-$	+0.652	+0.652
$I_2 + 2e^- \rightarrow 2I^-$	+0.536	+0.536
$O_2 + 2H^+ + 2e^- \rightarrow H_2O_2$		+0.30
Cytochrome-a $Fe^{+++} + e^- \rightarrow$ cytochrome-a Fe^{++}	+0.290	+0.29
Cytochrome-c $Fe^{+++} + e^- \rightarrow$ cytochrome-c Fe^{++}		+0.25
2,6-dichlorophenolindophenol + $2H^+ + 2e^- \rightarrow$ reduced 2,6-dichlorophenolindophenol		+0.22
Methemoglobin $Fe^{+++} + e^- \rightarrow$ hemoglobin Fe^{++}		+0.14
Ubiquinone + $2H^+ + 2e^- \rightarrow$ ubiquinone H_2		+0.10
Cytochrome-b $Fe^{+++} + e^- \rightarrow$ cytochrome-b Fe^{++}		+0.08
Dehydroascorbate + $2H^+ + 2e^- \rightarrow$ ascorbate	+0.390	+0.060
Fumarate + $2H^+ + 2e^- \rightarrow$ succinate	+0.433	+0.031
Methylene blue$^+$ + $2e^- + 2H^+ \rightarrow$ leuco-methylene blue H^+	+0.532	+0.011
FAD + $2H^+ + 2e^- \rightarrow$ FAD2H		−0.06
Oxalacetate + $2H^+ + 2e^- \rightarrow$ malate	+0.318	−0.102
Acetaldehyde + $2H^+ + 2e^- \rightarrow$ ethanol		−0.163
Pyruvate + $2H^+ + 2e^- \rightarrow$ lactate	+0.224	−0.190
(Glutathione-S)$_2$ + $2e^- + 2H^+ \rightarrow$ 2 glutathione-SH		−0.23
Sufranine-T + $2e^- \rightarrow$ leucosafranine-T		−0.289
1,3-diphosphoglycerate + $2H^+ + 2e^- \rightarrow$ 3-phosphoglyceraldehyde + phosphate		−0.29
Lipoic acid $\langle{}^{S}_{S}\vert$ + $2H^+ + 2e^- \rightarrow$ lipoic acid $(SH)_2$		−0.29
Acetoacetate + $2H^+ + 2e^- \rightarrow$ L-β-hydroxybutyrate		−0.293
$(C_6H_5S)_2 + 2H^+ + 2e^- \rightarrow 2C_6H_5SH$		−0.30
$NAD^+ + 2H^+ + 2e^- \rightarrow NADH + H^+$		−0.320
$NADP^+ + 2H^+ + 2e^- \rightarrow NADPH + H^+$		−0.32
(Cyst-S)$_2$ + $2H^+ + 2e^- \rightarrow$ 2-cysteine-SH		−0.34
Uric acid + $2H^+ + 2e^- \rightarrow$ xanthine		−0.36
Acetyl CoA + $2H^+ + 2e^- \rightarrow$ acetaldehyde + CoA		−0.41
$H^+ + e^- \rightarrow \frac{1}{2}H_2$	+0.000	−0.420
Gluconate + $2H^+ + 2e^- \rightarrow$ glucose + H_2O		−0.45

$$\text{redox equiv. wt.} = \frac{\text{g-mol. wt.}}{1}$$

$$H_2O_2 + 2e^- \rightleftharpoons H_2O + O^{--} \qquad (9.15)$$

$$\text{redox equiv. wt.} = \frac{\text{g-mol. wt.}}{2}$$

From these examples one could write the reaction involving the oxidation of lactic acid with either Fe^{+++} or H_2O_2. In the first case the 2-electron change of reaction (9.13) requires two moles of Fe^{+++}. Therefore, writing the two equations with the proper redox equivalents, reversing (9.13) to (9.16), and adding (9.14), we have

$$\begin{array}{l} CH_3 \\ | \\ CHOH \\ | \\ COOH \end{array} \rightleftharpoons \begin{array}{l} CH_3 \\ | \\ C{=}O + 2H^+ + 2e^- \\ | \\ COOH \end{array} \qquad (9.16)$$

$$2Fe^{+++} + 2e^- \rightleftharpoons 2Fe^{++}$$

Sum:

$$\begin{array}{l} CH_3 \\ | \\ CHOH + 2Fe^{+++} \\ | \\ COOH \end{array} \rightleftharpoons \begin{array}{l} CH_3 \\ | \\ C{=}O + 2H^+ + 2Fe^{++} \\ | \\ COOH \end{array} \qquad (9.17)$$

Similarly, from (9.16) and (9.15),

$$\begin{array}{l} CH_3 \\ | \\ CHOH \\ | \\ COOH \end{array} \rightleftharpoons \begin{array}{l} CH_3 \\ | \\ C{=}O + 2H^+ + 2e^- \\ | \\ COOH \end{array}$$

$$H_2O_2 + 2e^- \rightleftharpoons H_2O + O^{--}$$

Sum:

$$\begin{array}{l} CH_3 \\ | \\ CHOH + H_2O_2 \\ | \\ COOH \end{array} \rightleftharpoons \begin{array}{l} CH_3 \\ | \\ C{=}O + H_2O + \underbrace{(2H^+ + O^{--})}_{H_2O} \\ | \\ COOH \end{array} \qquad (9.18)$$

The fact that such pairs of half-reactions can be written does not predict that the overall reaction will take place, nor does it indicate the equilibrium position of the reaction.

REDOX POTENTIAL

The tendency for reduction to occur differs for each half-reaction and, as is appreciated from the electromotive series in inorganic chemistry, the half-reactions can be placed in an order of decreasing standard reduction potential (see Table 9.1). In other words, the oxidant of the higher half-reaction will oxidize the reduced form of any half-reaction below it. Thus, the half-reaction of oxygen, written as a reduction,

$$\tfrac{1}{2}O_2 + 2e^- \rightleftharpoons O^{--} \tag{9.19}$$

has a greater positive potential than the half-reaction of hydrogen, written in the same direction,

$$2H^+ + 2e^- \rightleftharpoons H_2 \tag{9.20}$$

and oxygen will oxidize H^+ and itself be reduced:

$$\begin{array}{c} \tfrac{1}{2}O_2 + H_2 \rightleftharpoons 2H^+ + O^{--} \\ \qquad\qquad\qquad\quad \Updownarrow \\ \qquad\qquad\qquad\quad H_2O \end{array} \tag{9.21}$$

We return now to the way in which half-reactions are tabulated. The reduction potentials are expressed in volts compared with a standard hydrogen electrode taken by definition as having zero potential. In these quantitative terms the half-reaction with the more positive (least negative) reduction potential will oxidize any half-reaction less positive in reduction potential. To put this in another way, there is a greater tendency to be reduced in the half-reactions with more positive potentials so that a higher half-reaction will take electrons from any lower half-reaction, thereby causing it to become oxidized.

As we have seen earlier, to construct an electrical battery two half-reactions are connected together. The potential so generated can be measured on a potentiometer. In the case of the reduction of Fe^{+++} by H_2, the available electrical energy could be generated in the system in Figure 9.3. The greater tendency for Fe^{+++} to be reduced by the hydrogen half-reaction causes electrons to flow from the hydrogen half-cell toward the $Fe^{+++}|Fe^{++}$ half-cell. In other words, the $Fe^{+++}|Fe^{++}$ half-cell has the more positive electrical potential, equal in fact to +0.771 volt compared to the standard $H^+|H_2$ half-cell and when $[Fe^{+++}]$ and $[Fe^{++}]$ are equal.

Some practical observations can be summarized from the above comments.

1. Tables of redox potentials may be given as either oxidation potentials or reduction potentials. From our definition of oxidation and reduction it should be easy to recognize the difference by seeing whether the

Figure 9.3

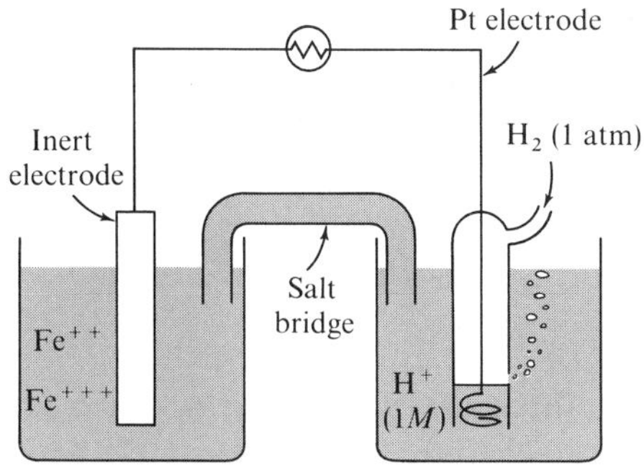

half-reaction is written in the direction of reduction or oxidation. Unfortunately it has been a common practice, particularly in textbooks, to represent the half-reaction as a ratio, for example, $\frac{1}{2}O_2/H_2O$, pyruvate/lactate, so that it was not clear whether these were indicating direction of reaction, such as

$$NAD^+ + 2H^+ + 2e^- \rightleftharpoons NADH + H^+ \tag{9.22}$$

$$\text{pyruvate} + 2H^+ + 2e^- \rightleftharpoons \text{lactate} \tag{9.23}$$

or whether they represented product/reactant, in which case the reactions above would be reversed. Since it is well known that oxygen is a good oxidizing agent, if the half-reaction potential for oxygen-water in the table is positive then the reactions are written as reductions and the values are reduction potentials. For these reasons, and following a convention of electrochemists, a half-reaction is represented in this treatment by separating the oxidized form by a vertical line from the reduced form: Ox|Red.

2. The oxidation potential is equal but opposite in sign to the reduction potential. This is important when summating the potentials of two half-reactions after having reversed the direction of one to an oxidation.

Problem 9.1: Write the redox reaction of ferridoxin and cytochrome-b from the half-reactions

cytochrome-b $Fe^{+++} + e^- \rightleftharpoons$ cytochrome-b Fe^{++}	$E_0 - 0.040$ volt	(i)
ferridoxin $Fe^{+++} + e^- \rightleftharpoons$ ferridoxin Fe^{++}	$E_0 - 0.432$ volt	(ii)

Solution: Since the least positive half-reaction (ii) will be oxidized by (i), the half-reaction (ii) should be rewritten in the direction of oxidation:

ferridoxin $Fe^{++} \rightleftharpoons$ ferridoxin $Fe^{+++} + e^-$	$E_0 + 0.432$ volt	(ii′)

Note that E_0 is changed in sign.

The total reaction (i) + (ii′) is therefore

cytochrome-b Fe^{+++} + ferridoxin $Fe^{++} \rightleftharpoons$
cytochrome-b Fe^{++} + ferridoxin Fe^{+++}

$$\Delta E_0 = (-0.040) + (+0.432) \text{ volt}$$
$$= +0.392 \text{ volt}$$

Further observations on half-reactions follow.

3. The potential difference between two half-reactions, ΔE_0 without regard to the direction of electron flow, is obviously the algebraic difference between the two half-reaction reduction potentials. In the above example, the two reduction potentials were −0.040 volt and −0.432 volt, so the potential gap between the two is (0.432 − 0.040) volt or 0.392 volt.

4. To write a balanced redox reaction, the electron change in each half-reaction must be the same. It is useful to consider a redox reaction to be composed of two reversible reactions coupled through the electrons. This may involve doubling or tripling one of the half-reactions, but such a manipulation *does not* double or triple the half-reaction potential, since potential is an intensive property:

$$Fe^{+++} + e^- \rightleftharpoons Fe^{++} \qquad E_0 + 0.771 \text{ volt}$$

$$2Fe^{+++} + 2e^- \rightleftharpoons 2Fe^{++} \qquad E_0 + 0.771 \text{ volt}$$

Problem 9.2: Write the reaction between ethanol and Cu^{++} in 1 N H_2SO_4.

Solution:

$$\text{acetaldehyde} + 2H^+ + 2e^- \rightleftharpoons \text{ethanol} \qquad E_0 - 0.163 \text{ volt} \qquad \text{(i)}$$

$$Cu^{++} + e^- \rightleftharpoons Cu^+ \qquad E_0 + 0.153 \text{ volt} \qquad \text{(ii)}$$

Since (i) has the least positive potential, it should be reversed to an oxidation reaction,

$$\text{ethanol} \rightleftharpoons \text{acetaldehyde} + 2H^+ + 2e^- \qquad E_0 + 0.163 \text{ volt} \qquad \text{(i′)}$$

To balance this equation, (ii) needs to be doubled

$$2Cu^{++} + 2e^- \rightleftharpoons 2Cu^+ \qquad E_0 + 0.153 \text{ volt} \qquad \text{(ii′)}$$

Note E_0 for (ii) and (ii′) is the same. Add (i′) and (ii′):

$$\text{ethanol} + 2Cu^{++} \rightleftharpoons \text{acetaldehyde} + 2Cu^+ + 2H^+ \qquad \Delta E_0 + 0.316 \text{ volt}$$

Problem 9.3: Calculate the potential (in volts) between $I_3^-|3I^-$ and $Fe^{+++}|Fe^{++}$ when all components are one molar. E_0: +0.536 volt for $I_3^-|3I^-$; +0.771 volt for $Fe^{+++}|Fe^{++}$.

Solution: The ΔE_0 for the total reaction will be appreciated most completely in the general electrochemical sense by proceeding by steps as follows.

(a) Write the half-reactions as reductions so that the half-potentials and their signs will be properly represented:

$$Fe^{+++} + e^- \rightleftharpoons Fe^{++} \qquad E_0 + 0.771 \text{ volt} \tag{i}$$

$$I_3^- + 2e^- \rightleftharpoons 3I^- \qquad E_0 + 0.536 \text{ volt} \tag{ii}$$

(b) Reverse the direction of writing the reaction with the least positive or more negative reaction potential. Reaction (ii) is thus written as an oxidation reaction so that the sign of the potential will be changed:

$$3I^- \rightleftharpoons I_3^- + 2e^- \qquad E_0 - 0.536 \text{ volt} \tag{ii'}$$

(c) Balance equations (i) and (ii′) so that the same number of electrons is involved in each:

$$2Fe^{+++} + 2e^- \rightleftharpoons 2Fe^{++} \qquad E_0 + 0.771 \text{ volt} \tag{i'}$$

$$3I^- \rightleftharpoons I_3^- + 2e^- \qquad E_0 - 0.536 \text{ volt} \tag{ii'}$$

$$\text{Sum: } 2Fe^{+++} + 3I^- \rightleftharpoons 2Fe^{++} + I_3^- \quad \Delta E_0 = +0.771 \text{ volt} + (-0.536) \text{ volt}$$

$$= +0.235 \text{ volt}$$

As a point of interest, it will be remembered that the components of the half-reactions were all equimolar. However, although a one molar solution of Fe^{+++} or Fe^{++} ions in reaction (i) is also a normal redox solution, one molar I_3^- is 2 normal and therefore is present in double excess. The consequences of this in terms of the equilibrium will be considered later, but since only the potential difference of half-cells in the given initial concentrations was asked for, the calculation above is sufficient.

RELATION OF REDOX POTENTIAL AND CONCENTRATION

In all of the above examples concerning the redox potentials it has been understood that the reactants and the products in each half-reaction were in their standard states—that is, one molar. At any other ratio the observed electrode potential E_{obs}, which is related to E_0, the standard electrode potential, is calculated in the following way:

$$E_{obs} = E_0 + \frac{RT}{nF} \ln \frac{[\text{oxidized form}]}{[\text{reduced form}]} \tag{9.24}$$

$$= E_0 + 2.303 \frac{RT}{nF} \log \frac{[\text{oxidized form}]}{[\text{reduced form}]} \tag{9.25}$$

where $R = 8.314$ joules degree^{-1} mole^{-1},

n = number of moles of electron change in the half-reaction per mole of reactant,

F = Faraday, which is 96,494 coulombs on the chemical scale,

T = degrees Kelvin.

It will be noted that the unit of RT/nF is the volt.

$$\frac{RT}{nF} = \frac{R\text{ joules}}{\text{degree} \cdot \text{mole of reactant}} \times T\text{ degrees} \times \frac{\text{moles of reactant}}{n\text{ moles of electrons}}$$

$$\times \frac{\text{moles of electrons}}{F\text{ coulombs}} = \text{joule} \cdot \text{coulombs}^{-1} = \text{volt}$$

Selecting a temperature of 30°C (or 303°K), Equation (9.25) reduces to (9.26) with logs to base 10:

$$E_{\text{obs}} = E_0 + \frac{2.303 \times 8.314 \times 303}{n \times 96{,}494} \log \frac{[\text{Ox}]}{[\text{Red}]} \tag{9.26}$$

or

$$E_{\text{obs}} = E_0 + \frac{0.06}{n} \log \frac{[\text{Ox}]}{[\text{Red}]} \tag{9.27}$$

At any other temperature the constant, 0.06 volt per mole of electrons, will be different. It will be realized that in any redox system it is the potential, E_{obs}, for each half-reaction that directs which will be oxidized or reduced.

Problem 9.4: Determine the direction of the reaction between a solution containing initially 10^{-4} M acetaldehyde, 0.1 M ethanol, 0.1 M pyruvate, and 10^{-2} M lactate.

Solution:

$$\text{acetaldehyde} + 2\text{H}^+ + 2e^- \rightleftharpoons \text{ethanol} \qquad E_0 - 0.163 \text{ volt} \tag{i}$$

$$\text{pyruvate} + 2\text{H}^+ + 2e^- \rightleftharpoons \text{lactate} \qquad E_0 - 0.190 \text{ volt} \tag{ii}$$

For (i),

$$E_{\text{obs}} = -0.163 + \frac{0.06}{2} \log \frac{10^{-4}}{10^{-1}}$$

$$= -0.163 + 0.03 \times (-3.00) = -0.253 \text{ volt}$$

For (ii),

$$E_{\text{obs}} = -0.190 + \frac{0.06}{2} \log \frac{10^{-1}}{10^{-2}}$$

$$= -0.190 + 0.03 \times (+1.00) = -0.160 \text{ volt}$$

The most positive half-reaction is pyruvate|lactate, so that the pyruvate will be reduced to lactate and the ethanol oxidized to acetaldehyde until such time as the amounts of these materials give equal E_{obs} values.

The following steps summarize the general approach to the solution of elementary redox problems.

1. Determine the nature of the reactants and products in the reaction in question.
2. Identify the half-reactions with the appropriate electron changes.
3. Write the half-reactions as reductions:

$$\text{oxidized form} + ne^- \rightleftharpoons \text{reduced form}$$

4. Assign to each half-reaction its appropriate standard potential, making particular note of the sign of the potential and remembering that this is the potential referred to the standard $H^+|H_2$ half-reaction which is at pH 0 and conventionally set at zero volt potential.
5. Use the number of electrons in each half-reaction to balance the overall reaction. This number of electrons is n in Equation (9.27).
6. From the initial concentrations of the oxidized and reduced forms in each half-reaction and their standard potentials, calculate from Equation (9.27) the initial reaction potential.
7. Rewrite the half-reaction with the least positive (or most negative) potential in the direction of oxidation, therefore reversing the sign of the potential.
8. Add the balanced half-reactions to obtain the overall reaction which should contain the same net ionic charges on each side.
9. Add the calculated half-reaction potentials (E_{obs}) to give the overall reaction potential (ΔE_{obs}).
10. By the nature of step 7 the reaction so obtained will be thermodynamically possible because ΔE_{obs} will be positive and will proceed spontaneously as written. This point will be amplified later.

Problem 9.5: What reaction will take place if 0.1 mole Fe^{+++}, 0.01 mole Fe^{++}, and 0.11 mole Br^- are added to 100 ml of water saturated with Br_2 in the presence of excess liquid Br_2?

Solution:

Step 1. If any reaction of a redox nature were to occur, the Br_2 would react with Fe^{++} and Br^- with Fe^{+++}.

Steps 2, 3, and 4. Half-reactions would refer to $Br_2|Br^-$ and $Fe^{+++}|Fe^{++}$:

$$Fe^{+++} + e^- \rightleftharpoons Fe^{++} \qquad E_0 + 0.771 \text{ volt} \tag{i}$$

$$Br_2 + 2e^- \rightleftharpoons 2Br^- \qquad E_0 + 1.087 \text{ volts} \tag{ii}$$

Step 5. To balance the electrons, (i) must be written

$$2Fe^{+++} + 2e^- \rightleftharpoons 2Fe^{++} \qquad E_0 + 0.771 \text{ volt} \qquad \text{(iii)}$$

Step 6. For the $Fe^{+++}|Fe^{++}$, the molar concentration is $0.10 \times (1000/100)$ for Fe^{+++} and $0.01 \times (1000/100)$ for Fe^{++}. From equation (iii),

$$\begin{aligned} E_{obs} &= +0.771 + \frac{0.06}{2} \log \frac{\left[0.10 \times \frac{1000}{100}\right]^2}{\left[0.01 \times \frac{1000}{100}\right]^2} \\ &= +0.771 + \frac{0.06}{2} \log 100 \\ &= +0.771 + 0.03 \times 2 \\ &= +0.831 \text{ volt} \end{aligned}$$

For the $Br_2|Br^-$ reaction, the Br_2 concentration is 1 (pure element in excess) and Br^- is $0.11 \times (1000/100)\ M$

$$\begin{aligned} E_{obs} &= +1.087 + \frac{0.06}{2} \log \frac{1}{\left[0.11 \times \frac{1000}{100}\right]^2} \\ &= +1.087 + 0.03 \log 0.827 \\ &= +1.087 + 0.03\ (\bar{1}.918) \end{aligned}$$

and the log must be converted to a complete negative number

$$\begin{aligned} E_{obs} &= +1.087 + 0.03\ (-0.082) \\ &= +1.087 - 0.0025 \\ &= +1.084 \text{ volts} \end{aligned}$$

Steps 7 and 8. Equation (iii) is least positive and is reversed to an oxidation:

$$2Fe^{++} \rightleftharpoons 2Fe^{+++} + 2e^- \qquad E_{obs} - 0.831 \text{ volt} \qquad \text{(iii)}$$

$$Br_2 + 2e^- \rightleftharpoons 2Br^- \qquad E_{obs} + 1.084 \text{ volts} \qquad \text{(i)}$$

$$\text{Sum: } Br_2 + 2Fe^{++} \rightleftharpoons 2Fe^{+++} + 2Br^-$$

Step 9. Overall reaction potential (initial) is $(-0.831 + 1.084)$ volt, or

$$\Delta E_{obs} = +0.253 \text{ volt}$$

Problem 9.6: What is the potential difference in volts between the half-reaction composed of $0.10\ M\ Cu^{++}$ and $0.05\ M\ Cu^+$, and one of $0.01\ M\ Fe^{+++}$ and $0.10\ M\ Fe^{++}$? E_0: $Cu^{++}|Cu^+ + 0.153$ volt; $Fe^{+++}|Fe^{++} + 0.771$ volt.

Solution:

Step 1.

$$Fe^{+++} + e^- \rightleftharpoons Fe^{++},\ E_{obs} = +0.771 + \frac{0.06}{1} \log \frac{0.01}{0.10} \qquad \text{(i)}$$

$$= +0.771 + [0.06 \times (-1.0)]$$

$$= +0.711 \text{ volt}$$

$$Cu^{++} + e^- \rightleftharpoons Cu^+,\ E_{obs} = +0.153 + \frac{0.06}{1} \log \frac{0.10}{0.05} \qquad \text{(ii)}$$

$$= +0.153 + 0.06 \times 0.301$$

$$= +0.171 \text{ volt}$$

Step 2. As in above problem,

$$Cu^+ \rightleftharpoons Cu^{++} + e^- \qquad E_{obs} - 0.171 \text{ volt} \qquad \text{(ii')}$$

Step 3. Reactions (i) and (ii') are balanced:

$$\text{Sum: (i) + (ii')}\ Fe^{+++} + Cu^+ \rightleftharpoons Fe^{++} + Cu^{++}$$

$$\Delta E_{obs} = +0.711 + (-0.171) = +0.540 \text{ volt}$$

---

pH DEPENDENCE OF REDOX SYSTEMS

The potential of the $H^+|H_2$ half-reaction varies with the concentration of the components just as do the other half-reactions. From Equation (9.27) we could write for $H^+|H_2$:

$$E_{obs} = E_0 + \frac{0.06}{1} \log \frac{[H^+]}{[H_2]^{1/2}} \qquad (9.28)$$

As noted earlier, in the standard state of H^+, which is one molar, the E_0 is arbitrarily set at zero volts. Then

$$E_{obs} = 0.06 \log \frac{[H^+]}{[H_2]^{1/2}}$$

Since $[H_2]$ is 1 and $-\log\,[H^+] = \text{pH}$, the equation simplifies to

$$E_{obs} = -0.06\ \text{pH} \qquad (9.29)$$

This relationship is most important in biochemical systems, where the pH is close to 7.0. It is therefore reasonable to adopt a secondary standard by which standard reduction potentials (E_0') are expressed at pH 7.0.

Problem 9.7: What is the potential of a $H^+|H_2$ half-reaction at pH 7.0 with reference to the standard $H^+|H_2$ in which $[H^+]$ is 1 *N*?

Solution: From Equation (9.29),

$$\begin{aligned} E_0' &= -0.06 \times \text{pH volt} \\ &= -0.06 \times 7.0 \text{ volts} \\ &= -0.42 \text{ volt} \end{aligned}$$

It follows that any table of reduction potentials that have been adjusted to conditions of pH 7.0, or any other pH, will have certain potentials reduced by some value given by Equation (9.29), and this will be −0.42 volt if the pH was 7.0. The half-reactions in question are those which involve H^+ in the redox reaction. It must be noted that these half-reactions are still referred to $H^+|H_2$ standard half-reaction where $[H^+]$ is 1 *M*. A few of the important redox systems are collected in Table 9.1, with some standard potentials at the two pH values given. Since many biological systems are oxidized by a dehydrogenation they will frequently show potentials that are pH-dependent. Another form of Equation (9.29) would be

$$E_0' \text{ (at pH} = x) = -0.06x + E_0 \tag{9.30}$$

Problem 9.8: What is the reduction potential of the oxalacetate|malate half-reaction at pH 7.4, given $E_0 = +0.318$ volt?

Solution: From Equation (9.30),

$$\begin{aligned} E_0' &= -0.06 \times (7.4) + 0.318 \text{ volt (at pH 7.4)} \\ &= (-0.444 + 0.318) \text{ volt} \\ &= -0.126 \text{ volt} \end{aligned}$$

POTENTIOMETRIC TITRATION OF REDOX SYSTEMS

As will have been noted from Equation (9.27), the potential of a half-reaction gives a measure of the proportion of the oxidized and reduced forms. It is used therefore to follow the titration of these systems. Consider the titration of 10 ml of *N* Fe^{+++} with *N* I^-. The solution of Fe^{+++} ion in a container is connected by a salt bridge to a standard reference cell, which could be a $H^+|H_2$ electrode but more likely is a reference calomel electrode for convenience. The voltage between the reference cell and the solution to be titrated is read on a potentiometer. Aliquots of N I^- solution are added and the potential difference measured. After each addition, an equivalent amount of Fe^{+++} is converted to Fe^{++} by the I^-, which is completely oxidized to I_3^- so that there is no residual $I_3^-|I^-$ half-reaction potential to consider at the same time. The titration is equivalent to the titration of a weak acid with base except that in the latter case the potentiometer is read in terms of pH. Equation (9.27) is also very similar to the Henderson-Hasselbalch equation, so it is not surprising to find that the oxidation-reduction titration curves have a similar shape. As in the case of weak acid titrations, we can calculate several important points in the above titration (see Table 9.2). Other points could be calculated and other half-reaction components titrated against each other similarly. The results are shown in Figure 9.4. The curves will be parallel for all half-reactions with the same number of electrons in the equation and separated from each other by a voltage equal to the algebraic difference in their standard reduction potentials. (The slope of the curve at E_0 is related to the number of electrons in the reactions.)

Table 9.2
Potentiometric Titration of Fe^{+++}

N I^- added (ml)	Fe^{+++} reduced (%)	$\frac{[Fe^{+++}]}{[Fe^{++}]}$	E_{obs} volts (using $H^+\|H_2$ electrode)
0.91	9.1	$\frac{10.0-0.91}{0.91}$	$\left(+0.771+\frac{0.06}{1}\log 10\right)=0.831$
5.0	50.	$\frac{10.0-5.0}{5.0}$	$\left(+0.771+\frac{0.06}{1}\log 1.0\right)=0.771$
9.1	91.	$\frac{10.0-9.1}{9.1}$	$\left(+0.771+\frac{0.06}{1}\log 0.1\right)=0.711$

Figure 9.4

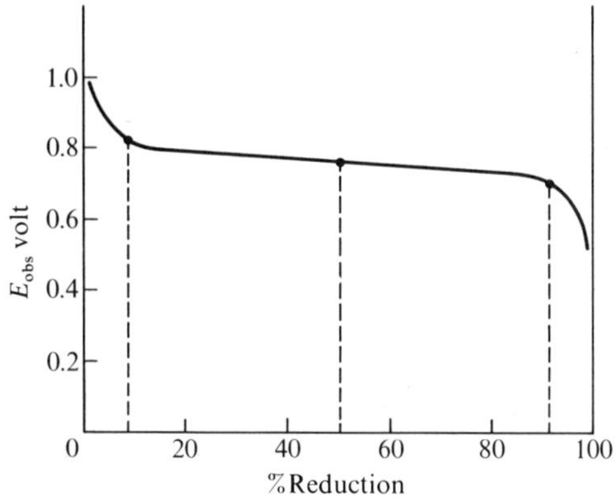

REDOX EQUILIBRIA

In reversible oxidation-reduction reactions an equilibrium will be reached in the normal way, at which time the concentrations of all the chemical species remain constant and the potentials of the corresponding half-reactions are equal. The situation is equivalent to an electrical dry cell running down completely. At equilibrium no energy can be obtained. In the general case, at equilibrium

$$A_{\text{red}}^{-n} + B_{\text{ox}} \rightleftharpoons A_{\text{ox}} + B_{\text{red}}^{-n}, \qquad \Delta E_{\text{obs}} = 0$$

We have the half-cells

$$A_{\text{ox}} + ne^- \rightleftharpoons A_{\text{red}}^{-n}, \qquad E_0^A \tag{9.31}$$

$$B_{\text{ox}} + ne^- \rightleftharpoons B_{\text{red}}^{-n}, \qquad E_0^B \tag{9.32}$$

Rewriting (9.31), we obtain

$$A_{\text{red}}^{-n} \rightleftharpoons A_{\text{ox}} + ne^-, \qquad -E_0^A \tag{9.33}$$

Since the overall potential is zero,

$$E^B + (-E^A) = 0 \qquad \text{and} \qquad E^B = E^A$$

but

$$E^A = E_0^A + \frac{RT}{nF} \ln \frac{[A_{\text{ox}}]_{\text{eq}}}{[A_{\text{red}}^{-n}]_{\text{eq}}}$$

and

$$E^B = E_0^B + \frac{RT}{nF} \ln \frac{[B_{\text{ox}}]_{\text{eq}}}{[B_{\text{red}}^{-n}]_{\text{eq}}}$$

Therefore

$$E_0^A + \frac{RT}{nF} \ln \frac{[A_{\text{ox}}]_{\text{eq}}}{[A_{\text{red}}^{-n}]_{\text{eq}}} = E_0^B + \frac{RT}{nF} \ln \frac{[B_{\text{ox}}]_{\text{eq}}}{[B_{\text{red}}^{-n}]_{\text{eq}}}$$

hence

$$E_0^B - E_0^A = \frac{RT}{nF}\left(\ln \frac{[A_{\text{ox}}]_{\text{eq}}}{[A_{\text{red}}^{-n}]_{\text{eq}}} - \ln \frac{[B_{\text{ox}}]_{\text{eq}}}{[B_{\text{red}}^{-n}]_{\text{eq}}}\right)$$

or

$$\Delta E_0 = \frac{RT}{nF} \ln \frac{[A_{\text{ox}}]_{\text{eq}}}{[A_{\text{red}}^{-n}]_{\text{eq}}} \frac{[B_{\text{red}}^{-n}]_{\text{eq}}}{[B_{\text{ox}}]_{\text{eq}}}$$

$$\Delta E_0 = \frac{RT}{nF} \ln K_{\text{eq}}$$

If we convert to logarithms to the base 10,

$$\Delta E_0 = \frac{2.303\ RT}{nF} \log K_{\text{eq}}$$

which at 30°C reduces to

$$\Delta E_0 = \frac{0.06}{n} \log K_{\text{eq}} \tag{9.34}$$

This relationship will be most important in a consideration of the energetics of biochemical systems.

Equation (9.34) refers only to redox systems at equilibrium, so that concentration terms are automatically expressed in K_{eq} and only the standard half-reaction potentials E_0 or E_0' are involved. The algebraic difference between the two E_0 values is sufficient to give a quantitative number to K_{eq}. The problem is to decide the direction for the equation. This is illustrated in Problem 9.9.

Problem 9.9: What is the equilibrium constant for the oxidation of acetaldehyde to acetate by NAD^+ at pH 7.0 and 30°C?

Solution:

Step 1. The half-reactions are

$$\text{acetate} + 2H^+ + 2e^- \rightleftharpoons \text{acetaldehyde} + H_2O \qquad E_0' - 0.60 \text{ volt} \qquad \text{(i)}$$

$$NAD^+ + 2H^+ + 2e^- \rightleftharpoons NADH + H^+ \qquad E_0' - 0.32 \text{ volt} \qquad \text{(ii)}$$

Step 2. Since we are asked to consider the oxidation of acetaldehyde, then (i) must be written in this form:

$$H_2O + \text{acetaldehyde} \rightleftharpoons \text{acetate} + 2H^+ + 2e^- \qquad E_0' + 0.60 \text{ volt} \qquad \text{(i}')$$

Step 3. Summation of the two half-reactions (i′) and (ii) gives

$$H_2O + \text{acetaldehyde} + NAD^+ \rightleftharpoons \text{acetate} + NADH + H^+, \quad \Delta E_0' + 0.28 \text{ volt}$$

Step 4. Since $\Delta E_0'$ is positive, the reaction will spontaneously go in the direction written and the equilibrium ratio K_r should be greater than one at pH 7.

Whenever an equilibrium constant is calculated from $\Delta E_0'$, the value refers to the ratio of [products]/[reactants] excluding the $[H^+]$, which is set by the standard state defining $\Delta E_0'$ (in this case $[H^+]$ is 10^{-7} *M*).
Step 5.

$$K_r = \frac{[\text{acetate}][\text{NADH}]}{[\text{acetaldehyde}][\text{NAD}^+]}$$

K_r is an equilibrium constant (see Chapter 6) that is true only for $[H^+] = 10^{-7}$ *M*, which is the condition for the measurement of half-cell potentials. The equilibrium constant, which is independent of pH, is given by

$$K_{eq} = \frac{[\text{acetate}][\text{NADH}][\text{H}^+]}{[\text{acetaldehyde}][\text{NAD}^+]}$$

where

$$\frac{K_{eq}}{10^{-7}} = K_r$$

Step 6. Calculate K_r from the $\Delta E_0'$ obtained in step 4. From Equation (9.34),

$$+0.28 = \frac{0.06}{2} \log K_r$$

or

$$\log K_r = \frac{0.56}{0.06}$$

$$= +9.333$$

$$K_r = \text{antilog } 9.333$$

$$= 2.15 \times 10^9$$

Therefore

$$K_{eq} = K_r \times 10^{-7}$$

$$= 2.15 \times 10^2$$

Any confusion in deciding which direction the equilibrium constant refers to, or the electron change, *n*, is avoided by proceeding logically through steps 1 to 6. The value for *n* in Equation (9.34) is the total number of electrons involved in each half-reaction when balanced prior to coupling. For example, if we wished to calculate K_{eq} for the reaction studied in Problem 9.2, *n* for Equation (9.34) is 2. The same is true for the reactions in Problems 9.4 and 9.5.

The problems of redox systems in which one or all of the components dissociate to give H^+ complicate the simple treatment given above. For more rigorous treatment the student is referred to advanced texts.

REFERENCES

W. M. Clark, *Oxidation-Reduction Potentials of Organic Systems*. Williams & Wilkins, Baltimore, 1960.

A. L. Lehninger, "Respiratory-energy transformation," *in* J. L. Oncley (ed.), *Biophysical Science—A Study Program*. Wiley, New York, 1959.

A. R. Patton, *Biochemical Energetics and Kinetics*. Saunders, Philadelphia, 1965.

9. PROBLEMS A

Answers on page 349

1. In the following oxidation-reduction reactions, what fraction of the molecular weight is the equivalent weight of
   (a) UDP D-glucose in the oxidation to UDP D-glucuronic acid,
   (b) 3-phospho-D-glyceraldehyde in the oxidative phosphorylation to 1,3-diphospho-D-glyceric acid,
   (c) crotonyl-coenzyme A in the reduction to butyryl CoA,
   (d) nitrite in the bacterial oxidation to nitrate,
   (e) cysteine in its oxidation to cysteic acid,
   (f) cystine in its oxidation to cysteic acid?

2. Write the balanced equation for the oxidation of
   (a) β-hydroxybutyrate to acetoacetate by NAD^+,
   (b) cytochrome-a by oxygen.

3. Compared to the standard hydrogen electrode, what is the half-cell potential of the $H^+|H_2$ reaction at pH 4, 5, 6, 7, and 8?

4. What is the half-reaction potential for the $\frac{1}{2}O_2|H_2O$ system at pH 6, 5, and 4.5 compared to the standard hydrogen electrode at pH 0?

5. (a) Will the following reaction occur spontaneously at pH 7.0 and 30°C with the appropriate catalyst?

$$\text{xanthine} + O_2 + H_2O \rightleftharpoons \text{uric acid} + H_2O_2$$

   (b) What is the equilibrium constant?
   (c) This reaction is catalyzed by xanthine oxidase, which is inhibited by allopurinol. What is the equilibrium constant of the reaction in (a) in the presence of xanthine oxidase, and a mixture of xanthine oxidase and allopurinol?

6. Compare, by calculating the ΔE_{obs}, the oxidation of lipoic acid by oxygen at pH 2.0 and pH 5.5.

7. (a) Will the following reaction occur spontaneously at pH 7.0 and 30°C?

$$\text{succinate} + 2\ \text{cytochrome-c}\ Fe^{+++} \rightleftharpoons \text{fumarate} + 2\ \text{cytochrome-c}\ Fe^{++}$$

(b) What is the apparent equilibrium constant for the reaction in the direction written?

8. (a) Write the balanced equation, and the component half-reactions, for the oxidation of FAD2H by cytochrome-c.
(b) What is the equilibrium constant for the reaction?
(c) What is the equilibrium constant for the reaction at pH 5.5?

9. (a) What is the direction of the oxidation-reduction reaction between equimolar amounts of lactate and pyruvate, coupled with $NAD^+|NADH$, also in equimolar amounts? Calculate the apparent equilibrium constant and $\Delta E_0'$ at pH 7.0 and 30°C.
(b) What ratio of $[NAD^+]/[NADH]$ must exist to oxidize an equimolar mixture of lactate and pyruvate?
(c) If the above reactions were performed at pH 5.5 and 30°C, what would be the ΔE_{obs}?

10. D-Glucose may be oxidized by O_2 with the formation of H_2O_2, the reaction being catalyzed by glucose oxidase. Alternatively, the oxidation can be achieved by the use of Br_2. Compare the two reactions, writing the component half-reactions and calculating the $\Delta E_0'$ at pH 7.0 and 30°C.

11. The variation in redox potential with percentage oxidation at constant pH and 30°C of dyes may be expressed

$$E_h' = E_0' + k$$

Calculate k for 1–99% oxidation at intervals of approximately 10%.

12. A dye at full color at 99%–100% oxidation would lose color proportionate to the percent oxidation, being zero in the reduced form. To a biological system, pH 7.0, was added *o*-cresol indophenol and the color was reduced to 80% of its full color when read against a blank. What is the redox potential of the system, using the information generated in Problem 11 above, and E_0' 0.191 volt?

13. Write the balanced equation and the component half-reactions for the oxidation of lipoic acid by dithiothreitol. What is the equilibrium constant for the reaction at pH 7.0 and 5.0 (diothithreitol E_0' − 0.34 volt at 30°C and pH 7.0)?

14. (a) Write the equation for the reaction between $(C_6H_5S)_2|C_6H_5SH$ and cystine|cysteine and calculate $\Delta E_0'$ at pH 7.0 and 30°C.
(b) What ratio of $[(C_6H_5S)_2]/[C_6H_5SH]$ would be necessary to maintain 90% of cysteine in this reduced form?

15. In the chloroplast

$$\text{cytochrome-b}_6\ Fe^{++} + \text{cytochrome-f}\ Fe^{+++} \rightleftharpoons \text{cytochrome-b}_6\ Fe^{+++} + \text{cytochrome-f}\ Fe^{++}$$

$$\Delta E_0' = +0.42 \text{ volt at pH 7.0}$$

(a) What is the apparent equilibrium constant?
(b) What would the relative concentrations of cytochrome-f Fe^{++} and cytochrome-f Fe^{+++} have to be to reduce cytochrome-b_6 Fe^{+++} in a solution where [cytochrome-b_6 Fe^{+++}]/[cytochrome-b_6 Fe^{++}] is unity?

---

9. PROBLEMS B

No answers given

1. (a) The redox dye safranine T is red in the oxidized state and colorless when reduced. What would be the color of the solution at $\Delta E_{obs} - 0.26$ volt (pH 7 and 30°C) compared to the standard hydrogen electrode?
   (b) If equimolar amounts of the red-colored oxidized safranine T and blue-colored oxidized methylene blue were mixed in solution at pH 7.0 and 30°C, what would be the resulting color?

2. The photosynthetic unit of the chloroplast is said to contain 1 molecule of cytochrome-b_6, 1 molecule of cytochrome-f and 5 molecules of ferredoxin. They have standard reduction potentials of −0.06 volt, +0.37 volt, and −0.42 volt respectively, at pH 7 and 30°C. In what array would these be arranged for the transfer of electrons, and what are the equilibrium constants of the three coupled reactions? How does the photosynthetic unit differ from the mitochondrial respiratory chain, cytochrome-b → O_2?

3. (a) Write the balanced equation, and the component half-reactions, for the oxidation of NADH by FAD.
   (b) What mole ratio of [NAD^+]/[NADH] is required to maintain at least 90% of the acetoacetate|L-β-hydroxybutyrate half-reaction in the form of acetoacetate?
   (c) For the reaction in (b), what is the equilibrium constant?
   (d) If the reaction in (a) is coupled to the reaction in (b), write the overall reaction and calculate its equilibrium constant. Would this equilibrium constant vary with pH?

4. The determination of E_0 for epinephrine was made by the step-by-step oxidation of a 0.002 *M* solution with ceric sulfate, 0.002 *N* in 0.5 *M* sulfuric acid, pH 0.286, at 30.0°C ± 0.05°. The potential was read against a standard hydrogen electrode with the following results.

Reduction (%)	$E_{observed}$ (volt)
28.51	+0.8030
30.48	+0.8027
38.94	+0.7969
47.82	+0.7916
56.06	+0.7864
60.14	+0.7850
81.88	+0.7697

Calculate the average standard reduction potential E_0 and compare with the results of Ball and Chen.*

5. It was suggested that the oxidation of ferrodoxin ($E_0' - 0.42$ volt) is linked to the formation of NADPH ($E_0' - 0.32$ volt) through a flavoprotein. Discuss this in the light of the redox potential of FAD|FAD2H ($E_0' = 0.06$ volt).

6. (a) What is the direction of the reaction between the half-cells oxalacetate|malate and NAD^+|NADH, assuming equimolar amounts of the reactants? Calculate the equilibrium constant from $\Delta E_0'$ at pH 7.0 and 30°C.
   (b) What ratio of $[NAD^+]/[NADH]$ must exist to oxidize a mixture of [oxalacetate]/[malate] $= 10^{-4}\ M/2 \times 10^{-2}\ M$ at pH 5.0?

7. Sodium lactate is frequently administered to a patient to combat metabolic acidosis.
   (a) How would you prepare 1 liter of 0.15 M sodium lactate at pH 7.4 from solid sodium hydroxide and 100% lactic acid, pK_a 3.86?
   (b) The first metabolic reaction of lactate is the oxidation (in liver at pH 7.4) to pyruvate with NAD^+. If the mole ratio of $[NAD^+]/[NADH]$ in the cell is 1/300, what is the minimum ratio of [lactate]/[pyruvate] that will permit oxidation of lactate to occur?

8. (a) Write the balanced equation, and the component half-reactions, for the oxidation of lactic acid to pyruvic acid by the dye, methylene blue.
   (b) What ratio of [pyruvate]/[lactate] would be required at pH 7.0 and 30°C to convert a methylene-blue indicator into 99% of its leuco form?

9. (a) Write the balanced equation, and the component half-reactions, for the oxidation of ascorbic acid to dehydroascorbic acid by iodine.
   (b) Ascorbic acid can be determined using the redox indicator 2,6-dichlorophenolindophenol. Write the balanced equation and calculate the $\Delta E_0'$ and K_{eq} at pH 7.0 and 30°C.
   (c) The determination in (b) is usually carried out at around pH 5.0. How does this change the equilibrium constant?
   (d) Construct titration curves for the reduction of dehydroascorbic acid and the corresponding oxidation of 2,6-dichlorophenolindophenol.

*E. G. Ball and T. T. Chen, *J. Biol. Chem.* **102**:691 (1933).

10

Biochemical Energetics

The transformation of energy is necessary for living cells to carry out their characteristic functions, such as motion, respiration, growth, and reproduction. The ultimate source of energy is the nuclear reaction

$$4\,{}^{1}_{1}\mathrm{H}^{+} \rightarrow {}^{4}_{2}\mathrm{He}^{+2} + 2\,{}^{0}_{1}e^{+} + 3\gamma\ (\text{gamma}) + 2\nu\ (\text{neutrino}) + \text{energy} \qquad (10.1)$$

which takes place on the surface of the sun. Energy from the sun is transported to the earth in the form of light energy. The conversion of this light energy into chemical energy is accomplished in plants by the process of photosynthesis. This chemical energy is stored in carbohydrates and other substances produced by plants and is released in living systems primarily by oxidation. A large amount of it is utilized for the production of adenosine triphosphate (ATP), a readily convertible form of energy, which drives many of the life processes.

The flow of energy is the controlling factor in all of these reactions. In many ways the flow of energy from the sun is like the flow of water down a fall. If the water is allowed to drop to the bottom of the fall, the energy it possessed (equal to the energy required to return it to the top of the fall) is converted to kinetic energy (energy of motion) and then to heat energy, which is dissipated at the bottom of the fall. No useful work is accomplished. If instead we pass the falling water through a turbogenerator, some of the energy of motion can be converted into mechanical energy and then electrical energy for performing work. The electrical energy could be used to charge a battery (chemical energy) and thus be stored. This turbogenerator is then analogous to plants, which absorb some of the light energy from the sun and transform it into the chemical energy of carbohydrates and other plant components. The quantitative treatment of these energy conversions is within the framework of the science of thermodynamics, which is the same for reactions in the living system as for ordinary chemical systems. In practice, however, one needs more simplifying assumptions, since the molecules treated are complex and in a multicomponent system.

A thermodynamic study is usually conducted on some small isolated group of objects, called a system, of which two types are defined, open systems and closed systems. When a reaction (process) takes place in a closed system, only forms of energy can enter and leave. Open systems allow the exchange of both matter and energy. The living system definitely falls into the latter category, whereas the majority of in vitro studies are conducted on closed systems.

The state of the system is described in terms of the values of some experimentally measurable properties. This thermodynamic description is complete when the specific properties allow only one value for the remaining variables. Temperature, pressure, volume, and energy are some of the properties commonly used to describe the state of the system.

A closer look at the various properties shows that they are of two types. Energy and volume are examples of the first type. Their value depends on the extent of the system considered and they are thus called extensive properties. Another way of defining an extensive property is to say that it is additive, meaning that the total property is the sum of all the constituent parts.

Temperature and pressure exemplify the second type of property. These nonadditive properties of the system are called intensive properties. The temperature of a house, for example, is not the sum of the temperatures of the various rooms.

Biochemists are interested primarily in the direction a reaction will proceed and how this is related to the flow of energy. The major concepts of thermodynamics, as stated in the first and second laws, enable one to make such predictions for chemical reactions. These concepts are introduced briefly and then applied to biochemical systems.

ENERGY

Energy can appear in many forms, such as heat, light, chemical, mechanical, electrical, and osmotic energy. These forms of energy are interconvertible, with the exception of heat. Thus electrical energy can be converted completely to mechanical energy or chemical energy and vice versa. All forms of energy can be converted completely to heat, *but the reverse is not true.* This fraction of the heat energy, which is unavailable for the performance of useful work, is related to the randomness of the system at the molecular level. The quantity of heat energy is usually measured in calories, a calorie being defined as the amount of energy required to raise 1 g of water from 15° to 16°C. It is readily converted to various other equivalent energy units, such as joules (electrical) and ergs (mechanical) (see Chapter 4).

In summary, energy has two important properties: it is additive and it is conserved.

FIRST LAW: CONSERVATION OF ENERGY

The first principle of thermodynamics asserts that energy is conserved. We can by some process change the form of energy, but the quantity is unchanged. If an amount of heat, q, is added to a system it may perform work, w, or the components of the system may change their internal energy, E, by an amount, ΔE. A mathematical statement of this first principle is

$$\Delta E = q - w \qquad \text{or} \qquad q = \Delta E + w \tag{10.2}$$

Included in the term w is pressure-volume work, which at constant temperature and pressure is $P\Delta V$—the product of the pressure times the volume change. In biochemical and chemical systems this is rarely a useful form of work.

A new quantity, H, called enthalpy or heat content, is now introduced, a change in which ΔH, under conditions of constant temperature and pressure, is defined as

$$\Delta H = \Delta E + P\Delta V \tag{10.3}$$

Rearranging Equation (10.2) and separating work into $P\Delta V$ work and useful work w', we have

$$q = \Delta E + w' + P\Delta V \tag{10.4}$$

Substituting for ΔE in terms of the new function, ΔH, we obtain

$$q = \Delta H - P\Delta V + w' + P\Delta V \tag{10.5}$$

which is another statement of the first law:

$$\Delta H = q - w' \tag{10.6}$$

It describes an ideal reversible system. If heat energy is put into this system it can either do useful work, w', or increase the enthalpy. The first law thus asserts nothing about the direction of chemical reactions; according to it all processes are reversible.

SECOND LAW: ENTROPY

Practical experience indicates that processes are in general not reversible. Heat energy is released when water drops in a waterfall. However, heating the water at the bottom of the fall does not cause the process to reverse. When a bar of metal is heated at one end, the energy is dissipated throughout the bar, but a bar of uniform temperature will not spontaneously become hot at one end and cold at the other.

We seek some criteria to explain the unidirectional nature of these spontaneous changes and hope to apply them to chemical reactions. They can be understood in terms of the concept of equilibrium. All spontaneous processes tend toward the equilibrium state. Furthermore, in their approach to equilibrium they can release energy for the performance of useful work. Once the system is at equilibrium no useful work can be obtained.

The concept that all spontaneous processes tend toward equilibrium is a statement of the second law of thermodynamics. It is defined mathematically in terms of a quantity called entropy S, or ΔS, change in entropy. The entropy is the fraction of the enthalpy which is not available to do useful work. It is related to the random motions of the system at the molecular level, which depend on temperature. The energy lost in these random motions is the product $T \times S$, where T is the absolute temperature. This is the reason that heat energy cannot be converted completely to other energy types—some of it is lost as entropy. A system is at equilibrium when the entropy is a maximum, and q, the heat absorbed, is equal to $T\Delta S$. This equality is true for a reversible process. A spontaneous change occurs without the maximum absorption of heat thereby increasing the entropy of the system. This is true for an irreversible process, $q < T\Delta S$. If we form the term, $\Delta S - (q/T)$, we note that for a nonspontaneous (equilibrium) process it is zero (reversible). For the spontaneous process (nonequilibrium) it is greater than zero (irreversible). Thus we can use the difference term $\Delta S - (q/T)$ as a criterion of spontaneity. Practically, this is difficult, as one needs to consider both the changes in the working substance, ΔS, and the heat exchanged with the surroundings q (e.g., for a chemical reaction taking place in a beaker we would need to know the nature of the species in the beaker plus the amount of heat exchanged between the beaker and its surroundings).

In the next section we introduce the more useful quantity, free energy, which can be used as a criterion of spontaneity without this difficulty. It should be noted that the thermodynamics introduced in the following sections applies to reversible systems; thus the work obtained from them will be the maximum useful work.

FREE ENERGY

A combination of the first and second laws gives the relationship which relates the direction (spontaneity) to the energy produced in a chemical reaction. *For a system proceeding toward equilibrium, the maximum useful work produced (w' is positive) is derived from Equation (10.6):*

$$w' = -\Delta H + T\Delta S \tag{10.7}$$

We introduce a new function, the Gibbs free energy, G, or a change in Gibbs free energy, ΔG,* to treat the concept of useful work. It is related to maximum useful work by the equation

$$w' = -\Delta G \tag{10.8}$$

Substituting Equation (10.8) in Equation (10.7), we obtain the relationship between free energy, enthalpy, and entropy:

$$\Delta G = \Delta H - T\Delta S \tag{10.9}$$

It is now apparent from Equation (10.9) that the change in Gibbs free energy, ΔG, at constant temperature and pressure, must be negative for spontaneity and the production of useful work. For systems already at equilibrium, ΔG is 0. A system can be displaced from equilibrium by supplying energy to it.

Now consider the reaction

$$A \rightleftharpoons B \tag{10.10}$$

Thermodynamics indicates that the conversion of A to B occurs spontaneously if the free energy of A, G_A, is greater than the free energy of B, G_B. This is indicated pictorially in Figure 10.1. The quantity $G_B - G_A$ is negative and energy is released in going from the reactants A to the products B. The reverse reaction requires an equivalent amount of energy. If we know the free energies of the products and the reactants we can calculate the free energy change for a reaction and predict the direction of spontaneous change.

It is also important to note that all the thermodynamic properties, for example, ΔG, ΔH, and ΔS, like energy, are additive and thus their values do not depend on how we get from state 1 to state 2 (G_A to G_B). They depend only on the initial and final states.

---

*In some texts this is indicated by ΔF.

Thermodynamic quantities, such as the free energy, B, and enthalpy, H, cannot be measured in the absolute sense. Only the changes in these quantities, indicated by Δ, are measurable. To define these changes exactly, and for convenience, we choose some arbitrary reference states for our calculations on a system. These reference states are called standard states, some of which are indicated in Table 10.1. The thermodynamic quantity in the standard state is usually indicated by the degree symbol—for example, $\Delta H°$ and $\Delta G°$. Standard states are rigorously defined only in terms of activities (see chapter 5). Concentrations are used as expressions of activity for our discussion of solutions. In solution the standard state concentration is 1 mole liter^{-1} for all species.

Table 10.1
Standard states

State	*Standard state*	*Expression of activity*
Solid	Pure solid in most stable form at 25°C and 1 atm	mole fraction = 1
Pure liquid or solvent in dilute solution	Pure liquid in most stable form at 25°C and 1 atm	mole fraction = 1
Gas	Ideal gas at 25°C and 1 atm	Partial pressure in atm
Solution	Solute in solvent at a concentration of one mole per liter at 25°C and 1 atm	Molarity

It is important to recall that all the thermodynamic quantities—ΔG, ΔH, and ΔS—are extensive properties, and it is necessary to define the extent of the system. For our purposes this is usually 1 mole of the substance and if we calculate ΔG to be equivalent to 2000 calories it is assumed, if not indicated otherwise, to be 2000 cal mole^{-1}.

STANDARD FREE ENERGY OF FORMATION, ΔG_f°

By definition the standard free energy of formation is the energy change associated with the production of 1 mole of the compound from its elements in their standard states. Thus the standard free energy of formation of D-glucose at 25°C refers to the reaction

$$6C(s)(1\ \text{atm}) + 3O_2(g)(1\ \text{atm}) + 6H_2(g)(1\ \text{atm}) \rightarrow C_6H_{12}O_6(s)(1\ \text{atm}),\ \Delta G_f^\circ = -217.54\ \text{kcal mole}^{-1}$$

where (s) and (g) refer to solid and gaseous, respectively. It is a consequence of our definition that the ΔG_f° for any element is adjusted to zero.

Standard free energies of formation, ΔG_f°, can be used to evaluate the standard free energy, ΔG°, for a reaction if they are known for each component. By reference to Figure 10.1, we can define the standard free energy, ΔG°, for a reaction as the difference in energy between 1 mole of each of the reactants and 1 mole of each of the products in their standard states. It is stated mathematically as

$$\Delta G^\circ = \Sigma \Delta G_f^\circ(\text{products}) - \Sigma \Delta G_f^\circ(\text{reactants}) \tag{10.11}$$

Numerical values for the standard free energies of formation are given in most handbooks.

Figure 10.1

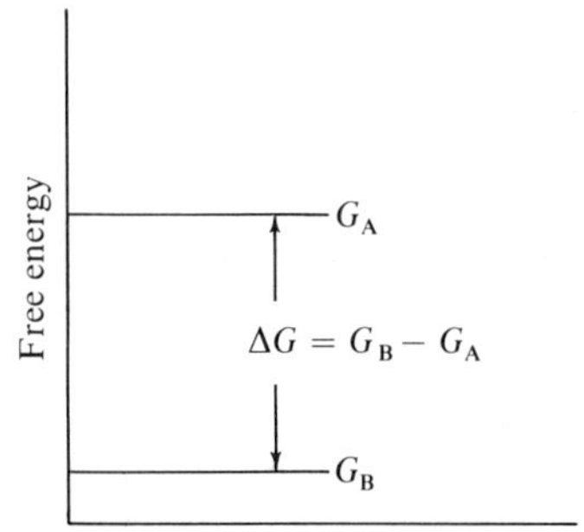

Problem 10.1: Calculate the standard free energy change for burning 1 mole of solid D-glucose to carbon dioxide gas and liquid water at 25°C and 1 atm from the standard free energies of formation.

$$\text{D-glucose}(s)\ \Delta G_f^\circ = -217.54 \text{ kcal mole}^{-1}$$

$$CO_2(g)\ \Delta G_f^\circ = -94.26 \text{ kcal mole}^{-1}$$

$$H_2O(l)\ \Delta G_f^\circ = -56.69 \text{ kcal mole}^{-1}$$

$$O_2(g)\ \Delta G_f^\circ = 0$$

Solution: Write the equation and balance:

$$C_6H_{12}O_6(s) + 6O_2(g) \rightleftharpoons 6CO_2(g) + 6H_2O(l)$$

The standard free energy of the reaction is the sum of the free energies of products minus the sum of free energies of the reactants, or Equation (10.11).

Substitute values to obtain

$$\Delta G_f^\circ \text{ products} = 6(-94.26) + 6(-56.69) = -905.70 \text{ kcal mole}^{-1}$$

$$\Delta G_f^\circ \text{ reactants} = (-217.54) + 6(0) = -217.54 \text{ kcal mole}^{-1}$$

$$\Delta G^\circ \text{ reaction} = -905.70 - (-217.54) = -688.16 \text{ kcal mole}^{-1} \text{ of D-glucose.}$$

This calculation indicates that the reaction should proceed spontaneously as written and produce energy.

The complete oxidation of 180 g (1 mole) of D-glucose produces 688 kcal of free energy. If glucose is burned, an irreversible process, this energy is released as heat and light and much of it would not be available to do useful work. Living systems convert their oxidative energy to other chemical energy in more nearly reversible steps at an efficiency of approximately 40%. For the most part this becomes the chemical energy of ATP.

Problem 10.2: The daily caloric requirement for an average adult is about 2000 Calories (Cal). In caloric studies the calorie referred to is the large calorie, Cal, which is the kilocalorie (1000 small calories). Kilocalorie is used here, as is customary in physics and chemistry.

(a) Calculate the grams of carbohydrate necessary to furnish the daily caloric requirement if the caloric equivalent of carbohydrate is 4 kcal g^{-1}.

(b) How many moles of ATP could be formed if the free energy required for the reaction

$$\text{ADP} + P_i \rightleftharpoons \text{ATP} + \text{H}_2\text{O}$$

is 8 kcal mole^{-1}?

Solution: (a) 2000 kcal is required and 4 kcal is produced per gram of carbohydrate

$$\frac{2000}{4}\frac{\text{kcal}}{\text{kcal g}^{-1}} = 500 \text{ g of carbohydrate}$$

(b) Assuming an efficiency of 40% for the synthesis of ATP,

$$0.4 \times 2000 \text{ kcal} = 800 \text{ kcal for ATP synthesis}$$

Since 8 kcal mole^{-1} is the free energy required for the synthesis of ATP in the above equation, the number of moles of ATP is

$$\frac{800 \text{ kcal}}{8 \text{ kcal}} \text{ mole ATP} = 100 \text{ moles of ATP}$$

The caloric calculation in Problem 10.2 was done in terms of free energy, which is the most generally useful thermodynamic function for biochemical systems. Caloric calculations involving foodstuffs are usually made in terms of enthalpy, $\Delta H°$, which is the heat produced by burning 1 mole of substance to CO_2 and H_2O in the presence of O_2 at standard conditions. The results calculated by the two methods differ by a few percent, the difference being due to the entropy contribution as indicated in Equation (10.9). In other reactions the entropy term may be larger, and thus the free energy and enthalpy will be very different.

FREE ENERGY AND CONCENTRATION

The free energy, ΔG, of a dissolved solute increases with solute concentration. This is apparent from studies of diffusion in solutions. The solute in a concentrated solution will diffuse into a dilute solution through a permeable junction until the concentrations are equal. This is a spontaneous process; thus the free energy of the concentrated solution must be greater than that of the dilute solution. The free energy of a dissolved solute, ΔG, in calories per mole is related to concentration by the equation

$$\Delta G = \Delta G_f^\circ + RT \ln c \tag{10.12}$$

where ΔG_f° is the standard free energy of formation in cal mole^{-1} in the standard state in solution, R is the gas constant in cal deg^{-1} mole^{-1}, T is temperature in degrees Kelvin, and c is the molar concentration.* This equation enables one to calculate the free energy in calories per mole for any concentration of solute. Using Equation (10.12) we can calculate the free energy per mole for diluting or concentrating a solution.

For saturated solutions there is equilibrium between the solid and solution, thus the two phases have the same free energy per mole.

Problem 10.3: The creatine concentration in blood and urine is 2 mg % and 75 mg %, respectively. Calculate the free energy per mole required for its transfer from blood to urine at 38°C.

Solution: Using Equation (10.12) we write the free energy expression for creatine in blood and urine. Note that the concentrations in mg % need not be converted to moles per liter, since we will eventually use them as a ratio.

$$\Delta G_{\text{blood}} = \Delta G_f^\circ + RT \ln [2] \tag{i}$$

$$\Delta G_{\text{urine}} = \Delta G_f^\circ + RT \ln [75] \tag{ii}$$

Since we are interested in the free energy difference per mole in going from the concentration in blood (initial) to the concentration in urine (final), we subtract to obtain the free energy per mole for the process:

$$\Delta G_{\text{urine}} - \Delta G_{\text{blood}} = \Delta G_{\text{transfer}} \tag{iii}$$

Substitute equation (i) and (ii) in (iii) to give

$$(\Delta G_f^\circ + RT \ln [75]) - (\Delta G_f^\circ + RT \ln [2]) = \Delta G_{\text{transfer}}$$

which simplifies to

$$RT \ln [75] - RT \ln [2] = \Delta G_{\text{transfer}}$$

---

*This equation is analogous to the usual expressions for partial molar free energy or chemical potential, $\overline{F} = \overline{F}^o + RT \ln a$ or $\mu = \mu^o + RT \ln a$. It is important to note that these are intensive free energies.

since ΔG_f° is the same for creatine in blood or urine. Combine to obtain

$$\Delta G_{\text{transfer}} = RT \ln \frac{[75]}{[2]}$$

Substituting for R and T and evaluating the log gives

$$\Delta G_{\text{transfer}} = 2.303(1.986)(311)(1.574) = 2230 \text{ cal mole}^{-1}$$

Since the free energy of the more concentrated solution is greater, energy must be expended in this process as is evident from the calculation. The free energy required for the production of 1 liter of urine can be calculated by finding the number of moles in a liter at the concentration given, 75 mg %, and multiplying this by the calculated free energy per mole.

From Problem 10.3 it is apparent that a general equation for free energy change and concentration can be derived. The free energy of dilution, as it is called, is

$$\Delta G = n_2 RT \ln \frac{c_2}{c_1} \qquad (10.13)$$

where n_2 is the number of moles of solute diluted and c_1 and c_2 are the initial and final concentrations, respectively. Equation (10.13) is also the free energy of concentrating, as this is just the reverse process.

FREE ENERGY AND EQUILIBRIUM CONSTANT, K_{eq}

The standard free energy, ΔG°, as calculated from ΔG_f°, can be directly related to the equilibrium constant, K_{eq}. This is an extremely important expression, for it enables one to get considerable insight into the thermodynamics of a system from a simple equilibrium measurement.

At equilibrium the energy of the products of a reaction must equal the energy of the reactants, since ΔG, the free energy change, must be zero. Since compounds inherently have different free energies of formation, ΔG_f°, how is this possible? It is accomplished by changing the concentration of the products and reactants from their standard state concentrations, one mole per liter, to some other concentrations whereby an energy balance is achieved. Since material balance is essential in the system, the energy of the products will increase and the reactants decrease, or vice versa. Free energy changes are related to concentration by Equation (10.12); in this equation it is the $RT \ln c$ term that is especially important to this discussion. Note that in Equation (10.12) all terms have the units of *calories per mole*.

In deriving the expression between ΔG° and K_{eq} we use as an example for discussion the isomerization

$$\alpha\text{-D-glucose} \rightleftharpoons \alpha\text{-D-galactose}$$
$$\text{A} \qquad\qquad\qquad \text{B}$$

The standard free energies of formation, ΔG_f°, for α-D-glucose and α-D-galactose, are, respectively, −219,220 and −220,730 cal mole^{-1} at 30°C. These are illustrated pictorially in Figure 10.2. Substituting the standard free energies in Equation (10.11), we obtain for the standard free energy

$$\Delta G^\circ = -220{,}730 - (-219{,}220) = -1{,}510 \text{ cal mole}^{-1}$$

For each component in the reaction at any concentration we can write

$$\Delta G_{\text{A}} = \Delta G_{f,\text{A}}^\circ + RT \ln [\text{A}] \tag{10.14}$$

$$\Delta G_{\text{B}} = \Delta G_{f,\text{B}}^\circ + RT \ln [\text{B}] \tag{10.15}$$

where the units of each term are calories per mole. For the log term the units are given by RT. At equilibrium, $[\text{B}]/[\text{A}] = K_{\text{eq}}$ and $\Delta G_{\text{A}} = \Delta G_{\text{B}}$, so equating we obtain

$$\Delta G_{f,\text{A}}^\circ + RT \ln [\text{A}] = \Delta G_{f,\text{B}}^\circ + RT \ln [\text{B}] \tag{10.16}$$

where $\Delta G_{f,\text{A}}^\circ$ and $\Delta G_{f,\text{B}}^\circ$ are the standard free energies of formation in calories per mole of α-D-glucose and α-D-galactose, respectively. Since $\Delta G_{f,\text{A}}^\circ$ and $\Delta G_{f,\text{B}}^\circ$ are constant the two log terms in the concentrations will be adjusted to satisfy the equation. Rearranging (10.16) we obtain

$$\Delta G_{f,\text{B}}^\circ - \Delta G_{f,\text{A}}^\circ = \Delta G^\circ = RT(\ln [\text{A}] - \ln [\text{B}]) \tag{10.17}$$

$$\Delta G^\circ = RT \ln \frac{[\text{A}]}{[\text{B}]}$$

The standard free energy, ΔG°, is −1510 cal mole^{-1} for our reaction; thus we can calculate the ratio of [A] to [B], which satisfies the equation

$$-1510 \text{ cal mole}^{-1} = RT \ln \frac{[\text{A}]}{[\text{B}]} \tag{10.18}$$

Solving for [A]/[B] we obtain

$$-1510 = (1.986)(303)(2.303) \log \frac{[\text{A}]}{[\text{B}]}$$

$$\log \frac{[\text{A}]}{[\text{B}]} = -\frac{1510}{1387} = -1.08$$

$$\log \frac{[\text{B}]}{[\text{A}]} = 1.08$$

$$\frac{[\text{B}]}{[\text{A}]} = 1.23 \times 10^1 = 12.3$$

Since for this derivation we specified equilibrium, [A] and [B] must be equilibrium concentrations and thus

$$K_{eq} = \frac{[B]}{[A]} \tag{10.19}$$

Equations (10.17) and (10.18) can be satisfied by any concentrations of A and B provided that the ratio [B]/[A] equals 12.3. In all the calculations involving $\Delta G°$ only the ratio K_{eq} is important.

We now will pick one set of concentrations and calculate ΔG_A and ΔG_B by Equations (10.14) and (10.15) to illustrate. Equation (10.19) is satisfied when the concentration of A is 1.9×10^{-3} M and the concentration of B is 2.4×10^{-2} M. Thus at 30°C

$$\Delta G_A = -219{,}220 + (1.986)(303)(2.303) \log 1.9 \times 10^{-3}$$
$$= -219{,}220 - 1387 \log \frac{1}{1.9 \times 10^{-3}} = -219{,}220 - 1387 \log 5.2 \times 10^2$$
$$= -219{,}220 - 1387(2.7) = -219{,}220 - 3740$$

That is, $-222{,}960$ cal mole^{-1} is the free energy of one mole of α-D-glucose at 1.9×10^{-3} M, and

$$\Delta G_B = -220{,}730 + (1.986)(303)(2.303) \log 2.4 \times 10^{-2}$$
$$= -220{,}730 - 1387 \log \frac{1}{2.4 \times 10^{-2}} = -220{,}730 - 1387 \log 4.17 \times 10^1$$
$$= -220{,}730 - 1387(1.61) = -220{,}730 - 2230$$

Or $-222{,}960$ cal mole^{-1} is the free energy of one mole of α-D-galactose at 2.4×10^{-2} M. The energies of α-D-glucose and α-D-galactose at equilibrium for these concentrations are shown in Figure 10.2. The free energies of

Figure 10.2

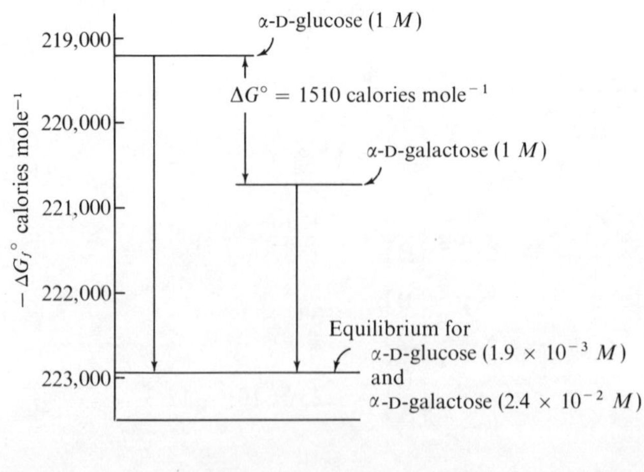

α-D-glucose and α-D-galactose are equal at equilibrium, $-222{,}960$ cal mole^{-1}. The concentrations of α-D-glucose and α-D-galactose are 1.9×10^{-3} M and 2.4×10^{-2} M, respectively, for this result; however, any concentrations which have a ratio of α-D-glucose/α-D-galactose equal to 12.3 make their free energies equal.

By rearranging (10.17) we can write the relationship between K_{eq} and $\Delta G°$

$$\Delta G° = -RT \ln K_{eq} \tag{10.20}$$

which is generally applicable.

FREE ENERGY, ΔG

The free energy change, ΔG, in calories per mole for the reaction

$$A + B \rightleftharpoons C + D$$

can also be derived from Equation (10.12). For each component we write

$$\Delta G_A = \Delta G°_{f,A} + RT \ln [A]$$
$$\Delta G_B = \Delta G°_{f,B} + RT \ln [B]$$
$$\Delta G_C = \Delta G°_{f,C} + RT \ln [C]$$
$$\Delta G_D = \Delta G°_{f,D} + RT \ln [D]$$

The free energy per mole for the reaction is the difference in the sum of the free energies of one mole of each product and the sum of the free energies of one mole of each reactant at the given concentrations. Subtracting the free energies of the reactants from the free energies of the products and collecting terms we obtain

$$\begin{aligned}(\Delta G_D + \Delta G_C) - (\Delta G_A + \Delta G_B) \\ = (\Delta G°_{f,C} + \Delta G°_{f,D}) - (\Delta G°_{f,A} + \Delta G°_{f,B}) \\ + RT[(\ln[C] + \ln[D]) - (\ln[A] + \ln[B])]\end{aligned}$$

where the logarithm term is a ratio of nonequilibrium concentrations. The term on the left side is ΔG for the reaction. On the right side the difference between the standard free energies of the products and the reactants is $\Delta G°$ for the reaction. When we substitute the quantities given

$$\Delta G = \Delta G° + RT \ln \frac{[C][D]}{[A][B]} \tag{10.21}$$

This equation is universally applicable and can be extended to include any number of components. Several points can be summarized concerning it.

1. ΔG is the energy difference per mole between the given concentrations and the equilibrium concentrations. This energy is equal to $\Delta G°$, the standard free energy, if the concentrations of each reactant and each product are one molar (substitute in Equation (10.21) 1 M concentration and $\Delta G = \Delta G°$).

2. At equilibrium $\Delta G = 0$ and Equation (10.21) becomes

$$0 = \Delta G° + RT \ln \frac{[C][D]}{[A][B]}$$

and [A], [B], [C], and [D] must be equilibrium concentrations. Thus we obtain Equation (10.20),

$$\Delta G° = -RT \ln K_{eq}$$

where $\Delta G°$ is the standard free energy in calories per mole. It is the difference in the free energy between one mole of each reactant and one mole of each product in their standard states. It is also defined as the energy released or absorbed when one mole of each reactant is converted to one mole of each product under standard conditions (1 molar concentration).

Problem 10.4: Calculate the free energy per mole for the following reaction in an open system

$$\alpha\text{-D-glucose} \rightleftharpoons \alpha\text{-D-galactose}$$

at 30°C where the concentrations of α-D-glucose (1×10^{-4} M) and α-D-galactose (0.1 M) are maintained.

Solution:

$$\Delta G = \Delta G° + RT \ln \frac{[\alpha\text{-D-galactose}]}{[\alpha\text{-D-glucose}]} \qquad \text{(i)}$$

We have already evaluated $\Delta G°$ for the reaction as -1510 cal mole^{-1}, so substituting in (i) we obtain

$$\Delta G = -1510 + (1.986)(303)(2.303) \log \frac{1 \times 10^{-1}}{1 \times 10^{-4}}$$

$$= -1510 + (1387) \log 10^3 = -1510 + (1387)(+3)$$

$$= -1510 + 4161 = 2651 \text{ cal mole}^{-1}$$

Thus forming 1 mole of α-D-galactose at a concentration of 10^{-1} M from 1 mole of α-D-glucose at a concentration of 10^{-3} M requires 2651 cal, in contrast to the standard state (1 molar) conditions, which could release 1510 cal.

Problem 10.5: For the reaction in Problem 10.4, calculate the free energy change in calories for the production of 1 liter of α-D-galactose solution from

α-D-glucose when the concentrations of α-D-glucose and α-D-galactose are $1 \times 10^{-2}\ M$ and $1.0 \times 10^{-1}\ M$, respectively.

Solution: First calculate the energy change per mole for the concentrations given ($\Delta G°$ is -1510 from Problem 10.4):

$$\Delta G = \Delta G° + RT \ln \frac{[\alpha\text{-D-galactose}]}{[\alpha\text{-D-glucose}]}$$

$$= -1510 + 1387 \log \frac{0.1}{1 \times 10^{-2}}$$

$$= -1510 + 1387 \log 10$$

$$= -1510 + 1387$$

$$= -123 \text{ cal mole}^{-1}$$

Since 1 liter of the α-D-galactose solution contains 0.1 mole, the free energy change is

$$\Delta G = (-123 \text{ cal mole}^{-1})\ 1.0 \times 10^{-1} \text{ mole} = -12.3 \text{ calories}$$

---

3. In biochemical systems we define a new standard state for convenience, which has all concentrations at 1 mole liter^{-1} except for hydrogen ion, which is 10^{-7} mole liter^{-1} and water, which is pure liquid. The standard free energy in this standard state is indicated by $\Delta G°'$. In Table 10.2 we have indicated the relationship between $\Delta G°$ (or $\Delta G°'$) and K_{eq} for three commonly used temperatures and a range for equilibrium constants.

Table 10.2
Relation between K_{eq} and $\Delta G°$ (or $\Delta G°'$)*

K_{eq}	$Log\ K_{eq}$	$\Delta G_{25} = -1364\ log\ K_{eq}$	$\Delta G_{30} = -1387\ log\ K_{eq}$	$\Delta G_{38} = -1420\ log\ K_{eq}$
0.001	−3	4089	4161	4260
0.01	−2	2726	2774	2840
0.1	−1	1364	1387	1420
1.0	0	9	0	0
10.	1	−1364	−1387	−1420
100.	2	−2726	−2774	−2840
1000.	3	−4089	−4161	−4260

*If the equilibrium constant contains hydrogen ion, note that the standard state concentration, $10^{-7}\ M$, must be used for the calculation of K_{eq} from $\Delta G°'$ (see Chapter 9).

Some clarification may be useful for calculations involving this standard state. When calculating free energies using Equation (10.21), the energy we obtain is the result of adding or subtracting to the standard state free energy,

$\Delta G°$. Recall that $\Delta G°$ is the energy obtained, ΔG per mole, when the products and reactants are at a concentration of one molar, for the general reaction

$$A + B \rightleftharpoons C + D$$

Equation (10.21) therefore becomes

$$\Delta G = \Delta G° + RT \ln \frac{[1][1]}{[1][1]}$$

For any other concentrations, since all are compared to the standard state concentrations, the equation is

$$\Delta G = \Delta G° + \left(RT \ln \frac{[C][D]}{[A][B]} - RT \ln \frac{[1][1]}{[1][1]} \right)$$

In the usual standard state the logarithm term is zero and drops out, giving Equation (10.21); if, however, some other standard state, such as $\Delta G'$, is used, this term will not be zero. For example, if A in the general equation were H^+, the equation would be

$$\Delta G = \Delta G°' + RT \ln \frac{[C][D]}{[H^+][B]} - RT \ln \frac{[1][1]}{[10^{-7}][1]}$$

or, combining terms,

$$\Delta G = \Delta G°' + RT \ln \frac{\frac{[C]}{1}\frac{[D]}{1}}{\frac{[H^+]}{10^{-7}}\frac{[B]}{1}}$$

In this equation we have not canceled factors to illustrate that each concentration term can be thought of as being divided by its respective standard state concentration. For this equation, ΔG equals $\Delta G°'$ when all the reactants and products are one molar except for H^+ ion, which must be 10^{-7} M.

These same ideas must be used in the calculation of K_{eq} from $\Delta G°'$ or $\Delta E_0'$ or vice versa. They are illustrated in Problem 10.6.

Problem 10.6: (a) Calculate the standard free energy, $\Delta G°'$, for the reduction of pyruvate to lactate by NADH from the equilibrium constant which at 38°C is 1.68×10^{11}.

(b) Calculate the free energy change per mole when the concentrations are as follows: $[NAD^+]$:[NADH] 10^1, [pyruvate] 1×10^{-4} M, [lactate] 2×10^{-3} M; the pH is 5.0.

Solution: Write the equation

$$H^+ + NADH + \text{pyruvate} \rightleftharpoons NAD^+ + \text{lactate}$$

(a) From Equation (10.20) and the standard state, $\Delta G^{o\prime}$, pH 7.0, $[H^+] = 10^{-7}$, we can write

$$\Delta G^{o\prime} = -1420 \log \frac{[NAD^+][\text{lactate}]}{[NADH][\text{pyruvate}] \dfrac{[H^+]}{[10^{-7}]}}$$

$$= -1420 \log K_{eq}\,[10^{-7}] = -1420 \log 1.68 \times 10^{11}\,(10^{-7})$$

$$= -6000 \text{ cal mole}^{-1}$$

(b) Using Equation (10.21) and the new standard state (pH 7.0),

$$\Delta G = \Delta G^{o\prime} + 1420 \log \frac{[NAD^+][\text{lactate}][10^{-7}]}{[NADH][\text{pyruvate}][H^+]}$$

Substituting concentrations and $\Delta G^{o\prime}$,

$$\Delta G = -6000 + 1420 \log\,[10^1]\,\frac{[2 \times 10^{-3}][10^{-7}]}{[1 \times 10^{-4}][10^{-5}]}$$

$$= -6000 + 1420 \log 2 = -6000 + 430$$

$$= -5570 \text{ cal mole}^{-1}$$

Note that had we calculated this ΔG for pH 7.0 the hydrogen ion concentration would have canceled. In general if a reactant concentration is the same as its standard state it can be neglected in the calculation. Note also that if the standard state ΔG° based on K_{eq} were used, the calculated ΔG would be the same.

FREE ENERGY AND LIVING SYSTEMS

The free energy change per mole which we have been calculating can be thought of as corresponding to a hypothetical process whereby the concentrations of the reactants and the products are maintained at some specified value while one mole of each reactant is converted to one mole of each product. The steady-state in living cells closely resembles this process. The biochemical reactions, which keep cells viable, are functioning as close to equilibrium as possible, so that the free energy is at a maximum and in a steady state by virtue of the fact that reactants are being fed in and products are removed. Thus, if the concentrations of the reactants and products are known and we assume the energy obtained to be the maximum, then we can make a crude approximation to the energy yield of a particular cellular reaction.

Problem 10.7: Estimate the free energy change per mole for the conversion of glucose at its concentration in interstitial fluid to CO_2 at its concentration in venous blood. Neglect the effect of H_2O formed on solution concentrations. Data are as follows.

Interstitial fluid	Cell	Venous blood
Glucose 5 mM	Glucose 10^{-4} M	CO_2 1.5×10^{-1} mM
Oxygen 4.2×10^{-5} M	Pyruvate 10^{-4} M	
	Carbon dioxide 10^{-4} M	
	Oxygen 1.4×10^{-5} M	

Standard free energies of formation, ΔG_f° in kcal mole^{-1} at 38°C and 1 atm:

Pyruvate (1 M)	−113.44
Glucose (1 M)	−219.22
Carbon dioxide (1 M)	−92.31
Water (1)	−56.69

Solution: The problem can be divided into three parts.

1. Free energy of bringing glucose and oxygen into the cell.
2. Free energy for the conversion of glucose to carbon dioxide and water.
3. Free energy of removing carbon dioxide from the cell.

Step 1. The free energy of dilution is, by Equation (10.13),

$$\Delta G = n_2 RT \ln \frac{c_2}{c_1}$$

with c_1 and c_2 equal to 5×10^{-3} M and 1×10^{-4} M, respectively, at 38°C; since the question calls for the free energy change per mole, n_2 equals one. Thus

$$\Delta G = 1420 \log \frac{1 \times 10^{-4}}{5 \times 10^{-3}} = 1420 \log 2 \times 10^{-2}$$

$$= 1420\ (-1.699)$$

$$= -2410 \text{ cal mole}^{-1}$$

For oxygen, the free energy of dilution is

$$\Delta G = 1420 \log \frac{1.4 \times 10^{-5}}{4.2 \times 10^{-5}} = -1420 \log 3$$

$$= -680 \text{ cal mole}^{-1}$$

Step 2. Write the reaction

$$C_6H_{12}O_6 + 6O_2 \rightleftharpoons 6CO_2 + 6H_2O \qquad \text{(i)}$$

Pyruvate is an intermediate in this process, but need not be considered. The calculation is carried out only for initial and final states. From Equation (10.11) the standard free energy is

$$\Delta G^\circ = [6\Delta G_f^\circ(CO_2) + 6\Delta G_f^\circ(H_2O)] - [\Delta G_f^\circ(C_6H_{12}O_6) + 6\Delta G_f^\circ(O_2)]$$
$$= [6(-92.31) + 6(-56.69)] - [(-219.22) + 6(0)]$$
$$= -553.86 - 340.14 + 219.22$$
$$= -674.78 \text{ kcal mole}^{-1}$$

At the concentration given, ΔG is from Equation (10.21):

$$\Delta G = -674{,}780 + 1420 \log \frac{[CO_2]^6[H_2O]^6}{[C_6H_{12}O_6][O_2]^6}$$
$$= -674{,}780 + 1420 \log \frac{[10^{-4}]^6[1]^6}{[10^{-4}][1.4 \times 10^{-5}]^6}$$

(Note that pure liquid water is the standard state so $[H_2O]$ is 1 for dilute solution.) Continuing the computation, we obtain

$$\Delta G = -674{,}780 + 1420 \log 1.33 \times 10^9$$
$$= -674{,}780 + 1420\ (9.124) = -674{,}780 + 12{,}960$$
$$= -661{,}820 \text{ cal mole}^{-1}$$

Step 3. The free energy of dilution for CO_2 is calculated as in part one for glucose. Substituting CO_2 concentration gives

$$\Delta G = 1420 \log \frac{1.5 \times 10^{-4}}{1 \times 10^{-4}} = 1420 \log 1.5 = 1420\ (0.176)$$
$$= 250 \text{ cal mole}^{-1}$$

The overall free energy for the process is the sum of the parts:

$$\Delta G = -2410 - 4080 - 661{,}820 + 1500 \text{ cal mole}^{-1} = -666{,}810 \text{ cal mole}^{-1}$$
$$= -6.67 \times 10^5 \text{ cal mole}^{-1}$$

TEMPERATURE DEPENDENCE OF K_{eq}

If the equilibrium constant can be measured as a function of temperature, then ΔH°, the standard enthalpy, and ΔS°, the standard entropy, can be calculated for the process. Using Equations (10.9) and (10.20), the following relationship can be derived between the equilibrium constant, K_{eq}, and ΔH°:

$$\log K_{eq} = -\frac{\Delta H^\circ}{2.303\ RT} + I \qquad (10.22)$$

where I is an integration constant. The derivation of this equation assumes that ΔH° is constant over the range of temperatures used. This being true,

$\Delta H°$ can be determined from a plot of log K_{eq} versus $1/T$, the slope of which is equal to $-\Delta H°/2.303R$. The standard entropy at a given temperature is calculated from Equation (10.9) for the standard state as

$$\Delta G° = \Delta H° - T\Delta S° \tag{10.23}$$

which gives

$$\Delta S° = \frac{\Delta H° - \Delta G°}{T}$$

Note that although $\Delta H°$ and $\Delta G°$ have units of calories per mole, $\Delta S°$ has units of calories per degree per mole, commonly called entropy units, e.u.

Alternatively, $\Delta H°$ can be obtained from calorimetric data and combined with an equilibrium constant measurement to give all the thermodynamic parameters without having to vary the experimental temperature.

Problem 10.8: The equilibrium constant for the reaction

$$\text{fumarate} + H_2O \rightleftharpoons \text{L-malate}$$

has been measured* as a function of temperature. From the data given for T and K_{eq}, calculate $\Delta H°$, $\Delta G°$, and $\Delta S°$ for the reaction at 25°C.

T(°C)	K_{eq}	log K_{eq}	$1/T \times 10^4$
14.3	4.78	0.68	34.8
20.2	4.46	0.65	34.1
25.0	3.98	0.60	33.6
30.0	3.55	0.55	33.0
34.6	3.27	0.52	32.5
40.0	3.09	0.49	32.0
44.4	2.75	0.44	31.5
49.6	2.43	0.38	31.0

Solution: Calculate to obtain the values of $1/T$ and find the logs of the K_{eq}'s. Plot log K_{eq} versus $1/T$ by Equation (10.22) as shown in Figure 10.3 and determine the slope. Using the point x and the intercept with the y-axis, the slope, m, is

$$m = \frac{0.70 - 0.37}{(34.9 - 30.5) \times 10^{-4}} = \frac{0.33}{4.4 \times 10^{-4}} = \frac{33 \times 10^{-2}}{4.4 \times 10^{-4}}$$

$$= 7.5 \times 10^2$$

Equate this to $-\Delta H°/2.3R$

$$\Delta H° = -2.3(1.986)(7.5 \times 10^2)$$

$$= -3.43 \times 10^3 \text{ cal mole}^{-1}$$

---

*E. M. Scott and R. Powell, *J. Amer. Chem. Soc.* **70**:1104 (1948).

To calculate $\Delta G°$, substitute in Equation (10.20) (see Table 10.2):

$$\Delta G° = -1364 \log 3.98 = -1364\ (0.600)$$
$$= -818 \text{ cal mole}^{-1}$$

Figure 10.3

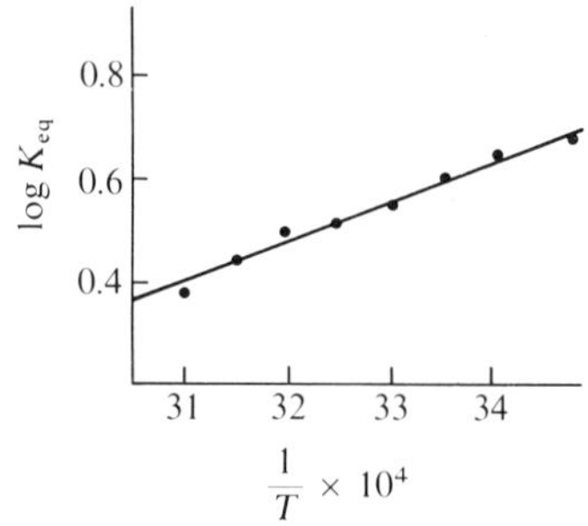

The standard entropy is from Equation (10.23):

$$\Delta S° = \frac{\Delta H° - \Delta G°}{T} = \frac{-3430 - (-818)}{298}$$
$$= -8.76 \text{ e.u. (cal deg}^{-1} \text{ mole}^{-1})$$

RELATION BETWEEN $\Delta G^{o\prime}$ AND $\Delta E_0'$

Oxidation reactions are the primary source of energy in biochemical systems. The standard redox potentials for these reactions enable us to calculate the free energy for the process without a knowledge of K_{eq} as used in the previous energy calculations. The relationship between standard free energy and standard electrode potential can be developed as follows (see also Chapter 9):

$$\Delta E_0' = 2.303 \frac{RT}{nF} \log K_{eq} \qquad (10.24)$$

Rearrange:

$$nF\Delta E_0' = 2.303\ RT \log K_{eq} \qquad (10.25)$$

Now

$$\Delta G^{o\prime} = -2.303\ RT \log K_{eq} \qquad (10.26)$$

$$-\Delta G^{o\prime} = 2.303\ RT \log K_{eq} \qquad (10.27)$$

and thus

$$-\Delta G^{o\prime} = nF\Delta E_0' \qquad (10.28)$$

$$\Delta G^{o\prime} = -nF\Delta E_0' \qquad (10.29)$$

where $\Delta G^{o\prime}$ is the standard free energy, n is the number of moles of electrons transferred in the reaction, F is the number of coulombs per mole of electrons, and $\Delta E_0'$ is the standard electrode potential in joules coulomb^{-1}. With a conversion factor to change joules to calories, the equation is

$$\Delta G^{o\prime} = -n(23{,}063)\Delta E_0' \tag{10.30}$$

Problem 10.9: Calculate the standard free energy for the oxidation of nicotinamide adenine dinucleotide (NAD^+) by flavin adenine dinucleotide (FAD) from the reduction potentials at 30°C:

$$NAD^+ + 2H^+ + 2e^- \rightleftharpoons NADH + H^+ \qquad E_0' - 0.320 \text{ volt}$$

$$FAD + 2H^+ + 2e^- \rightleftharpoons FADH_2 \qquad E_0' - 0.060 \text{ volt}$$

Solution: The $\Delta E_0'$ for the reaction is calculated by the procedure given in Chapter 9:

$$\Delta E_0' = +0.260 \text{ volt}$$

Substitute the $\Delta E_0'$ in Equation (10.29):

$$\Delta G^{o\prime} = -nF\Delta E_0' = -(2)(23{,}063)(0.260) \text{ cal mole}^{-1}$$

Evaluate to obtain $\Delta G^{o\prime}$ at 30°C and pH 7.0:

$$\Delta G^{o\prime} = -12{,}000 \text{ cal mole}^{-1}$$

Problem 10.10: Calculate the standard free energy for the reduction of acetate to acetaldehyde by NADH.

Solution: Use the procedure described in Chapter 9 to obtain $\Delta E_0'$ for the reaction

$$\text{acetaldehyde} + NAD^+ \rightleftharpoons NADH + H^+ + \text{acetate}, \qquad \Delta E_0' \ 0.148 \text{ volt}$$

We want to reverse the reaction so $\Delta E_0'$ is -0.148 volt. Calculate $\Delta G^{o\prime}$, using Equation (10.29):

$$\Delta G^{o\prime} = -n(23{,}063)(\Delta E_0')$$

Substitute:

$$\Delta G^{o\prime} = -(2)(23{,}063)(-0.148)$$

Evaluate:

$$\Delta G^{o\prime} = +6830 \text{ cal mole}^{-1}$$

COUPLED REACTIONS

A highly endergonic reaction can be driven to the right by coupling it with a highly exergonic one. Consider the two reactions

$$A \rightleftharpoons B \qquad \Delta G^{o\prime} + 3000 \text{ cal mole}^{-1} \tag{10.31}$$

$$B \rightleftharpoons C \qquad \Delta G^{o\prime} - 4000 \text{ cal mole}^{-1} \tag{10.32}$$

These reactions couple through the common intermediate B, which is a product in reaction (10.31) and a reactant in reaction (10.32). An overall version of A to C is achieved by the coupling. The $\Delta G^{o\prime}$ for this conversion is the sum of the $\Delta G^{o\prime}$'s for the two reactions:

$$A \rightleftharpoons C \qquad \Delta G^{o\prime} - 1000 \text{ cal mole}^{-1} \tag{10.33}$$

Coupling of reactions was discussed in Chapter 6 in terms of equilibrium constants: the overall equilibrium constant is the product of the constants. Standard free energy is related to the logarithm of the equilibrium constant; thus the overall free energy is the sum of the free energies of the constituent reactions.

Problem 10.11: Calculate the free energy change at 30°C and pH 7.0 when the following reactions couple to give alanine and oxalacetate at the given concentration of reactants and products: [pyruvate] = [aspartate] = 10^{-2} M, [alanine] = 10^{-4} M, [oxalacetate] = 10^{-5} M.

$$\text{glutamate} + \text{pyruvate} \rightleftharpoons \text{ketoglutarate} + \text{alanine} \tag{i}$$

$$\Delta G^{o\prime}_{30^\circ} - 240 \text{ cal mole}^{-1}$$

$$\text{glutamate} + \text{oxalacetate} \rightleftharpoons \text{ketoglutarate} + \text{asparate} \tag{ii}$$

$$\Delta G^{o\prime}_{30^\circ} - 1150 \text{ cal mole}^{-1}$$

Solution: Write the coupled reaction and calculate the standard free energy:

$$\text{pyruvate} + \text{aspartate} \rightleftharpoons \text{alanine} + \text{oxalacetate}$$

The standard free energy is obtained by adding the component free energies after reversing reaction (ii):

$$\Delta G^{o\prime} = -240 \text{ cal mole}^{-1} + (+1150) \text{ cal mole}^{-1} = 910 \text{ cal mole}^{-1}$$

The free energy change in terms of the given concentrations is calculated by using Equation (10.18), corrected for a temperature of 30°C (see Table 10.2). All terms have units of calories per mole:

$$\Delta G = \Delta G^{o\prime} + 1387 \log \frac{[\text{alanine}][\text{oxalacetate}]}{[\text{pyruvate}][\text{aspartate}]}$$

Substituting numbers, we obtain

$$\Delta G = 910 + 1387 \log \frac{10^{-4} \times 10^{-5}}{10^{-2} \times 10^{-2}}$$

$$= 910 + 1387 \log 10^{-5} = 910 + 1387\,(-5) = 910 - 6935$$

$$= -6025 \text{ cal mole}^{-1}$$

Note that the coupled reaction in the standard state is endergonic but at the concentrations used here it is exergonic.

The relationships between $\Delta G^{\circ\prime}$, $\Delta E_0'$, and K_{eq} are now apparent. An equilibrium constant can be expressed in terms of $\Delta G^{\circ\prime}$ or $\Delta E_0'$, and vice versa, whichever is convenient and relevant for the reaction being considered.

Problem 10.12: Couple the reactions given below in terms of free energy per mole to produce ATP at 30°C and pH 7.0.

$$NAD^+ + 2H^+ + 2e^- \rightleftharpoons NADH + H^+ \qquad E_0' - 0.32 \text{ volt} \qquad \text{(i)}$$

$$\text{1,3-DPG} + 2H^+ + 2e^- \rightleftharpoons P_i + \text{3-PG} \qquad E_0' - 0.29 \text{ volt} \qquad \text{(ii)}$$

$$ADP + P_i \rightleftharpoons ATP \qquad \Delta G^{\circ\prime} + 8000 \text{ cal mole}^{-1} \qquad \text{(iii)}$$

3-PG is 3-phosphoglyceraldehyde, and 1,3-DPG is 1,3-diphosphoglyceric acid.

Solution: First couple reactions (i) and (ii) to give the redox reaction (iv). By the methods described in Chapter 9 we obtain

$$\text{1,3-DPG} + NADH + H^+ \rightleftharpoons NAD^+ + \text{3-PG} + P_i \qquad \Delta E_0' \; 0.03 \text{ volt} \qquad \text{(iv)}$$

To obtain the standard free energy first convert $\Delta E_0'$ to $\Delta G^{\circ\prime}$ by Equation (10.30):

$$\Delta G^{\circ\prime} = -n(23{,}063)(\Delta E_0') = -(2)(23{,}063)(0.03) = -1380 \text{ cal mole}^{-1} \quad \text{(v)}$$

Now couple equations (iv) and (iii) to give the overall reaction,

$$\text{1,3-DPG} + NADH + H^+ + ADP \rightleftharpoons ATP + NAD^+ + \text{3-PG} \qquad \text{(vi)}$$

and add the standard free energies from (iii) and (v) to give

$$\Delta G^{\circ\prime} = +8000 \text{ cal mole}^{-1} - 1{,}380 \text{ cal mole}^{-1} = +6620 \text{ cal mole}^{-1}$$

The equilibrium constant can be calculated for reaction (vi) or any of the other reactions from either Equation (10.24) or (10.26). For the overall reaction (vi) the equilibrium constant is obtained from Equation (10.24) (see Problems 9.9 and 10.6):

$$\Delta G^{\circ\prime} = -1387 \log K_{eq}[10^{-7}]$$

Substitute for $\Delta G^{\circ\prime}$:

$$+6620 = -1387 \log K_{eq}[10^{-7}]$$

$$-\frac{6620}{1387} = \log K_{eq}[10^{-7}]$$

$$-4.77 = \log K_{eq}[10^{-7}]$$

$$\text{antilog } K_{eq} = 1.7 \times 10^2$$

The equilibrium constant, 1.7×10^2, indicates that the formation of the products is favored, the standard free energy, $\Delta G^{o\prime} = +6620$ cal mole^{-1}, indicates that the reactants are favored. It is, of course, the standard state for the hydrogen ion, 10^{-7} *M*, which makes $\Delta G^{o\prime}$ positive.

HIGH-ENERGY COMPOUNDS

Coupling of reactions requires that we have a reactant that partakes in a highly exergonic reaction and is simultaneously a product in the endergonic reaction. The most common "high-energy compound" is adenosine triphosphate. It is not a high-energy compound, but its hydrolysis is a high-energy reaction, which can couple with endergonic reactions:

$$ATP^{-4} + H_2O \rightleftharpoons ADP^{-3} + P_i^{-2} + H^+ \qquad \Delta G^{o\prime}_{37^\circ} - 8000 \text{ cal mole}^{-1} \quad (10.34)$$

Hydrogen ion is directly involved in the reaction and upon hydrolysis the medium becomes acidic. In living systems the tissue-buffers neutralize these hydrogen ions to form water,

$$H^+ + OH^- \rightleftharpoons H_2O \qquad (10.35)$$

thereby forcing the reaction to the products. The hydrolysis of ATP is thus already a coupled reaction, and the large negative $\Delta G^{o\prime}$ reflects the coupling of hydrolysis and neutralization.

The hydrolysis energies of certain other organic phosphates are also large and negative—for example, creatine phosphate, guanosine triphosphate, uridine triphosphate. These compounds are formed from ATP hydrolysis and may be used in endergonic metabolic reactions. If excess ATP is present, the energy may be stored as creatine phosphate.

Problem 10.13: Calculate the minimum ratio of ATP to ADP required to form creatine PO_4 from creatine at pH 7.0 at 30°C, with [creatine] $= 2 \times 10^{-3}$ *M*, [creatine PO_4] $= 1 \times 10^{-3}$ *M*,

$$ATP \rightleftharpoons ADP + P_i \qquad \Delta G^{o\prime} - 8000 \text{ cal mole}^{-1} \quad \text{(i)}$$

$$\text{creatine phosphate} \rightleftharpoons \text{creatine} + P_i \qquad \Delta G^{o\prime} - 10{,}500 \text{ cal mole}^{-1} \quad \text{(ii)}$$

Solution: Couple the two reactions and obtain the standard free energy. Since we want to form creatine phosphate, the equation is

$$ATP + \text{creatine} \rightleftharpoons ADP + \text{creatine } PO_4 \qquad \text{(iii)}$$

The standard free energy for reaction (iii) is obtained by summing the standard free energies for reactions (ii) and (iii) after reversing reaction (iii):

$$\Delta G^{o\prime} = -8000 + 10{,}500 = +2500 \text{ cal mole}^{-1}$$

Now calculate the ATP to ADP ratio by substituting in Equation (10.16):

$$\Delta G^{o\prime} = -1387 \log \frac{[\text{ADP}][\text{creatine PO}_4]}{[\text{ATP}][\text{creatine}]}$$

The minimum ratio will just maintain equilibrium:

$$2500 = -1387 \log \frac{[\text{ADP}][1 \times 10^{-3}]}{[\text{ATP}][2 \times 10^{-3}]}$$

$$\frac{2500}{1387} = \log \frac{[\text{ATP}][2]}{[\text{ADP}]}$$

$$1.802 = \log \frac{[\text{ATP}][2]}{[\text{ADP}]}$$

Use the antilog to give

$$\frac{[2][\text{ATP}]}{[\text{ADP}]} = 6.35 \times 10^{1}$$

$$\frac{[\text{ATP}]}{[\text{ADP}]} = 31.8$$

Thus any ratio of concentrations of ATP and ADP greater than 31.8 to 1 would produce creatine phosphate.

ACTIVATION ENERGY

In our discussion of free energy we have been concerned primarily with the equilibrium state and the energy difference between the reactants and the products. Using the principles of thermodynamics, we are able to predict which reactions should be spontaneous: such reactions have a negative free energy change. Many reactions for which we calculate a negative free energy, however, do not form the products spontaneously. For example, the oxidation of glucose (Problem 10.1) was found to be exergonic (energy-producing) and thus should be spontaneous. It is common knowledge, however, that glucose in the presence of O_2 (air) does not oxidize spontaneously. To initiate this reaction some energy must be expended, as is the case for most chemical reactions; that is, an energy barrier must exist between the reactants and the products. This is represented pictorially in Figure 10.4.

The conversion of the reactants to the products is proportional to the number of molecules in the activated state. In any population of molecules there exists a distribution of energies. The number of molecules with sufficient energy to reach the activated state can be increased by the addition of heat and results in an increase in the reaction velocity. Homeotherms obviously do not control the reaction rate by changing the temperature.

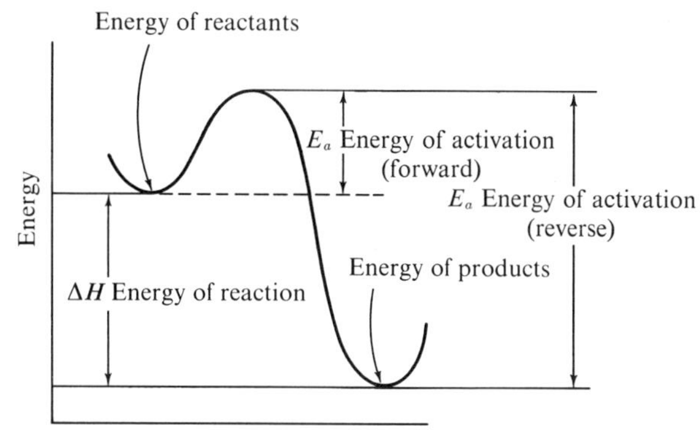

Figure 10.4

Instead, efficient catalysts, enzymes, are involved, which function to lower the activation energy. This method of increasing the rate also gives a measure of control over the reaction. The rate of the reaction depends in part on the amount of enzyme present.

The activation energy, E_a, is related to the reaction velocity, v, by the empirical expression

$$v = Ae^{-E_a/RT} \tag{10.36}$$

first used by Arrhenius, R and T are the gas constant and absolute temperature, respectively. E_a can be determined experimentally by measuring the reaction velocity as a function of temperature. If log v is plotted versus $1/T$, the slope of the line is $-E_a/2.3\ R$.

It is apparent from Figure 10.4 that $\Delta H°$ can be calculated from the measurement of the activation energy for the forward and reverse reactions.

$$\Delta H° = E_a(\text{reverse}) - E_a(\text{forward})$$

The equilibrium constant can be obtained from the ratio of forward and reverse rate constants and $\Delta G°$ can be calculated from Equation (10.20).

Problem 10.14: Calculate the activation energy for the hydrolysis of sucrose catalyzed by the enzyme invertase. The solution is given in part in Table 10.3 with these data.

Table 10.3
Data for Problem 10.14

v (*mole min*$^{-1}$)	T(°C)	T(°K)	$(1/T) \times 10^3$	*Log v*
2.82	20	293	3.413	0.450
6.25	35	308	3.247	0.796
10.20	45	318	3.145	1.008
13.10	55	328	3.049	1.116

Solution: Calculate $1/T$ after converting to absolute temperature. Obtain the logarithms of the velocities. Plot log v versus $1/T$. This data is plotted in Figure 10.5. Draw a line through the points and extrapolate to the x- and y-axes. Determine the slope of the line from the values at these two points (see Figure 10.5):

$$\text{slope} = \frac{1.240 - 0.100}{(3.00 - 3.60) \times 10^{-3}\ \text{deg}^{-1}}$$

Evaluate to give

$$\text{slope} = -2.10 \times 10^{3}\ \text{deg}$$

Equate this to the function from the Arrhenius equation (10.36) to give

$$-\frac{E_a}{2.30R} = -2.10 \times 10^{3}\ \text{deg}$$

Solve for E_a (R is a constant 1.986 cal deg^{-1} mole^{-1}):

$$E_a = (1.986\ \text{cal deg}^{-1}\ \text{mole}^{-1})(2.30)(2.10 \times 10^{3}\ \text{deg})$$

Thus

$$E_a = 9450\ \text{cal mole}^{-1}$$

Activation energies are always positive, meaning that energy must be supplied to keep the reaction going isothermically.

Figure 10.5

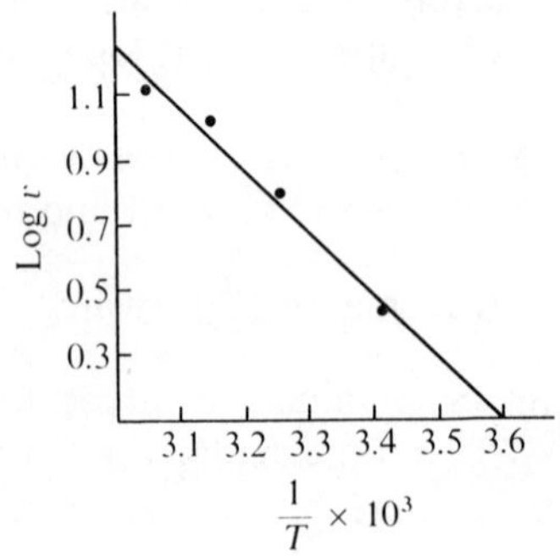

REFERENCES

I. M. Klotz, *Introduction to Chemical Thermodynamics.* Benjamin, New York, 1965.

I. M. Klotz, *Energy Changes in Biochemical Reactions.* Academic Press, New York, 1967.

A. L. Lehninger, *Bioenergetics.* Benjamin, New York, 1965.

B. H. Mahan, *Elementary Chemical Thermodynamics.* Benjamin, New York, 1962.

A. R. Patton, *Biochemical Energetics and Kinetics*. Saunders, Philadelphia, 1965.
A. B. Pardee and L. L. Ingraham, "Free energy and entropy in metabolism," *in* D. M. Greenberg (ed.), *Metabolic Pathways*, Vol. 1, Chapter 1. Academic Press, New York, 1966.
L. E. Strong and W. J. Stratton, *Chemical Energy*. Reinhold, New York, 1965.
K. E. van Holde, *Physical Biochemistry*. Prentice-Hall, New Jersey, 1971.

---

10. PROBLEMS A

Answers on page 351

1. Calculate the standard free energy change for the formation of isocitrate in the following coupled system at 38°C and pH 7.0:

$$\text{citrate} \rightleftharpoons \text{oxalacetate} + \text{acetate}, \qquad K_{eq} = 2.0$$

$$\textit{cis}\text{-aconitate} + H_2O \rightleftharpoons \text{citrate}, \qquad K_{eq} = 0.07$$

$$\textit{iso}\text{-citrate} \rightleftharpoons \textit{cis}\text{-aconitate}, \qquad K_{eq} = 0.5$$

2. The following reactions are occurring together in solution:

$$\text{ATP} + \text{3-phosphoglycerate} \rightleftharpoons \text{ADP} + \text{1,3-diphosphoglycerate}$$

$$K_{eq} = 3.1 \times 10^{-4} \text{ at pH 7.0 and 38°C}$$

$$\text{3-phosphoglycerate} \rightleftharpoons \text{2-phosphoglycerate}$$

$$K_{eq} = 0.24 \text{ at pH 7.0 and 38°C}$$

(a) Calculate the standard free energy for the formation of ADP and 1,3-diphosphoglycerate from ATP and 2-phosphoglycerate.
(b) Assuming the concentration of 2-phosphoglycerate to be twice that of 1,3-diphosphoglycerate, calculate the minimum ratio of ATP and ADP required to form 1,3-diphosphoglycerate from 2-phosphoglycerate.

3. The equilibrium constant for the system

$$\textit{iso}\text{-citrate} \rightleftharpoons \textit{cis}\text{-aconitate}$$

is 0.5. Calculate the free energy change per mole at 38°C for a solution which is 10^{-2} *M* in *iso*-citrate and 3×10^{-2} *M* in *cis*-aconitate.

4. Calculate the standard free energy for the reaction

$$2 \text{ lactate } (1\ M) + O_2(g) \rightleftharpoons 2 \text{ pyruvate } (1\ M) + 2H_2O(l)$$

The values of the standard free energies, ΔG_f°, for these compounds at 38°C formed from the elements and in their standard state in solution are

$$\text{lactate } (1\ M) = -117.9 \text{ kcal mole}^{-1}$$

$$\text{pyruvate } (1\ M) = -106.4 \text{ kcal mole}^{-1}$$

$$H_2O(l) = -56.2 \text{ kcal mole}^{-1}$$

5. Calculate K_{eq} for the following reaction at 38°C and pH 7.0:

$$\text{L-glutamate} + \text{pyruvate} \rightleftharpoons \alpha\text{-ketoglutarate} + \text{L-alanine}$$

$\Delta G^{\circ\prime} = -250$ cal mole^{-1}.

6. Given the reaction

$$\text{uridine} + P_i \rightleftharpoons \text{uracil} + \text{D-ribose-1-P}$$

The equilibrium constant is 0.91 at pH 7.0 and 38°C.

(a) Calculate the standard free energy.
(b) Calculate the ratio of products to reactants required to produce 3000 cal mole^{-1} of free energy.

7. What is the standard free energy for the oxidation of FAD2H by NAD^+ at pH 7.0 and 38°C?

8. Calculate $\Delta G^{\circ\prime}$ for the oxidation of succinate to fumarate by NAD^+:

$$NAD^+ + 2e^- + 2H^+ \rightleftharpoons NAD2H \qquad E_0' = -0.320 \text{ volt}$$

$$\text{fumarate} + 2e^- + 2H^+ \rightleftharpoons \text{succinate} \qquad E_0' = +.031 \text{ volt}$$

9. If the concentrations of the products and reactants are [ADP]/[ATP] = 4, [pyruvate] = 3×10^{-2} M, and [phosphoenolpyruvate] = 1×10^{-5} M, what is the free energy in cal mole^{-1} in the following reaction?

$$\text{phosphoenolpyruvate} + \text{ADP} \rightleftharpoons \text{pyruvate} + \text{ATP}$$

$$K_{eq} = 2 \times 10^3 \text{ at pH 7.0 and 38°C.}$$

10. Glucose dehydrogenase combines with the inhibitor urethane at 38°C with no enthalpy change and an entropic change of +2.75 e.u.* What is $\Delta G^{\circ\prime}$ at 37°C for the reaction?

11. The hydrolysis of urea catalyzed by urease has a heat of activation of 12.5 kcal mole^{-1}.† What is the reaction velocity at 38°C? Assume A, the Arrhenius constant, to be 1×10^{13} sec^{-1}.

12. Calculate the standard free energy and entropy of formation for ethanol (l) from the following data at 25°C.

$$CH_3CH_2OH(l) + 3O_2(g) \rightleftharpoons 2CO_2(g) + 3H_2O(l), \qquad \Delta G^{\circ} = -306.82 \text{ kcal mole}^{-1}$$

$CO_2(g)$	$\Delta G_f^{\circ} = -94.26$ kcal mole^{-1}
$H_2O(l)$	$\Delta G_f^{\circ} = -56.69$ kcal mole^{-1}
$O_2(g)$	$\Delta G_f^{\circ} = 0$
$CH_3CH_2OH(l)$	$\Delta H_f^{\circ} = -66.356$ kcal mole^{-1}

*Y. Ogura, *J. Biochem.* (*Tokyo*), **41**:351 (1954)
†E. J. Casey and K. J. Laidler, *J. Amer. Chem. Soc.*, **72**:2159 (1950).

13. (a) What is the free energy required to concentrate 10 liters of 12% (w/v) of aqueous ethanol to 42% (w/v) at 38°C?
    (b) The caloric value of ethanol is given as 7 kcal g^{-1}. How does this compare with the $\Delta G°$ from Problem 12?

---

10. PROBLEMS B No answers given

1. Calculate the standard free energy and the equilibrium constant for the reaction

$$\alpha\text{-ketoglutarate} + \text{alanine} \rightleftharpoons \text{pyruvate} + \text{glutamate}$$

The values of the standard free energies of formation, ΔG_f°, for these compounds formed from the elements in their standard states at 25°C in solution are

α-ketoglutarate, −190.62 kcal $mole^{-1}$

alanine, −88.75 kcal $mole^{-1}$

pyruvate, −113.44 kcal $mole^{-1}$

glutamate, −171.76 kcal $mole^{-1}$

2. Calculate the standard free energy for the reaction

$$\text{oxalacetate} + \text{acetate} \rightleftharpoons \text{citrate}$$

$K_r = 0.5$ at pH 7.0 and 38°C.

3. The standard free energy change for the conversion of uridine diphosphate glucuronic acid to uridine diphosphate galacturonic acid at 38°C and pH 7.0 is 60 cal $mole^{-1}$. Calculate the equilibrium constant.

4. For the following reaction the apparent equilibrium constant at 38°C and pH 7.0 is 2.33×10^{-5}:

$$NAD^+ + \text{malate} \rightleftharpoons \text{oxalacetate} + NADH + H^+$$

   (a) Calculate the standard free energies $\Delta G°$ and $\Delta G°'$.
   (b) Calculate the free energy change in calories per mole when the concentrations are 0.1 M in malate, 0.01 in NAD^+, 0.001 M in NADH, 0.0002 M in oxalacetate, and 10^{-7} M in H^+.

5. What is the standard free energy for the formation of ribose-5-P at 37°C if the two component reactions are

$$\text{ribulose-5-P} \rightleftharpoons \text{ribose-5-P} \qquad K_{eq} = 2.30$$

$$\text{ribulose-5-P} \rightleftharpoons \text{xylulose-5-P} \qquad K_{eq} = 0.67$$

6. Couple the following reactions:

$$\text{1,3-diphosphoglycerate} + ADP \rightleftharpoons ATP + \text{3-phosphoglycerate}$$

$$K_{eq} = 3.2 \times 10^3 \text{ at pH 7.0 and 38°C}$$

$$\text{phosphoenolpyruvate} + \text{ADP} \rightleftharpoons \text{ATP} + \text{pyruvate}$$

$$K_{eq} = 2.0 \times 10^3 \text{ at pH 7.0 and 38°C}$$

(a) Calculate $\Delta G^{\circ\prime}$ for each reaction.
(b) Assuming the concentration of 1,3-diphosphoglycerate and 3-phosphoglycerate to be equal, what ratio of pyruvate to phosphoenolpyruvate would be required to maintain equilibrium?

7. Calculate the standard free energy, $\Delta G^{\circ\prime}$, for the oxidation of reduced flavin adenine dinucleotide, FAD_2H, by cytochrome-c at pH 7.0 and 38°C.

8. Calculate the free energy change for the oxidation of cytochrome-a by oxygen at 38°C and pH 7.0 when the ratio of products to reactants is 3×10^{10}.

9. What is the work done by the kidneys in producing 1 liter of urine, when the only solute considered is urea, which is 0.005 *M* in plasma and 0.333 *M* in urine?

10. In the reaction between inorganic phosphate and glucose to form glucose-6-phosphate, the $\Delta G^{\circ\prime}$ is +3000 cal mole^{-1} at pH 7.0 and 38°C; when the same product is formed from glucose and ATP with glucokinase at pH 7.0 and 38°C, the $\Delta G^{\circ\prime}$ is −4000 cal mole^{-1}.
(a) Calculate the equilibrium constant for each reaction.
(b) Calculate the standard free energy of hydrolysis of ATP.

11. The second pK_a of phosphoric acid varies with temperature according to the equation

$$pK_a = \frac{A}{T} - D + CT$$

where A is 1979.5, D is 5.3541, and C is 0.01980. Calculate ΔH°, ΔG°, and ΔS° at 25°C. As the temperature is increased would the pH of a phosphate buffer (pH 7.4) increase or decrease?

12. The parietal cells of the gastric glands secrete a solution of 0.16 *M* HCl and 0.007 *M* KCl. Assuming plasma to be the source of the H^+, Cl^-, and K^+ ions, calculate the energy required to produce 450 ml of this solution. For plasma, pH 7.4, the levels of Cl^- and K^+ are 100 meq liter^{-1} and 4.2 meq liter^{-1}, respectively.

13. For the reaction

$$\text{isocitrate} + \tfrac{1}{2}O_2 + H^+ \rightleftharpoons \alpha\text{-ketoglutarate} + CO_2 + H_2O$$

in aqueous solution, $\Delta G^{\circ\prime}$ is given as −54.9 kcal mole^{-1} at 38°C, with O_2 and CO_2 at 1 atm, $[H^+]$ 10^{-7} *M*, and the other reactants are 1 *M*. What is the free energy change, ΔG, in calories per mole at 38°C if α-ketoglutarate and isocitrate are 0.01 *M*, O_2 is 0.2 atm, CO_2 is 0.05 atm, and $[H^+]$ is 10^{-7} *M*?

14. What is the free energy change for removing 0.3 mmole of asparagine from the mitochondria at 1.0×10^{-4} *M* concentration to the cytoplasm at a concentration of 1.7×10^{-4} *M*?

15. The intermediates of the tricarboxylic acid cycle do not normally accumulate in cells. The average concentrations of some of the intermediates in fasted-rat liver are succinate 1.5×10^{-4} *M*, fumarate 7.3×10^{-4} *M*, malate 1.1×10^{-4} *M*, and oxalacetate 1.0×10^{-5} *M*. The $[NAD^+]/[NADH]$ ratio is approximately 0.59. Assume that the total $FAD + FAD_2H$ concentration is 10^{-4} *M*. Calculate the free energy change per mole for the conversion of succinate to oxalacetate at pH 7.0 and 38°C. The equilibrium constants for the reaction sequence succinate $\rightleftharpoons$ fumarate $\rightleftharpoons$ malate $\rightleftharpoons$ oxalacetate are, respectively, 1.0, 3.20, and 2.33×10^{-12}.

16. In the conversion of succinate to fumarate under the conditions described in Problem 15, what is the ΔG (cal $mole^{-1}$) for the reaction when the succinate concentration is reduced from 1.5×10^{-4} *M* to 1.3×10^{-4} *M*? Assume that the fumarate and FAD concentrations remain constant.

11

Enzyme Kinetics

MICHAELIS-MENTEN EQUATION

It is now generally accepted that the catalysis of reactions by enzymes involves the formation of a complex between the enzyme, E, and the substrate, S. This enzyme-substrate complex, ES, may then either dissociate to reform E and S, or it may undergo some rearrangement and form E and products P. This simplest of all mechanisms may be represented as

$$\mathrm{E} + \mathrm{S} \underset{k_2}{\overset{k_1}{\rightleftharpoons}} \mathrm{ES} \underset{k_4}{\overset{k_3}{\rightleftharpoons}} \mathrm{E} + \mathrm{P} \tag{11.1}$$

When the concentration of ES is constant, then the rate at which it is being formed is equal to the rate of its decomposition. Thus, from the Law of Mass Action, for this steady state we have

$$k_1[\mathrm{E}][\mathrm{S}] + k_4[\mathrm{E}][\mathrm{P}] = k_2[\mathrm{ES}] + k_3[\mathrm{ES}] \tag{11.2}$$

Considering only the *initial* stage of the reaction, when [P] is insignificant and [S] is virtually constant, then the term involving [P] is nearly zero and we can write (11.2) as

$$k_1[\mathrm{E}][\mathrm{S}] = [\mathrm{ES}](k_2 + k_3)$$

or

$$\frac{[\mathrm{E}][\mathrm{S}]}{[\mathrm{ES}]} = \frac{k_2 + k_3}{k_1} \equiv K_m \tag{11.3}$$

The maximum initial velocity, V_{max}, will be achieved when *all* the enzyme present is in the complex, active form ES, so that

$$V_{max} = k_3[E_{tot}] \tag{11.4}$$

At any other stage of enzyme saturation the initial velocity, v, for the formation of P will be

$$v = k_3[ES] \tag{11.5}$$

Since $[E_{tot}] = [E] + [ES]$,

$$[E] = [E_{tot}] - [ES] \tag{11.6}$$

and substituting (11.4) and (11.5) in (11.6), we obtain

$$[E] = \frac{V_{max}}{k_3} - \frac{v}{k_3} \tag{11.7}$$

Substituting [E] from (11.7) in (11.3), we have after rearranging terms

$$\frac{[S]}{k_3[ES]} \cdot (V_{max} - v) = K_m$$

and from (11.5)

$$\frac{[S]}{v} \cdot (V_{max} - v) = K_m$$

or, in the more usual form of the Michaelis-Menten equation,

$$v = \frac{[S]V_{max}}{[S] + K_m} \tag{11.8}$$

where K_m is constant for the particular enzyme and substrate under discussion. Before considering the practical aspects of the parts of this equation, it may be well to recall the assumptions on which it is based or which are implied. Three assumptions are made.

1. The reaction is in a steady state and the concentration of ES is constant. The free enzyme is thus in an equilibrium with the complex ES.

2. [P] is insignificant compared to [S] and the substrate concentration is constant. Such a situation exists at the initiation of the reaction, and v, as we have noted above, is only the initial rate of the reaction.

3. $[S] >>> [E]$ and the analysis applies to a single substrate system. Reactions involving two or more substrates have been treated by evaluating the kinetic parameters of each substrate in turn while maintaining the other reactants at fixed concentrations. We will consider these more complex systems later.

The practical significance of the terms in Equation (11.8) is now treated.

INITIAL VELOCITY, *v*

The velocity of a reaction is measured by following the disappearance of reactant (substrate) or the formation of product with time. The velocity will differ depending upon the order of the reaction, but for the classical single substrate-enzyme system discussed above, the rate of change of substrate concentration with time will be linear at first, and then decrease as the substrate is used, until equilibrium is established between substrate and product. It should be remembered that the enzyme only changes the rate at which this equilibrium is reached and not its position.

For purposes of this discussion, a time versus substrate course is shown in Figure 11.1, where $[^1S] < [^2S] < [^3S] < [^4S] < [^5S] < [^6S]$. The initial velocity is literally the velocity of the reaction at zero time before any product is formed (see assumption 2 above). It is determined from the slope of the substrate versus time plot at this initial point in time. Such a velocity is difficult to determine for many practical reasons, such as the time required for mixing the reactants together and then measuring the substrate concentration rapidly. Perhaps the most convenient procedure determines the change in concentration continuously by spectrophotometry, but many kinetic studies require that aliquots of the reaction mixture be taken for analysis at specified time intervals. It is assumed that the reaction velocity during this early stage of the reaction is constant. Initial velocities are determined frequently in the literature on the basis of the amount of substrate changed after a fixed time. It is seen in Figure 11.1 that unless these first measurements can be made before t_1 minutes, then the calculated initial velocities of the reactions with $[^5S]$ and $[^6S]$ will be incorrect. Similarly, t_2 would be satisfactory for $[^2S]$ and $[^1S]$ but for no other substrate concentration in that figure. In other words, fixed-time assays, where the initial velocities are taken as linear in the time span chosen, must be suspect until

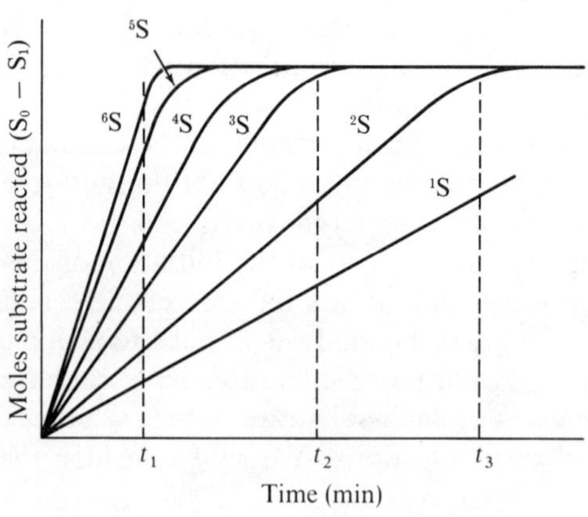

Figure 11.1

it is established that the [S] versus t graph is linear within experimental error up to that time.

Problem 11.1: A substrate, S, was cleaved by an enzyme and the rate of disappearance measured every 30 sec for 3 min. A series of six test tubes was arranged with 1.5 μg of enzyme (mol. wt. 30,000) being added to the same volume of substrate solution, but each at a different concentration. The results are summarized in Table 11.1. What are the initial velocities at each substrate concentration?

Table 11.1
Data for Problem 11.1

Time (min)	*Substrate reacted (μmole)*						
	^0S	^1S	^2S	^3S	^4S	^5S	^6S
0.5	0.0	0.49	0.56	0.71	1.10	1.36	1.45
1.0	0.1	1.05	1.23	1.52	2.24	2.74	3.04
1.5	0.2	1.66	1.93	2.36	3.48	–	–
2.0	0.3	2.23	2.61	3.18	–	–	–
2.5	0.4	2.81	3.30	4.00	–	–	–
3.0	0.5	3.40	4.00	–	–	–	–
Initial conc. (μmole ml^{-1})	0.0	2.32	3.00	4.55	12.66	38.50	200.00

Solution:

Step 1. Correct each reading for the amount of apparent change in the blank solution [^{0}S].
Step 2. Plot micromoles of substrate reacted at each concentration against time (Figure 11.2). Draw the best fit to each series of points.

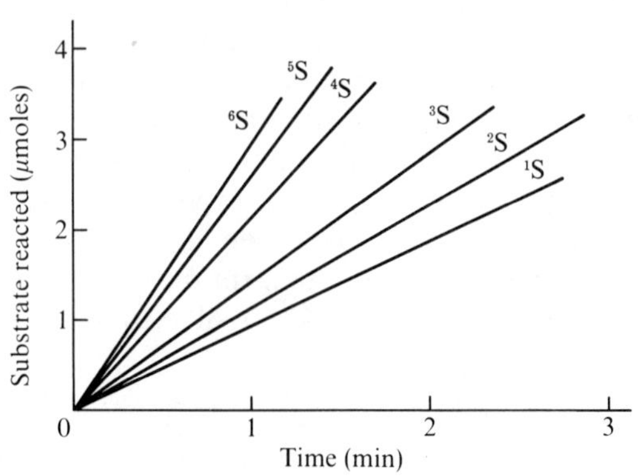

Figure 11.2

Step 3. From Figure 11.2 determine the rate of substrate reacted per minute for each concentration in the linear portion of each curve. In this case the reading of each can be taken at 1 min. These values will be the slopes of the linear portions and will be the initial velocities, v.

At $[^1S]$ $2.32 \times 10^{-3}\ M$	$^1v = 0.95\ \mu$mole min^{-1}
$[^2S]$ $3.00 \times 10^{-3}\ M$	$^2v = 1.13\ \mu$mole min^{-1}
$[^3S]$ $4.52 \times 10^{-3}\ M$	$^3v = 1.42\ \mu$mole min^{-1}
$[^4S]$ $12.66 \times 10^{-3}\ M$	$^4v = 2.14\ \mu$mole min^{-1}
$[^5S]$ $38.5 \times 10^{-3}\ M$	$^5v = 2.64\ \mu$mole min^{-1}
$[^6S]$ $200.0 \times 10^{-3}\ M$	$^6v = 2.94\ \mu$mole min^{-1}

These values are plotted against each other in Figure 11.3.

Figure 11.3

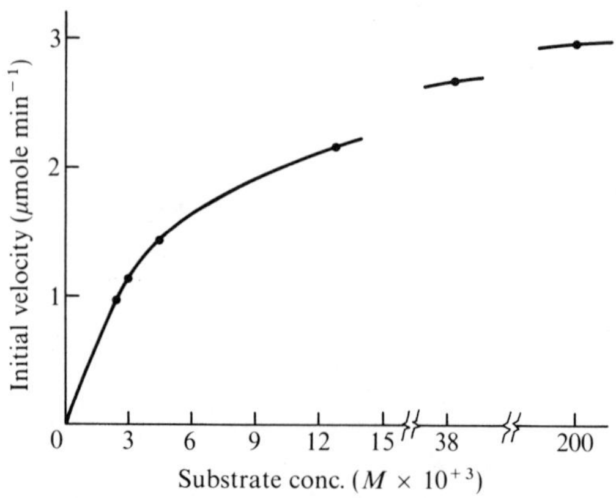

It will be appreciated that the change with time of the concentration of the substrate or product during the reaction may be followed continuously by measuring change in absorbance of the reaction solution in a spectrophotometer, change in optical activity in a polarimeter, change in redox potential with a potentiometer, change in pH, and so on. The readings from these various instruments may be plotted directly to give a measure of the velocity of the reaction, knowing that at all times the reading can be converted to the concentration of the material in question.

It is possible, of course, for the linear portion to be so short in time that either an approximation must be accepted or, preferably, a continuous measurement must be made and a tangent drawn to the curve at zero time.

In such instances it may also be possible to use the integrated form of the Michaelis-Menton equation*

*H. J. Lee and I. B. Wilson, *Biochem. Biophys. Acta* **242**:519 (1971).

$$\frac{2.303}{t}\log_{10}\frac{[S_0]}{[S_t]} = -\frac{1}{K_m} \times \frac{[S_0]-[S_t]}{t} + \frac{V_{max}}{K_m}$$

(in the form

$$y = m \cdot x + b)$$

$[S_0] - [S_t]$ is the amount of substrate converted in time t. By plotting

$$\frac{2.303}{t}\log_{10}\frac{[S_0]}{[S_t]} \text{ versus } \frac{[S_0 - S_t]}{t}$$

the line has slope $-1/K_m$ and an intercept of the y-axis equal to V_{max}/K_m.

SUBSTRATE CONCENTRATION, [S]

As will be noted from the Michaelis-Menten equation, the units of the constant K_m must be those of [S], and since such constants are conventionally expressed as moles per liter it follows that [S] should be in these units.

MICHAELIS-MENTEN CONSTANT, K_m

An analysis of the Michaelis-Menten equation (11.8) will show that the constant, K_m, has the following properties.

1. Like S it has units of concentration, mole liter^{-1}.
2. It is that concentration of substrate where the initial velocity is half the maximum rate possible under the conditions of the experiment. This relationship is independent of enzyme concentration. For the enzymic reaction in Problem 11.1 we see that a substrate concentration of 5.0×10^{-3} M gives an initial velocity of 1.5 μmole substrate reacted per min, which is $\frac{1}{2}$ V_{max} as calculated in Problem 11.6. Substituting these values in Equation (11.8), we obtain

$$1.5\ \mu\text{mole min}^{-1} = \frac{5.0 \times 10^{-3}\ M \times 3.0\ \mu\text{mole min}^{-1}}{5.0 \times 10^{-3}\ M + K_m}$$

When we cancel common units (μmole min^{-1}) on each side of the equation and cross-multiply,

$$1.5(5.0 \times 10^{-3}\ M + K_m) = 3.0 \times 5.0 \times 10^{-3}\ M$$

Collecting terms, we obtain

$$1.5K_m = (3.0 \times 5.0 \times 10^{-3} - 1.5 \times 5.0 \times 10^{-3})M$$
$$K_m = 5.0 \times 10^{-3}\ M$$

which is the substrate concentration for $\frac{1}{2}\ V_{max}$.

3. Since Equation (11.3) gives $K_m = (k_2 + k_3)/k_1$, the identity of K_m will vary according to the relative sizes of the constants k_1, k_2, and k_3. We take two examples.

(a) If k_3 is very small compared to k_2 then K_m is more identifiable as the dissociation constant (stability constant) of the complex ES. This condition occurs with some enzymes and substrates.

(b) For most enzymes, k_3 is not small compared to k_2, and K_m is better considered as a kinetic constant.

MAXIMUM INITIAL VELOCITY, V_{max}

It is consistent with the concept of an enzyme-substrate intermediate in enzymic catalysis that for any given amount of enzyme in the solution there will be a maximum attainable velocity. This velocity will be approached asymptotically by relatively large increases in substrate concentration and in some cases it may not be experimentally attained. The units, however, will be exactly those used to express the velocity of the reaction and v. It has been noted above that v can be measured in many ways, but for comparative purposes V_{max} may be expressed as moles per minute so that at least at this point in the analysis of the kinetics the units of v should be converted. For example, the hydrolysis of sucrose by invertase is conveniently followed directly by measuring the change in the optical activity of the reaction solution with time. The initial velocity may, therefore, be expressed as the change in optical rotation per minute. At the maximum velocity for this experiment the value could then be converted to moles of sucrose hydrolyzed per minute, from a knowledge of the extent of inversion of sucrose to produce this change in optical rotation.

TURNOVER NUMBER AND MOLECULAR ACTIVITY

The V_{max} depends upon the amount of enzyme in the reaction. The theoretical V_{max} in the presence of one mole of enzyme is called the *turnover number* or *molecular activity* of that enzyme for the substrate in question. The turnover number, the moles of substrate reacted per minute per mole of enzyme, therefore has units of reciprocal minutes (min^{-1}) or seconds (sec^{-1}). It is sometimes expressed in terms of an arbitrarily chosen molecular weight, such as 10^5 g, in those cases where the molecular weight of the enzyme is not known. The understanding of turnover number, which is a quantitation of the speed with which the active sites of the enzyme react, is complicated by the fact that some enzyme molecules are composed of subunits and may have more than one active site per molecule. The classical example is catalase, which catalyzes the decomposition of 5×10^6 moles of hydrogen peroxide per min per mole (2.5×10^5 g), but has four catalytic sites per mole so that the turnover number is better given as $1.25 \times 10^6\ \text{min}^{-1}$ to be comparable to another enzyme with one active center per molecule.

Problem 11.2: In Problem 11.1 it could be determined graphically (see Problem 11.5) that the V_{max} is 3.0 μmole of substrate reacted per min in the presence of 1.5 μg enzyme (mol. wt. 30,000). What is the turnover number?

Solution:

$$\text{Moles of substrate reacting per min} = 3.0 \times 10^{-6}$$

$$\text{Moles of enzyme} = \frac{1.5 \times 10^{-6}}{30{,}000}$$

$$\text{Turnover number} = \frac{3.0 \times 10^{-6}}{\dfrac{1.5 \times 10^{-6}}{3.0 \times 10^{4}}}\ \text{min}^{-1}$$

$$= \frac{3.0 \times 10^{-6} \times 3.0 \times 10^{+4}}{1.5 \times 10^{-6}}$$

$$= 6.0 \times 10^{+4} = 60{,}000\ \text{min}^{-1}$$

A problem arises for those reactions in which a polymer is being degraded or synthesized. A polymer molecule in solution does not change its molar concentration as a result of being reduced or increased in molecular size. The change in molecular weight does not change the number of molecules. The catalytic activity is then referred to the number of chemical bonds hydrolyzed or synthesized per minute.

Problem 11.3: A 1% (w/v) solution of starch at pH 6.7 is digested by 15 μg of β-amylase (mol. wt. 152,000). The rate of maltose liberation was determined to have a maximal initial velocity of 8.5 mg formed per min. What is the turnover number?

Solution: For every maltose residue liberated, one glucosidic bond in the starch molecule is hydrolyzed. The rate of maltose formation in moles per minute is, therefore, a measure of the bonds hydrolyzed per minute:

$$8.5 \text{ mg maltose (mol. wt. 342)} = \frac{8.5}{342} \text{ mmole}$$

$$= \frac{8.5 \times 10^{3}}{342} \ \mu\text{mole}$$

$$= 24.8 \ \mu\text{mole}$$

that is, the bonds hydrolyzed per minute = 24.8 μmole

$$\text{Amount of enzyme} = 15 \ \mu\text{g}$$

$$= \frac{15}{1.52 \times 10^{+5}} \ \mu\text{mole}$$

$$= 9.88 \times 10^{-5} \ \mu\text{mole}$$

$$\text{Turnover number} = \frac{24.8 \times 10^{-6} \text{ mole}}{9.88 \times 10^{-5} \times 10^{-6} \text{ mole}} \text{ min}^{-1}$$

$$= 2.5 \times 10^{+5} \text{ min}^{-1}$$

ENZYMATIC ACTIVITY

Another method of expressing enzyme activity relates the micromoles of substrate reacted or product formed per minute to the weight of protein in the enzyme preparation. Therefore, one unit of enzyme is equivalent to the micromoles of product formed per minute. *Specific enzyme activity* gives the micromoles of product per min per mg of protein.

Although these definitions are suggested by the Enzyme Commission (1961),* they are not always followed, particularly in the case of crude preparations of enzymes. Different investigators apply their own definition of enzyme unitage and careful interconversion may be necessary to compare different preparations of the same enzyme. Also, the Enzyme Commission (1972)† recommends that the unit of activity be moles per second.

Problem 11.4: What is the specific activity of the enzymes in Problems 11.1 and 11.3?

Solution: (a) In Problem 11.1, 1.5 μg of enzyme catalyzed the reaction of 3 μmole of substrate per min under maximum conditions:

*M. Florkin and E. H. Stotz (eds.), *Comparative Biochemistry,* Vol. 13. Elsevier, New York, 1964.

†*Enzyme Nomenclature,* Commission of I.U.P.A.C. and I.U.B., Elsevier Scientific Publishing Co., Amsterdam, 1973.

1 mg of enzyme would catalyze the reaction of

$$3 \times \frac{1.0}{1.5 \times 10^{-3}} \mu\text{mole substrate}$$

or

$$2 \times 10^{3} \mu\text{mole}$$

$$\text{specific enzyme activity} = 2 \times 10^{3} \mu\text{mole substrate/minute}$$

or

$$33 \mu\text{mole substrate/second}$$

(b) In Problem 11.3 15 μg of β-amylase "hydrolyzed 24.8 μmoles bonds" per min:

$$1 \text{ mg } \beta\text{-amylase would hydrolyze } 24.8 \times \frac{1.0}{15 \times 10^{-3}} \text{ bonds per minute}$$

or

$$16.5 \times 10^{+3} \text{ bonds}$$

$$\text{specific enzyme activity} = 1.65 \times 10^{4} \text{ bonds per minute}$$

or

$$= 275 \quad \text{bonds per second}$$

GRAPHICAL SOLUTIONS OF THE MICHAELIS-MENTEN EQUATION

Evaluation of V_{max} and K_m is difficult to do by simple inspection or extrapolation of the raw data or by using the hyperbolic relationship of Equation (11.8) because high substrate levels are required to reach the value of V_{max}. The equation can be rectified in two ways.

1. The inverse of the equation can be written

$$\frac{1}{v} = \frac{K_m}{V_{max}} \frac{1}{[S]} + \frac{1}{V_{max}} \tag{11.9}$$

$$\frac{[S]}{v} = \frac{1}{V_{max}} [S] + \frac{K_m}{V_{max}} \tag{11.10}$$

(in the form $y = m \cdot x + b$)

with both forms representing straight lines. The form (11.9) is the Lineweaver-Burk plot (Figure 11.4, next page), of which (11.10) is a simple modification (Figure 11.5).

2. Equation (11.8) may be rearranged to the linear form,

$$v = -K_m \frac{v}{[S]} + V_{max}$$

which is shown in Figure 11.6.

Figure 11.4

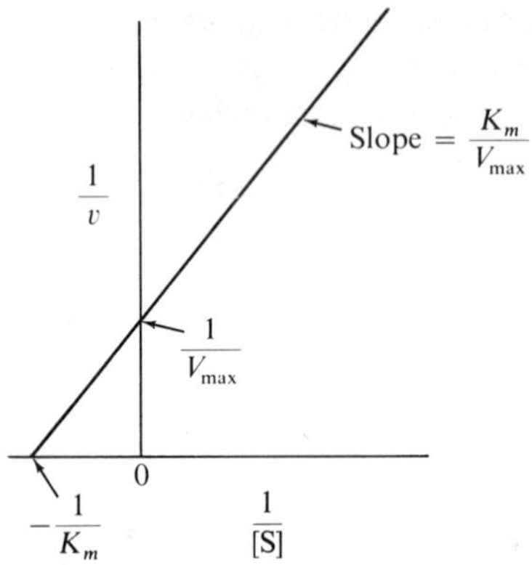

Figure 11.5

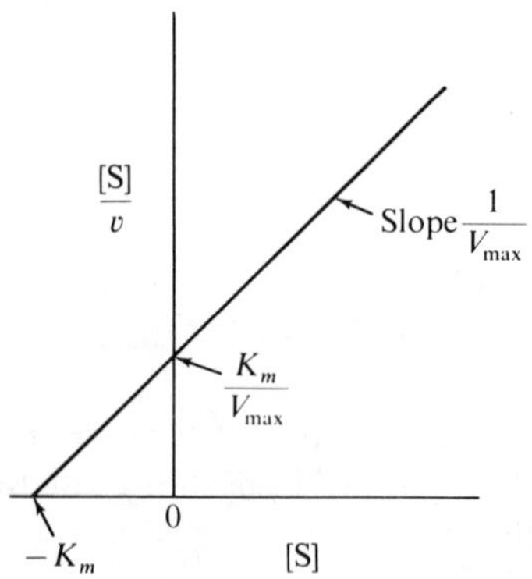

Figure 11.6

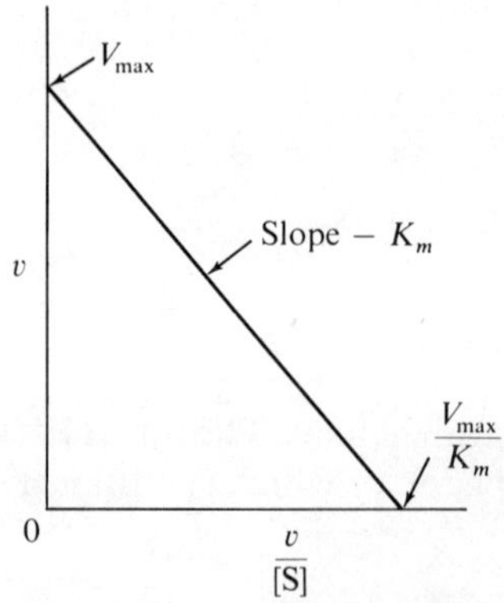

Arguments may be applied for the adoption of one or another of these graphical solutions, but basically the limitations derive from the determination of experimental values with points so spread that a line can be fitted to them. Clear deviations from linearity may be due to impure substrate or enzyme, more than one enzyme acting on the substrate, or to the various inhibitions of the system that are discussed later. It is claimed* that v versus $v/[S]$ plots provide the most sensitive method for recognizing deviations from the Michaelis-Menten relationships. Computer programs† are written for testing the fit of data to the equations.

Problem 11.5: Analyze the data from Problem 11.1 by the three plots given above.

Solution: Since we are concerned with the initial velocities of the reaction, v, and the corresponding substrate concentration [S], the values from Problem 11.1 are summarized in Table 11.2.

Table 11.2
Data for Problem 11.5

$\frac{v}{[S]}\left(\frac{\mu mole\ liter}{mole\ min}\right)$	$[S]\,(M)$	$v\left(\frac{\mu mole}{min}\right)$	$\frac{1}{v}\left(\frac{min}{\mu mole}\right)$	$\frac{1}{[S]}\left(\frac{1}{M}\right)$	$\frac{[S]}{v}\left(\frac{mole\ min}{\mu mole\ liter}\right)$
4.09×10^{2}	2.32×10^{-3}	0.95	1.052	4.3×10^{2}	2.44×10^{-3}
3.73×10^{2}	3.00×10^{-3}	1.13	0.885	3.3×10^{2}	2.66×10^{-3}
3.12×10^{2}	4.52×10^{-3}	1.42	0.706	2.2×10^{2}	3.19×10^{-3}
1.69×10^{2}	12.66×10^{-3}	2.14	0.466	7.9×10^{1}	5.90×10^{-3}
6.86×10^{1}	38.5×10^{-3}	2.64	0.379	2.6×10^{1}	1.46×10^{-2}
1.47	200.0×10^{-3}	2.94	0.340	0.5	0.68×10^{-1}

Figure 11.7 on the next page shows $1/v$ versus $1/[S]$. (a) When $1/[S]=0$, $1/v = 1/V_{max}$. By inspection,

$$1/V_{max} = 0.333 \text{ min } \mu\text{mole}^{-1}$$

$$\therefore V_{max} = 3.00\ \mu\text{mole min}^{-1}$$

(b) When $1/v = 0$, $1/[S] = -1/K_m$. By inspection,

$$-\frac{1}{K_m} = -2.0 \times 10^{2}$$

$$K_m = 5.0 \times 10^{-3}\ M$$

---

*J. E. Dowd and D. S. Ruggs, *J. Biol. Chem.* **240**:863 (1965); C. Walter, *J. Biol. Chem.* **249**:699 (1974).

†See, W. W. Cleland, *Advances in Enzymol,* F. F. Nord (ed.), Vol. 29, Wiley, New York, 1967.

Figure 11.7

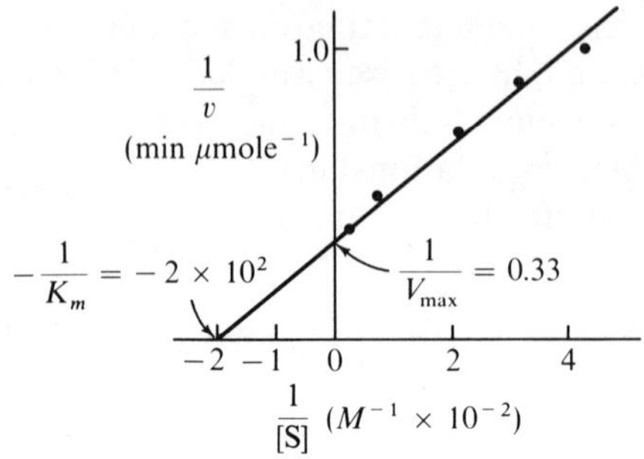

Figure 11.8 shows $[S]/v$ versus $[S]$. It is seen that $K_m = 5.0 \times 10^{-3}\ M$, and

$$\frac{K_m}{V_{\max}} = 1.65 \times 10^{-3}$$

Therefore

$$V_{\max} = \frac{5.0 \times 10^{-3}}{1.65 \times 10^{-3}}\ \mu\text{mole min}^{-1} = 3.03\ \mu\text{mole min}^{-1}$$

Figure 11.8

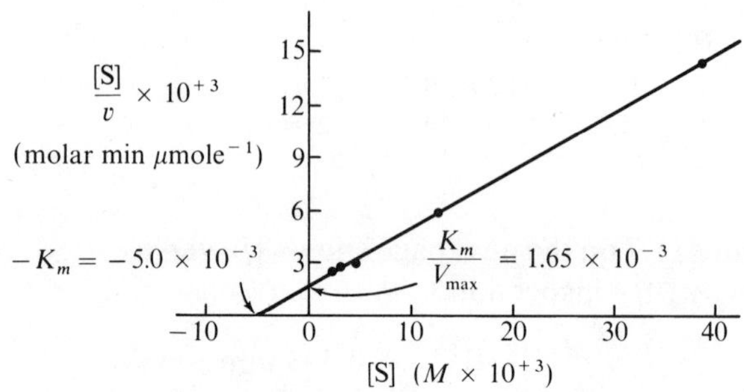

Note, however, that three of the points occur close together in this plot so that the slope of the line when judged by best fit is influenced greatly by the one point at $38.5 \times 10^{-3}\ M$.

Figure 11.9 shows v versus $v/[S]$. It is seen that

$$V_{\max} = 2.95\ \mu\text{mole min}^{-1}$$

$$\frac{V_{\max}}{K_m} = 607\ \mu\text{mole mole}^{-1}\ \text{min}^{-1}\ \text{liter}$$

$$K_m = \frac{2.95}{607}\,\frac{\text{mole}}{\text{liter}} = 4.86 \times 10^{-3}\ M$$

Figure 11.9

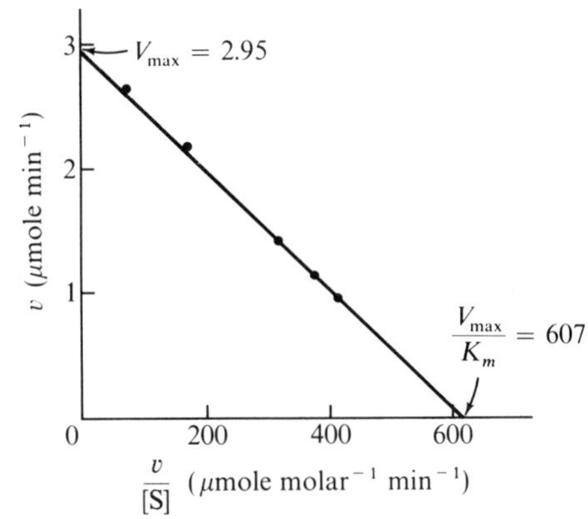

The points in Figures 11.7 and 11.9 are better for determining K_m and V_{max} with these particular results. It could be different for other experimental values and it is, therefore, worthwhile to plot the results in at least two ways.

Problem 11.6: The nicotinamide mononucleotide (NMN) adenyl transferase* catalyzes the reaction

$$\text{NMN} + \text{ATP} \rightleftharpoons \text{NAD}^+ + P \sim P_i$$

The rate of formation of NAD^+ was followed by coupling the above reaction with that catalyzed by alcohol dehydrogenase,

$$CH_3CH_2OH + NAD^+ \rightleftharpoons NADH + CH_3CHO + H^+$$

and the formation of NADH was measured spectrophotometrically at 340 nm. At pH 4.95 the results, after correction for the blank values in the parallel control experiments, were the following.

[S]	v
$M \times 10^{+3}$ of NMN	μmole NAD^+ formed in 3 min with 1 mg of enzyme
0.138	0.148
0.220	0.171
0.291	0.234
0.560	0.324
0.766	0.390

Determine the apparent values of K_m and V_{max} by several graphical methods. This is an example of a two-substrate enzyme reaction being analyzed with the Michaelis-Menten equation.

---

*M. R. Atkinson, J. F. Jackson, and R. K. Morton, *Biochem. J.* **80**:318 (1961).

Solution: The necessary reciprocal values for the graphical solution are calculated directly, since the values in the literature were corrected (Table 11.3).

Table 11.3
Data for graphical solution of Problem 11.6

$\frac{1}{[S]}\left(\frac{1}{mM}\right)$	$\frac{1}{v}\left(\frac{3\ min}{\mu mole}\right)$	$\frac{v}{[S]}\left(\frac{\mu mole}{M \cdot 3\ min}\right) \times 10^{-3}$	$\frac{[S]}{v}\left(\frac{M \cdot 3\ min}{\mu mole}\right) \times 10^{+3}$
7.25	6.76	1.07	0.93
4.55	5.85	0.78	1.29
3.44	4.27	0.81	1.25
1.79	3.09	0.58	1.73
1.31	2.56	0.51	1.96
0.68	2.03	0.34	2.96

The appropriate ordinates are shown in Figures 11.10, 11.11, and 11.12, and two individuals (A and B) independently drew what they considered to be the best fit to the points. Two sets of values were also analyzed by the method of least squares. The appropriate values for K_m and V_{max} are summarized in Table 11.4. It is easy to see the variations in graphing by the two individuals. Check the original paper and also your own solution.

Figure 11.10

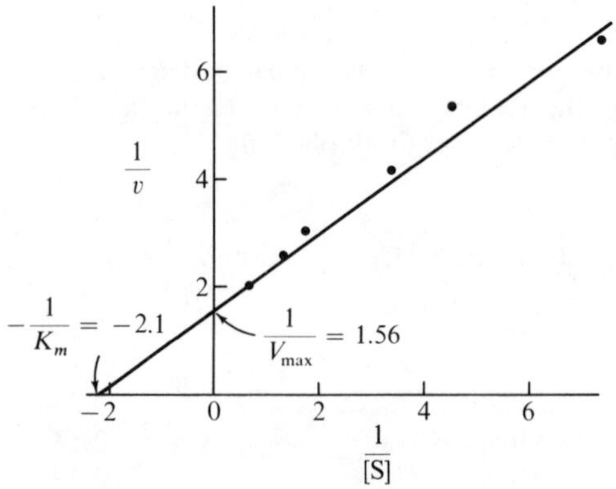

Figure 11.11

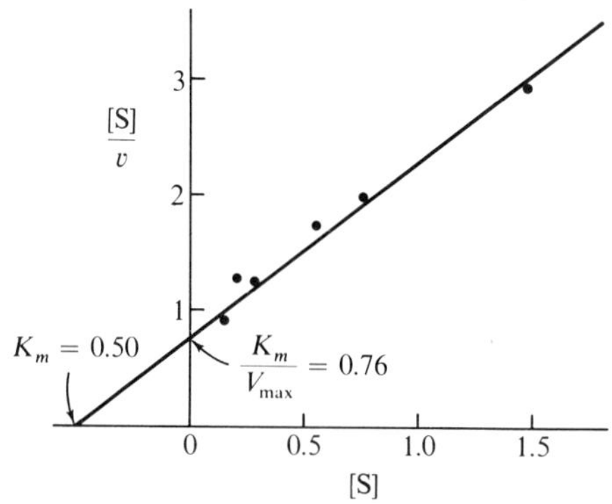

Figure 11.12

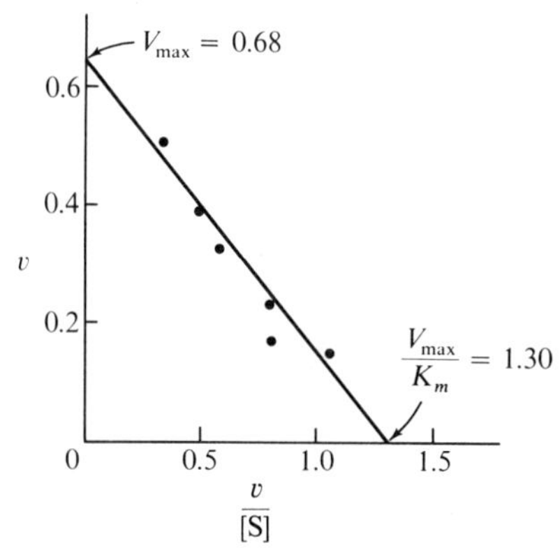

Table 11.4
Solutions of Problem 11.6

Graphical method	*Apparent K_m ($M \times 10^3$)*			*Apparent V_{max} ($\mu mole/3\ min$)*		
	A	*B*	*Least squares*	*A*	*B*	*Least squares*
$1/v$ versus $1/[S]$	0.50	0.48	0.46	0.65	0.64	0.61
$[S]/v$ versus $[S]$	0.54	0.50	0.58	0.68	0.66	0.69
v versus $v/[S]$	0.54	0.53	–	0.66	0.68	–

TWO-SUBSTRATE REACTIONS

Consider the general equation for a reversible reaction of the type

$$A + B \rightleftharpoons C + D$$

Several models can be proposed for the enzyme-catalyzed reaction mechanisms. The various mechanisms are identified in several ways:

Sequential Mechanisms. Both substrates bind to the enzyme before either of the products is released. For example

$$\mathrm{A + E \rightleftharpoons EA} \tag{11.11a}$$

$$\mathrm{EA + B \rightleftharpoons (EAB \rightleftharpoons ECD) \rightleftharpoons ED + C} \tag{11.11b}$$

$$\mathrm{ED \rightleftharpoons E + D} \tag{11.11c}$$

The kinetic equation for this mechanism is

$$v = \frac{V_{\max}[\mathrm{A}][\mathrm{B}]}{[\mathrm{A}][\mathrm{B}] + K_a[\mathrm{B}] + K_b[\mathrm{A}] + K_{ia}K_b} \tag{11.12}$$

where K_a is the Michaelis constant for A, K_b is the Michaelis constant for B, K_{ia} is a dissociation constant for the complex $\mathrm{EA \rightleftharpoons E + A}$, $V_{\max}$ is the maximum initial velocity for the reaction as written, v is the observed initial velocity for the reaction as written.

One type of sequential mechanism is that in which the substrates bind in a definite order to the enzyme, one example of which is an *Ordered Bi Bi mechanism*. This mechanism is given above in Equations (11.11a) to (11.11c) or, using the shorthand notation of Cleland (11.13), we see substrate A binding to the enzyme to form the binary complex EA, which then binds B to give the ternary complex EAB. The central complexes

$$\begin{array}{ccccccccc} & \mathrm{A} & & \mathrm{B} & & \mathrm{C} & & \mathrm{D} & \\ & \downarrow & & \downarrow & & \uparrow & & \uparrow & \\ \hline \mathrm{E} & & \mathrm{EA} & & \mathrm{(EAB \rightleftharpoons ECD)} & & \mathrm{ED} & & \mathrm{E} \end{array} \tag{11.13}$$

EAB and ECD interconvert and this is followed by the dissociation of product C in the forward direction, leaving the binary complex ED, which then dissociates to D and the regenerated enzyme E.

Many dehydrogenase enzymes, with NAD^+ or $NADP^+$ coenzymes, have Ordered Bi Bi mechanisms. For example, alcohol dehydrogenase catalyzes the reaction

$$\text{ethanol} + \mathrm{NAD^+} \rightleftharpoons \text{acetaldehyde} + \mathrm{NADH + H^+}$$

It is also possible, however, that the substrates of A and B do not bind in a definite order to the enzyme and the two binary complexes EA and EB bind the second substrate to give the same ternary complex EAB. This is a *random* reaction. Following interconversion to ECD, the dissociations of C and D are also without effect on each other. The mechanism may be represented*

*W. W. Cleland, *Biochem. Biophys. Acta* **67**:104–137 (1963).

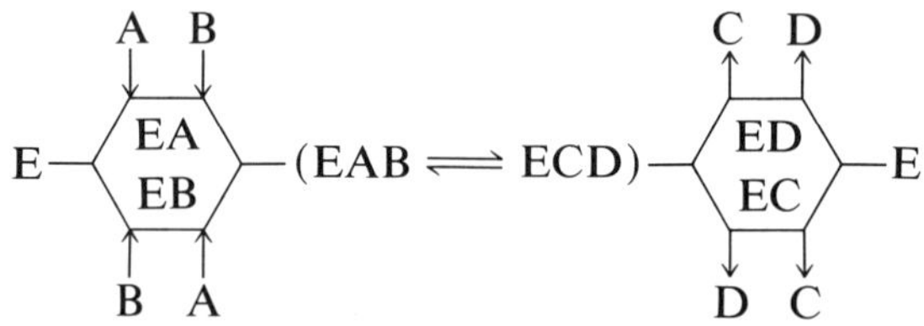

(11.14)

If the association and dissociation of substrates and products are very rapid compared to the interconversion of the central complexes, the mechanism is identified as Rapid Equilibrium Random and the kinetic equation is given by Equation (11.12). Creatine kinase appears to conform to the random sequential mechanism.

Nonsequential Mechanisms – Ping-Pong Mechanisms. A second group of mechanisms for enzyme-catalyzed reactions has a reaction sequence in which one of the products is released before all substrates are bound. One such system has what is called a *ping-pong* mechanism in which the first binary complex EA undergoes modification to FC from which one of the products dissociates. This is followed by the binding of substrate B to F, an altered form of the enzyme. The next process is then similar to the first and the graphical representation is

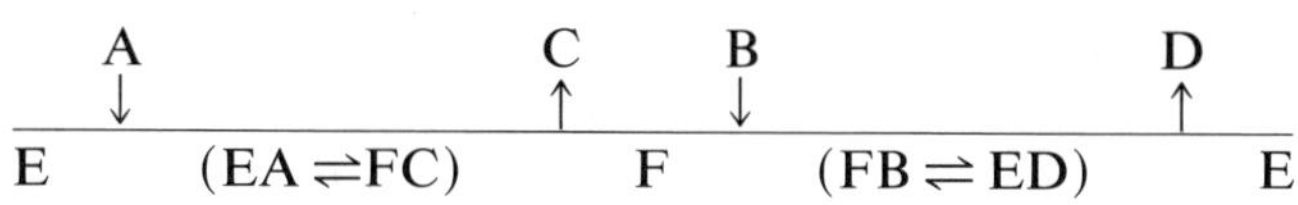

(11.15)

Many of the transaminases are best described by a ping-pong mechanism, the rate equation being

$$v = \frac{V_{\max}[\mathrm{A}][\mathrm{B}]}{[\mathrm{A}][\mathrm{B}] + K_a[\mathrm{B}] + K_b[\mathrm{A}]} \tag{11.16}$$

Note the absence of the $K_{ia}K_b$ term – see Equation (11.12).

GRAPHICAL SOLUTIONS FOR THE UNKNOWN PARAMETERS

Equations (11.12) and (11.16) can be linearized by transformation to their inverse forms, as in the case of the simple Michaelis-Menten equation (11.8). In each case the slopes of the lines or the intercepts on the axes or both are values that contain more than one parameter, so that the slopes or intercepts of a family of lines need to be replotted to give complete graphical solutions. To illustrate these general points let us consider the rate equation for the sequential mechanism. Equation (11.12) in the inverse form (11.17) is written

$$\frac{1}{v} = \frac{K_a}{V_{\max}}\left(1 + \frac{K_{ia}K_b}{K_a[\mathrm{B}]}\right)\frac{1}{[\mathrm{A}]} + \frac{1}{V_{\max}}\left(1 + \frac{K_b}{[\mathrm{B}]}\right) \tag{11.17}$$

$$(y = \qquad m \cdot x + b)$$

If the initial velocities of the reaction are measured with varying concentrations of both substrates A and B and in the absence of significant amounts of products C and D, then the double reciprocal plots of $1/v$ versus $1/[\mathrm{A}]$ will give one family of lines (Figure 11.13) and a similar family of lines will be generated for $1/v$ versus $1/[\mathrm{B}]$. On the primary plot (Figure 11.13) for differing values of [B] there are a corresponding set of intercepts $1/V_{\mathrm{app}}$ on the y-axis, where

$$\frac{1}{V_{app}} = \frac{1}{V_{max}}\left(1 + \frac{K_b}{[\mathrm{B}]}\right)$$

Figure 11.13

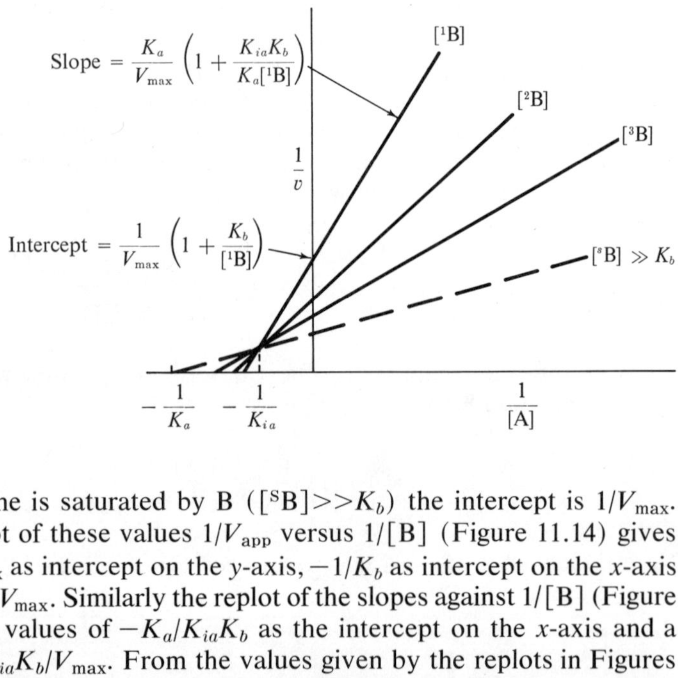

When the enzyme is saturated by B ($[^s\mathrm{B}] >> K_b$) the intercept is $1/V_{\max}$. A secondary plot of these values $1/V_{\mathrm{app}}$ versus $1/[\mathrm{B}]$ (Figure 11.14) gives values for $1/V_{\max}$ as intercept on the y-axis, $-1/K_b$ as intercept on the x-axis and a slope of $K_bV_{\max}$. Similarly the replot of the slopes against $1/[\mathrm{B}]$ (Figure 11.15) gives the values of $-K_a/K_{ia}K_b$ as the intercept on the x-axis and a slope equal to $K_{ia}K_b/V_{\max}$. From the values given by the replots in Figures 11.14 and 11.15 the quantities K_a, K_{ia}, K_b and $V_{\max}$ are obtained. A similar evaluation obtained from the primary plot of $1/v_0$ versus $1/[\mathrm{B}]$ should confirm the values of these constants.

It may be noted that the intersection of lines in the primary plots gives the value of $-1/K_{ia}$ and may be above or below the $1/[\mathrm{A}]$-axis depending upon the relative values of K_{ia} and K_a.* In the case of the random sequen-

---

*J. D. Lueck, W. R. Ellison, and H. J. Fromm, *FEBS Letters* **30**:321 (1973).

Figure 11.14

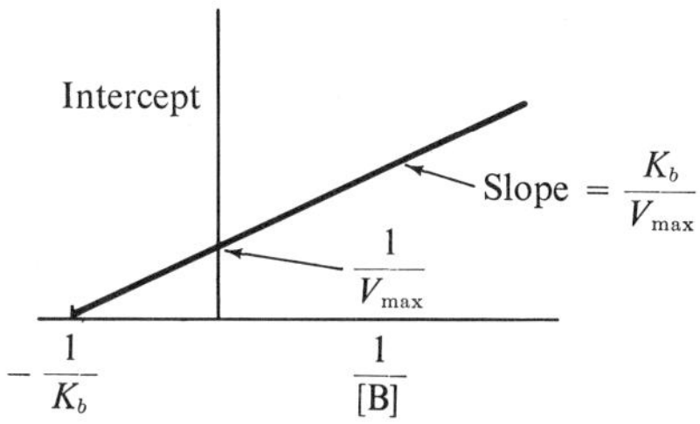

Figure 11.15

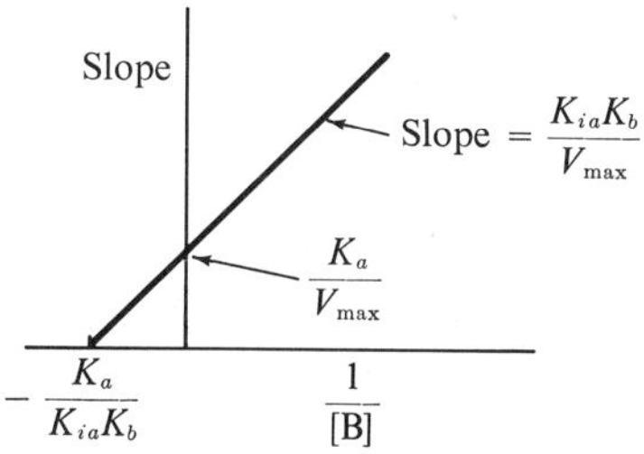

tial mechanism illustrated in Figure 11.13, if $K_{ia} = K_a$ then the intersection is on the $1/[A]$-axis.

The primary double reciprocal plots for the *ping-pong* mechanism

$$\frac{1}{v} = \frac{K_b}{V_{max}} \cdot \frac{1}{[B]} + \frac{1}{V_{max}}\left(1 + \frac{K_a}{[A]}\right) \tag{11.18}$$

or

$$\frac{1}{v} = \frac{K_a}{V_{max}} \cdot \frac{1}{[A]} + \frac{1}{V_{max}}\left(1 + \frac{K_b}{[B]}\right)$$

$$(y = m \quad \cdot \quad x \; + \; b)$$

give families of straight lines with parallel slopes (Figures 11.16 and 11.17). The secondary plots of the intercepts $1/V_{app}$ against $1/[\text{Substrate}]$ give

Figure 11.16

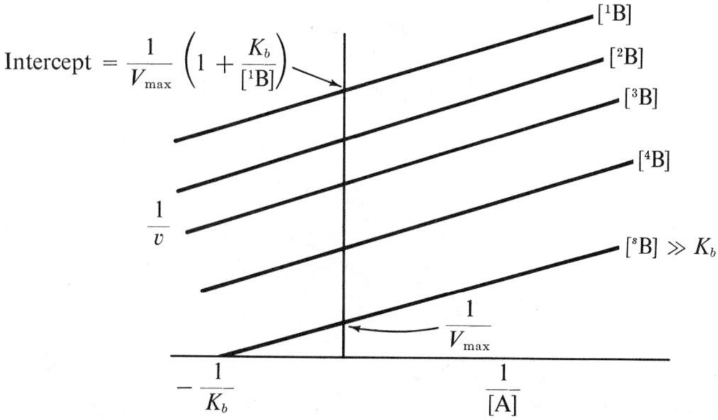

Figure 11.17

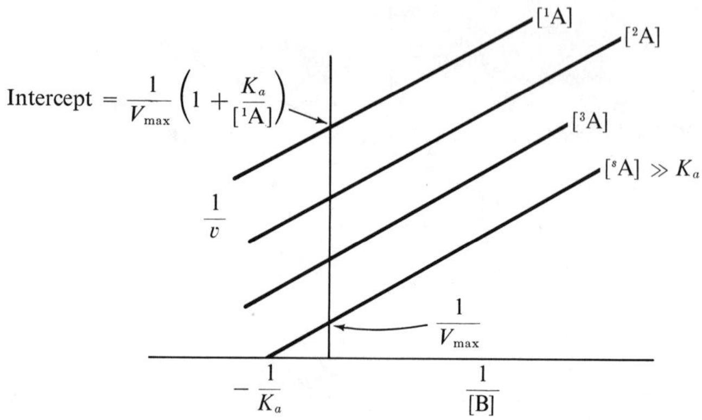

values for $-1/K_b$ or $-1/K_a$, $1/V_{\max}$ and $K_b/V_{\max}$ or $K_a/V_{\max}$ (Figures 11.18 and 11.19). Again, these constants can all be evaluated.

Figure 11.18

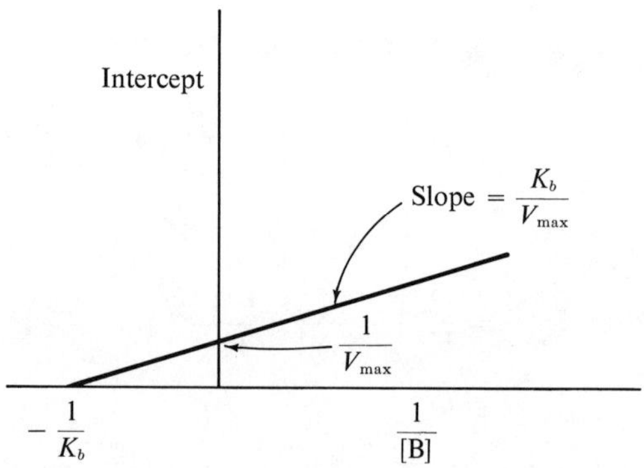

Figure 11.19

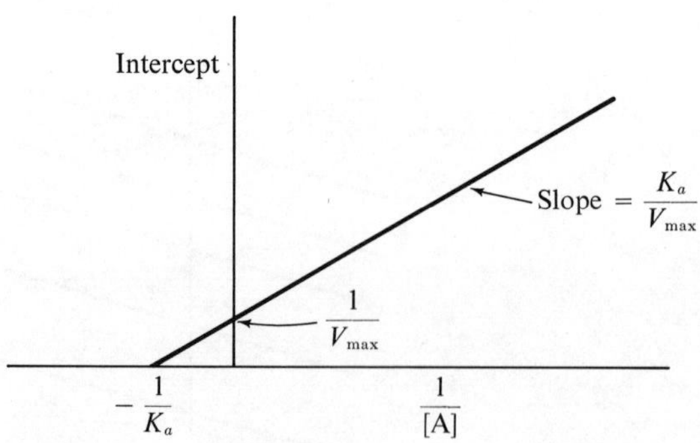

Problem 11.7: Initial velocity data were collected for alcohol dehydrogenase by varying the concentrations of both substrates (NAD^+ and ethanol) systematically in the absence of significant concentrations of products. The data are tabulated below. Make the appropriate double reciprocal plots (primary and secondary) and evaluate V_{max}, K_a, K_b, and K_{ia}. Is the mechanism of the enzyme sequential or ping-pong?

[A] m*M* NAD	[B] m*M* Ethanol	v ΔAbs 340/min	[A] m*M* NAD	[B] m*M* Ethanol	v ΔAbs 340/min
0.50	50	0.0551	0.125	16.5	0.0175
0.25	50	0.0460	0.100	16.5	0.0168
0.165	50	0.0360	0.50	12.5	0.0335
0.125	50	0.0325	0.25	12.5	0.0233
0.10	50	0.026	0.165	12.5	0.0190
0.50	25	0.0471	0.125	12.5	0.0149
0.25	25	0.0342	0.100	12.5	0.0138
0.165	25	0.0268	0.50	10	0.0288
0.125	25	0.0216	0.25	10	0.0214
0.100	25	0.0195	0.165	10	0.0154
0.50	16.5	0.039	0.125	10	0.0139
0.25	16.5	0.026	0.100	10	0.0120
0.165	16.5	0.0206			

Solution: The necessary reciprocal values for the graphical solutions are calculated in Table 11.5 on the next page.

The converging nature of the lines indicates a sequential mechanism.

For the replots of the intercepts and slopes against 1/[EtOH] the data is read from Figure 11.20 and tabulated in Table 11.6.

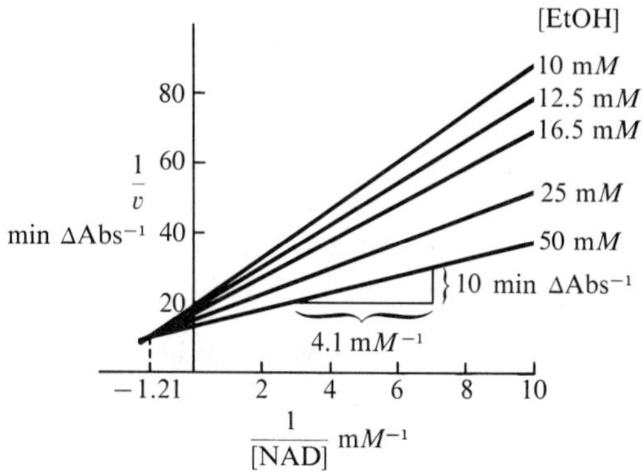

Figure 11.20

Table 11.5
Data for graphical solution of Problem 11.7

$\frac{1}{[NAD^+]}$ mM^{-1}	$\frac{1}{v_0}$ $(\Delta Abs_{340})^{-1}$ *min*	*EtOH* mM
2.0	18.15	
4.0	21.74	
6.1	27.78	50.0
8.0	30.77	
10.0	38.46	
2.0	21.23	
4.0	29.24	
6.1	37.31	25.0
8	46.30	
10	51.28	
2.0	25.64	
4.0	38.46	
6.1	48.54	16.5
8.0	57.14	
10.0	59.52	
2.0	29.85	
4.0	42.92	
6.1	52.63	12.5
8.0	67.11	
10.0	72.46	
2.0	34.72	
4.0	46.73	
6.1	64.94	10.0
8.0	71.94	
10.0	83.33	

It is important to note that in the calculation of the slope of a line the values of the numbers read from the graph must be in the correct units. For example in Table 11.6 the slope at 1/[EtOH] 20 M^{-1} is the ratio of the numbers selected from the primary graph (Figure 11.20), 10 min $\Delta Abs^{-1}/4 \cdot 1$ mM^{-1}. The ratio of the values chosen for determination of the slopes of the lines are given in parenthesis.

From the two secondary replots it is seen that

(a) Intercept versus 1/[EtOH] (Figure 11.21)

Table 11.6
Replot data for graphical solution of Problem 11.7

Intercept $\left(\frac{1}{V_{app}}\right)$	$\frac{1}{[EtOH]}$ M^{-1}	Slope
12.8	20	$2.4\left(=\frac{10}{4.1}\right)$
14.7	40	$3.8\left(=\frac{10}{2.65}\right)$
16.4	61	$5.3\left(=\frac{10}{1.9}\right)$
17.5	80	$6.1\left(=\frac{15}{2.46}\right)$
19.0	100	$6.9\left(=\frac{20}{2.9}\right)$

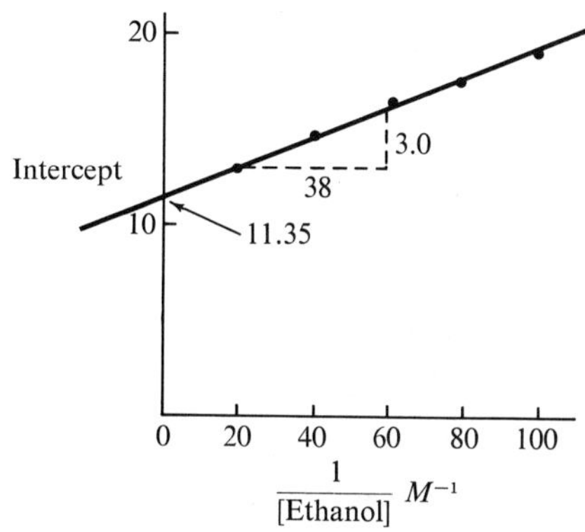

Figure 11.21

$$\text{Intercept} = \frac{1}{V_{max}}$$

$$= 11.35$$

$$\therefore \underline{V_{max} = 0.088} \text{ absorbance units at 340 nm/min}$$

$$\text{Slope} = \frac{K_b}{V_{max}}$$

that is

$$\frac{30}{38} = \frac{K_b}{0.088}$$

$$\therefore \underline{K_b = 6.95 \text{ m}M}$$

(b) Slope versus 1/[EtOH] (Figure 11.22)

Figure 11.22

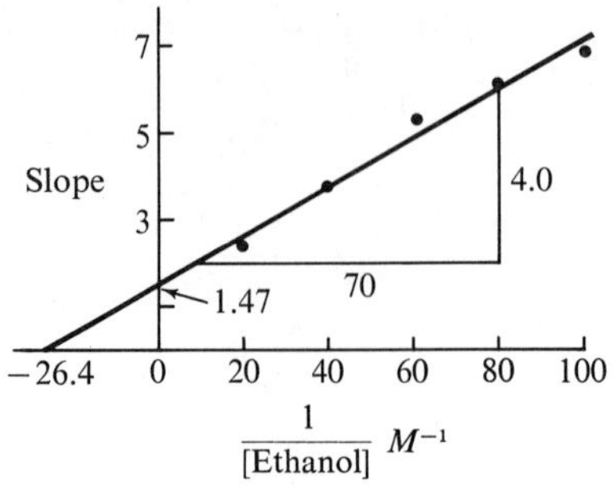

$$\text{Intercept} = \frac{K_a}{V_{max}}$$

that is

$$= 1.47$$

$$\therefore\ \underline{K_a = 0.13\ \text{m}M}$$

$$\text{Slope} = \frac{K_{ia}K_b}{V_{max}}$$

$$= \frac{4.0}{70}$$

$$\therefore\ K_{ia} = \frac{4.0}{70} \times \frac{V_{max}}{K_b}$$

$$\underline{= 0.72\ \text{m}M}$$

$$\text{Other intercept} = \frac{-K_a}{K_{ia}K_b}$$

$$= -26.4$$

$$\therefore\ K_{ia} = \frac{K_a}{26.4K_b}$$

$$\underline{= 0.71\ \text{m}M}$$

From the intersection of lines on the primary plot (Figure 11.20)

$$-\frac{1}{K_{ia}} = -1.21$$

$$\underline{K_{ia} = 0.83\ \text{m}M}$$

ENZYME INHIBITION

The presence of certain molecules in an enzyme-catalyzed reaction may reduce the reaction rate by interfering with the formation or breakdown of the enzyme-substrate complex. Such interfering molecules are called inhibitors. There are several ways in which the inhibitor can influence the kinetics of the reaction, but we will treat only the three simplest and most common types of inhibitors: competitive, noncompetitive, and uncompetitive.

The experimental procedure involves the determination of initial observed velocities for the reaction under conditions where the concentrations of all but one of the reactants are fixed while the concentrations of the inhibitor and the substrate under study are varied. The effect of inhibitors on the kinetics of single or multisubstrate enzyme-catalyzed reactions therefore are considered in a similar fashion so that the identification of the type of inhibition can be discussed in terms of the simple Michaelis-Menten equation (11.13).

Competitive Inhibition. If the inhibitor (I) competes with the substrate molecule such that the normal Michaelis-Menten reaction, Equation (11.1),

$$\mathrm{E + S \rightleftharpoons ES \rightarrow E + P}$$

is accompanied by the *reversible* reaction to form the dead-end complex EI

$$\mathrm{E + I \rightleftharpoons EI} \tag{11.19}$$

thereby essentially removing some of the enzyme molecules from the reaction with S, the situation of competitive inhibition applies. Then the initial velocity of the reaction is given by

$$v = \frac{V_{\max}}{1 + \dfrac{K_m}{[\mathrm{S}]}\left(1 + \dfrac{[\mathrm{I}]}{K_{is}}\right)} \tag{11.20}$$

where [I] is the concentration of the inhibitor and K_{is} has the dimension of concentration.

$$K_{is} = \frac{[\mathrm{E}][\mathrm{I}]}{[\mathrm{EI}]} \tag{11.21}$$

Equation (11.20) is a hyperbolic function, as was the uninhibited Equation (11.8), which again can be rectified in several ways. The Lineweaver-Burk plot of $1/v$ versus $1/[\mathrm{S}]$,

$$\frac{1}{v} = \frac{K_m}{V_{\max}} \cdot \left(1 + \frac{[\mathrm{I}]}{K_{is}}\right) \cdot \frac{1}{[\mathrm{S}]} + \frac{1}{V_{\max}} \qquad (11.22)$$

which is of the form

$$(y = \qquad m \qquad \cdot \quad x + b)$$

demonstrates that, whereas the intercept of the $1/v$-axis is the same as for the uninhibited reaction (11.10), both the slope and the K_m values are changed by a factor of

$$1 + \frac{[\mathrm{I}]}{K_{is}}$$

See Figure 11.23.

Figure 11.23

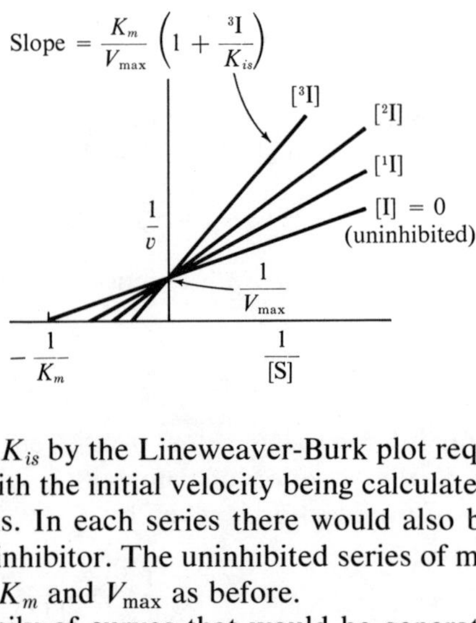

The determination of K_{is} by the Lineweaver-Burk plot requires several series of measurements, with the initial velocity being calculated for a range of substrate concentrations. In each series there would also be included a constant concentration of inhibitor. The uninhibited series of measurements would give the values for K_m and $V_{\max}$ as before.

The slopes of the family of curves that would be generated in Figure 11.23 contain both K_{is} and $V_{\max}$, and [I], from which each can be obtained by the secondary replot of the slopes against [I]. Since for competitive inhibition

$$\text{Slope} = \frac{K_m}{V_{\max}} + \frac{K_m}{V_{\max} K_{is}} [\mathrm{I}]$$

$$(y = \quad b \ + m \cdot x)$$

then for the replot of slope versus [I] (Figure 11.24) at $y = 0$

$$-\frac{K_m}{V_{\max}} = \frac{K_m}{V_{\max} K_{is}} \cdot [\mathrm{I}]$$

Figure 11.24

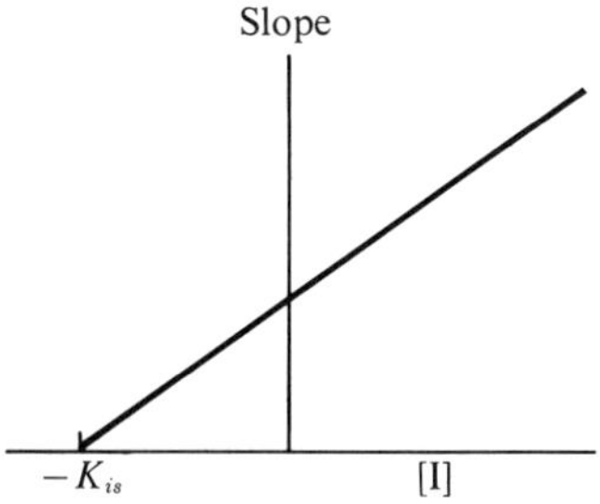

and

$$[I] = -K_{is}$$

This procedure for the evaluation of K_{is} is thought to be better than the method of Dixon* for differentiating between competitive and noncompetitive inhibition.†

Noncompetitive Inhibition. This type of inhibition results from the inhibitor complexing with the enzyme or the enzyme-substrate complex:

$$E + I \rightleftharpoons EI$$

which was given as Equation (11.12), or

$$ES + I \rightleftharpoons IES \tag{11.23}$$

but IES does *not* dissociate into IE + P. For these equations,

$$K_{is} = \frac{[E][I]}{[EI]} \tag{11.24}$$

and a similar constant applies to the second equation,

$$K_{ii} = \frac{[ES][I]}{[IES]} \tag{11.25}$$

The rate equation becomes

$$v = \frac{V_{max}[S]}{K_m\left(1 + \frac{[I]}{K_{is}}\right) + S\left(1 + \frac{[I]}{K_{ii}}\right)} \tag{11.26}$$

When $[I] = 0$, the initial velocity becomes that for the uninhibited reaction. However, unlike the rate Equation (11.20) for competitive inhibition, when

*M. Dixon, *Biochem. J.* **55**:170 (1953).
†D. L. Purich and H. J. Fromm, *Biochem. Biophys. Acta* **268**:1 (1972).

[S] is very large, V_{max} is reduced by a factor of $1 + ([I]/K_{ii})$. This is also evident from the Lineweaver-Burk rectification of the curve:

$$\frac{1}{v} = \left(1 + \frac{[I]}{K_{is}}\right) \frac{K_m}{V_{max}} \cdot \frac{1}{[S]} + \left(1 + \frac{[I]}{K_{ii}}\right) \frac{1}{V_{max}} \qquad (11.27)$$

$$(y = \quad m \quad \cdot x + \quad b \quad)$$

When $1/v = 0$, then $1/[S] = -1/K_m$ in the absence of any inhibitor.

The graphical solutions of the noncompetitive situation are the same as those used for competitive inhibition. They are summarized in Figures 11.25, 11.26, and 11.27.

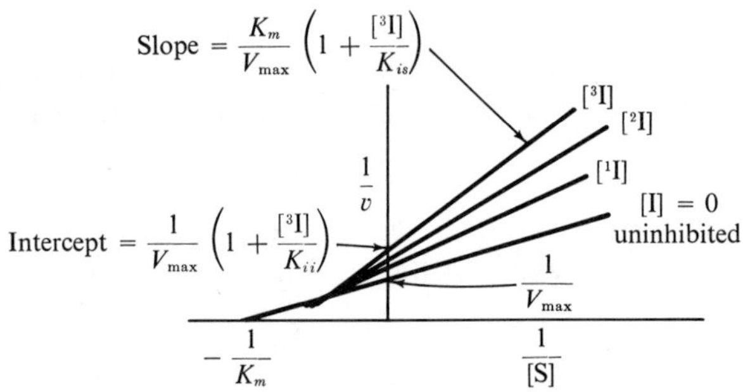

Figure 11.25

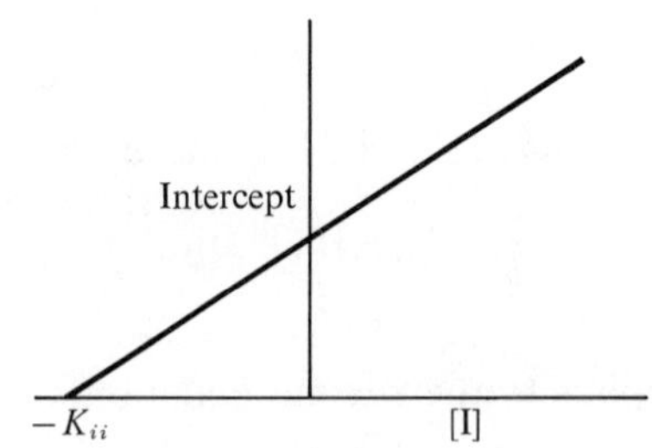

Figure 11.26

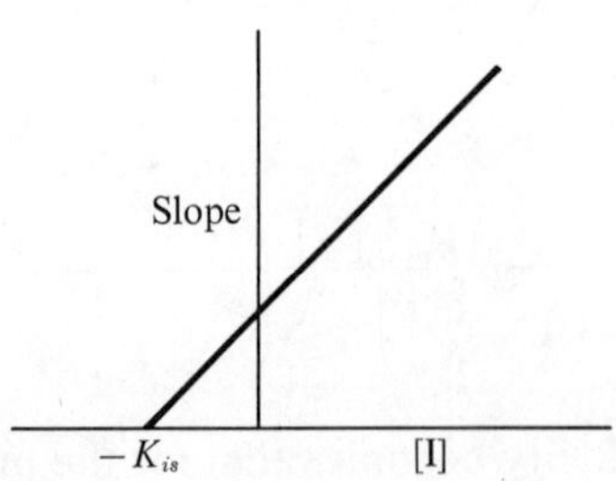

Figure 11.27

Uncompetitive Inhibition. In this type of inhibition the inhibitor complexes with the enzyme-substrate complex to form a dead-end complex

$$\mathrm{ES + I \rightleftharpoons ESI}$$

$$K_{ii} = \frac{[\mathrm{ES}][\mathrm{I}]}{[\mathrm{ESI}]}$$

The rate equation in its linearized form (Figure 11.28) becomes

$$\frac{1}{v} = \frac{K_m}{V_{\max}} \frac{1}{[\mathrm{S}]} + \frac{1}{V_{\max}} \left(1 + \frac{[\mathrm{I}]}{K_{ii}}\right) \tag{11.28}$$

Figure 11.28

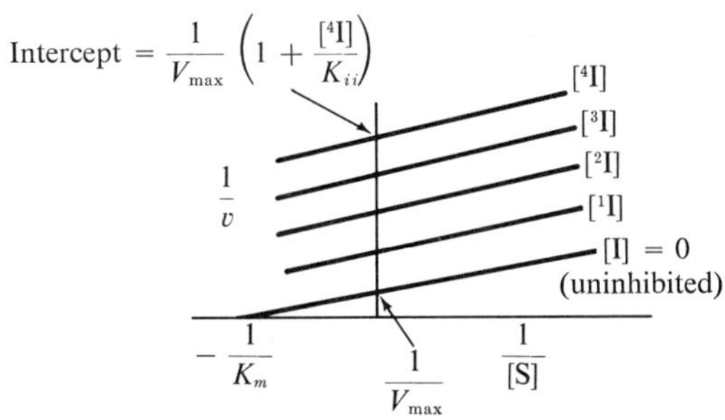

The slopes of the family of lines are the same, $K_m/V_{\max}$, and the intercepts may be replotted in the form

$$y(\text{intercept}) = \frac{1}{V_{\max}} + \frac{1}{V_{\max} K_{ii}} [I]$$

The intercept of the replot (Figure 11.29) on the [I]-axis is therefore $-K_{ii}$.

Figure 11.29

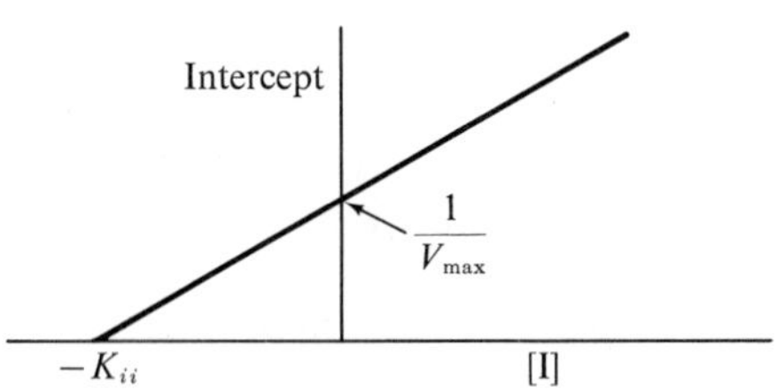

Problem 11.8: A product inhibition study on alcohol dehydrogenase was carried out in which the concentration of NAD^+ was fixed at 1.75 mM while the concentrations of the other substrate, ethanol, and the inhibitor, acetaldehyde, were varied. The data are collected below. Plot the results in double reciprocal form. What kind of inhibition pattern do you get? What are the apparent values for $V_{\max}$, K_b, and K_{is}?

v ΔAbs_{340}/min	[B] mM Ethanol	[I] mM Acetaldehyde
0.152	100	0
0.142	50	0
0.130	30	0
0.119	20	0
0.117	100	4.0
0.105	50	4.0
0.0897	30	4.0
0.0730	20	4.0
0.101	100	8.0
0.0831	50	8.0
0.0664	30	8.0
0.0538	20	8.0
0.0850	100	12.0
0.0655	50	12.0
0.0518	30	12.0
0.0431	20	12.0

Solution.* The necessary reciprocal values for the graphical solution of the above problem are summarized in Table 11.7.

Table 11.7
Data for graphical solution of Problem 11.8

$\frac{1}{[EtOH]}$ M^{-1}	$\frac{1}{v}(\Delta Abs)^{-1}$ *min*	[*Acetaldehyde*] mM
10	6.58	0.0
20	7.04	
30	7.69	
50	8.40	
10	8.54	
20	9.52	4.0
30	11.15	
50	13.70	
10	9.9	
20	12.03	8.0
30	15.06	
50	18.59	
10	11.76	
20	15.27	12.0
30	19.31	
50	23.20	

*B. V. Plapp: *J.Biol. Chem.* **245:**1727 (1970).

The experimental data fit the diagnostic characteristics of noncompetitive inhibition in which $K_{ii} \neq K_{is}$ and $K_{ii} > K_{is}$ as can be seen from the intersection of lines in the primary plot occurring above the 1/[S]-axis (Figure 11.30).

Figure 11.30

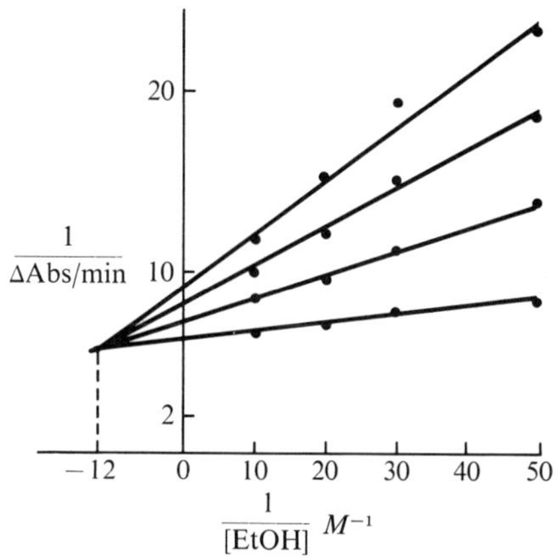

The slopes and intercepts of the primary plots are collected in Table 11.8. Where [I] = 0 on the replot (Figure 11.31) of the intercept verus [I], the value 6.21 is the reciprocal of $V_{\max}$. Similarly the intercept, 1.65, of the replot (Figure 11.32) of slope versus [I] gives $K_m/V_{\max}$ from which K_m can be calculated if we know $V_{\max}$:

$$V_{\max} = \frac{1}{6.21}$$

$$= 0.161 \text{ absorbance units at 340 nm per min}$$

$$K_b = 0.04 \times 0.161$$

$$= 6.4 \text{ m}M$$

$V_{\max}$ and K_b may also be calculated from Figure 11.30 when [I] = 0.

Table 11.8
Data for replotting of slopes and intercepts versus I

Slope	*Intercept*	*[I]mM*
$0.045\left(=\frac{1.8}{40}\right)$	6.20	0
$0.120\left(=\frac{5.0}{40}\right)$	7.15	4
$0.211\left(=\frac{5.7}{27}\right)$	8.10	8
$0.291\left(=\frac{8.0}{27.5}\right)$	9.05	12

Figure 11.31

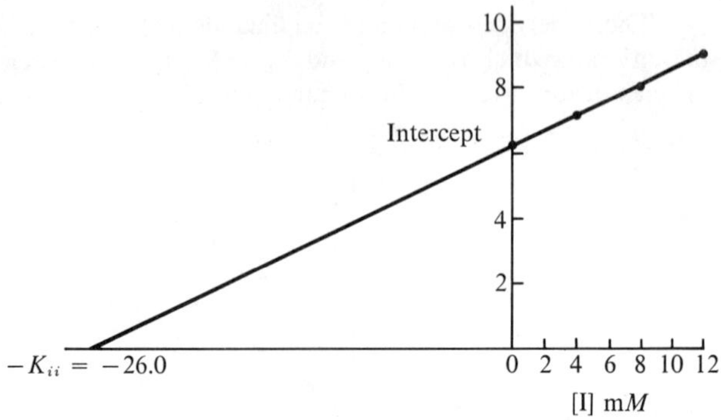

Figure 11.32

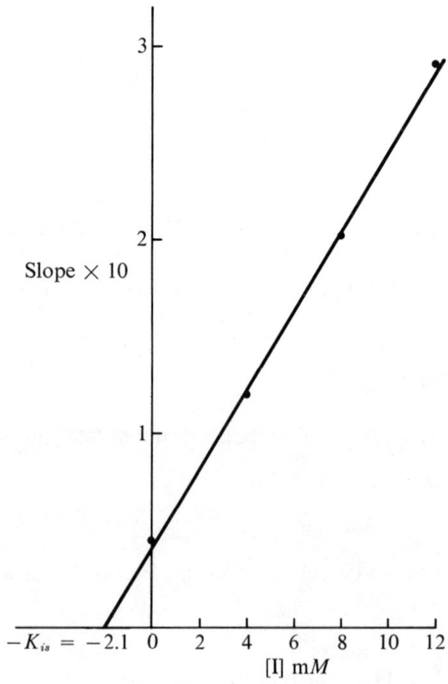

$$K_{ii} = 26.0 \text{ m}M$$

$$K_{is} = 2.1 \text{ m}M$$

Such graphical solutions are usually correct to within 20% of the "true" values obtained by computer fitting of the data.

Substrate Inhibition. There will be inhibited enzymic reactions that do not follow the course of the competitive and noncompetitive types. The data will not plot satisfactorily by any of the methods described above. For example, the Lineweaver-Burk plots of the uninhibited data will neither intersect at $1/[S]=0$ (competitive type) nor $1/v=0$ (noncompetitive type). Such an example may be that of substrate inhibition at high substrate concentration, giving a nonlinear Lineweaver-Burk plot (Figure 11.33).

Figure 11.33

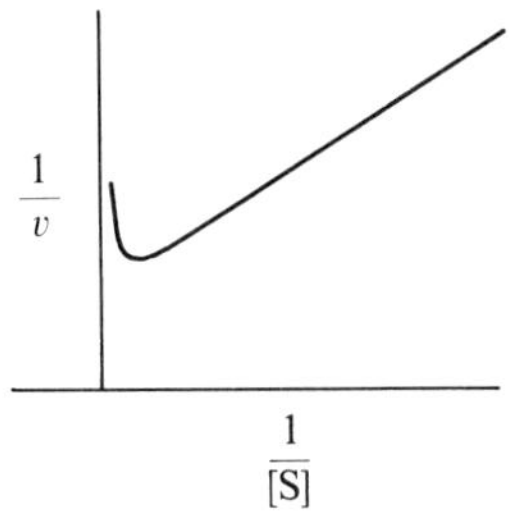

For studies of these types of inhibition the student is referred to more advanced texts.

SIGMOIDAL KINETICS

The rates of some enzymatic reactions do not follow the same patterns as those described above. Double reciprocal plots of $1/v$ versus $1/[S]$ are not linear. Instead it is found that a sigmoidal relationship between v and $[S]$ is followed (Figure 11.34), as was seen in the cooperative binding of ligands to macromolecules (page 152). The rate equation applying to these reactions is (11.29),

$$v = \frac{[S]^m V_{max}}{[S]^m + K'} \tag{11.29}$$

which can be rearranged to (11.30)

$$\frac{v}{V_{max}} = \frac{[S]^m}{[S]^m + K'} \tag{11.30}$$

or (11.31)

$$\log \frac{v}{V_{max} - v} = m \log [S] - \log K' \tag{11.31}$$

The similarity between Equations (11.31) and (8.29) is obvious. The term $v/(V_{max} - v)$ relates to amount of ES complex that is formed compared to that where the enzyme is saturated, just as $\theta/1 - \theta$ gives the ratio of binding sites filled to those unoccupied; θ is similar to $v/V_{max} - v$.

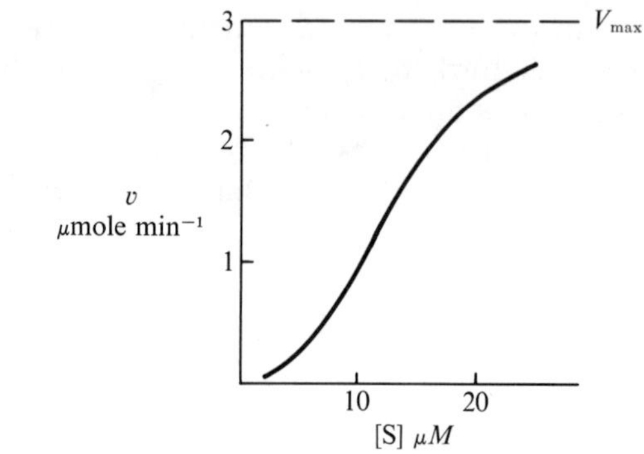

Figure 11.34

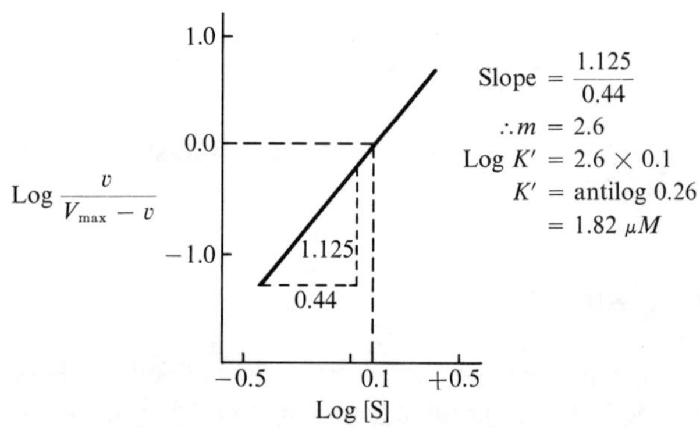

The affect of the binding of one molecule of substrate or effector molecule on the binding of the next substrate molecule is expressed in the exponent m, which is the Hill coefficient. If m is unity then Equation (11.30) collapses to the classical Michaelis-Menten equation (11.8).

The constant K' is not the concentration of substrate where the initial velocity of the reaction is equal to half the maximum velocity. If $[S_{1/2}]$ is the substrate concentration where $v = \frac{1}{2}V_{max}$, then $K' = [S_{1/2}]^m$ (Figure 11.34).

The value of K' changes as different concentrations of inhibitor or activator effector molecules are present. The positive effectors (activators) reduce the value of $[S_{1/2}]$ and the cooperativity until the enzyme behaves in a classical manner, giving Michaelis-Menten type kinetics. The reverse is true of the negative effectors (inhibitors) as is seen in Figure 11.35.

Figure 11.35

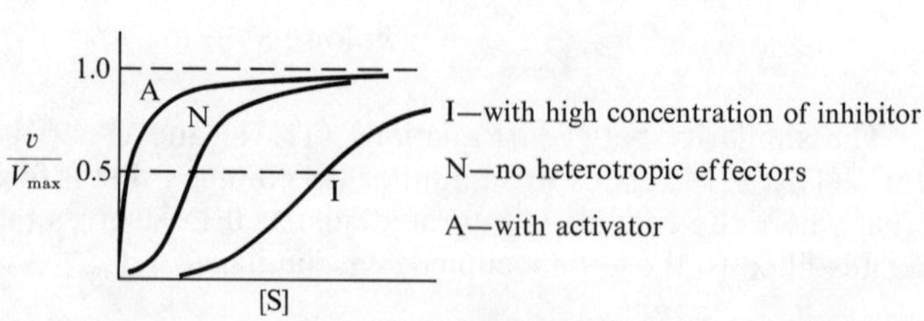

Problem 11.9: A dehydrogenase catalyzes the reduction of a substrate S by NADH. The initial velocity of the reaction is determined by the rate of change of absorbance at 340 nm. The following data were collected:

[S] μM	v μmole/min Effector concentration 0	 10 μM
1.0	1.67	6.25
2.0	4.44	7.69
3.0	6.43	8.33
6.0	8.78	9.09
8.0	9.28	9.30
10.0	9.52	
12.0	9.52	

What type of kinetics applies to these reactions? Calculate the appropriate kinetic constants.

Solution: A plot of v versus [S] shows the sigmoidal kinetics of the enzyme reaction in the absence of effector molecules (Figure 11.36). The allosteric behavior is modified by the effector molecules, activation by A and inhibition by B.

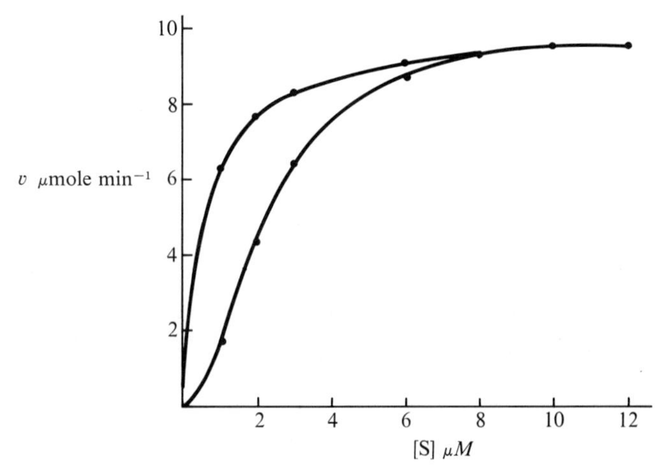

Figure 11.36

The kinetics are now studied by using a Hill plot, for which V_{max} is taken as 10.0 μmole min^{-1} from the plot in Figure 11.36. From this value the term $\log_{10} v/(V_{max} - v)$ can be calculated for each reaction mixture. These values together with those for $\log_{10}$ [S] are summarized in Table 11.9, next page.

The slopes of the Hill plots and the value of log [S] at $\log v/(V_{max} - v) = 0.$ permit the calculation of m and K', respectively, from Figure 11.37. These are summarized:

Table 11.9
Data for the solution of Problem 11.9

$Log_{10}[S]$	$Log \frac{v}{V_{max} - v}$			
	No effector		*Effector A*	
	$\frac{v}{V_m - v}$	$\log \frac{v}{V_m - v}$	$\frac{v}{V_m - v}$	$\log \frac{v}{V_m - v}$
0.00	$\frac{1.67}{10 - 1.67}$	−0.699	$\frac{6.25}{10 - 6.25}$	0.222
0.301	$\frac{4.44}{10 - 4.44}$	−0.098	$\frac{7.69}{10 - 7.69}$	0.522
0.477	$\frac{6.43}{10 - 6.43}$	0.254	$\frac{8.33}{10 - 8.33}$	0.698
0.778	$\frac{8.78}{10 - 8.78}$	0.857	$\frac{9.09}{10 - 9.09}$	1.000
0.903	$\frac{9.28}{10 - 9.28}$	1.110	$\frac{9.30}{10 - 9.30}$	1.124

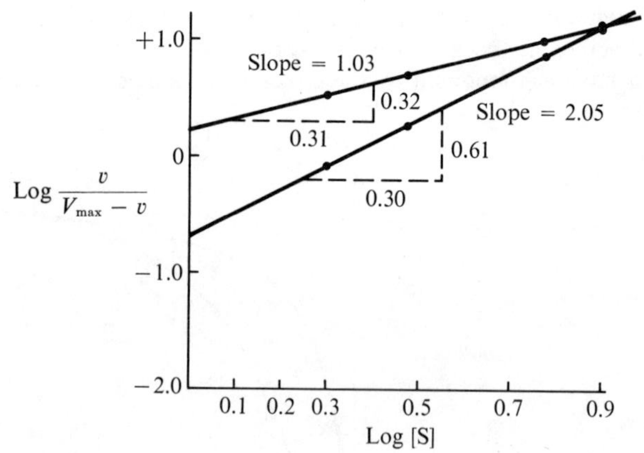

Figure 11.37

(a) With no effector present

$$\text{Slope} = \frac{0.61}{0.30}$$

$$\therefore m = 2.05$$

$$\text{At } v = \tfrac{1}{2}V_{max},\ \log \frac{v}{V_{max} - v} = 0 \text{ and}$$

$$\log [S] = 0.35$$

$$S = 2.24\ \mu\text{molar}$$

$$K' = (2.24)^{2.05}$$

$$= 5.1$$

(b) With effector A

$$\text{Slope} = \frac{0.32}{0.31}$$

$$\therefore m = 1.03$$

$$K' = (0.617)^{1.03}$$

$$= 0.62$$

REFERENCES

T. E. Barman, *Enzyme Handbook,* Vol. I and II. Springer-Verlag, New York, 1969; Supplement No. 1, 1974.

H. N. Christensen and G. A. Palmer, *Enzyme Kinetics,* Saunders, Philadelphia, 1967. (A programmed course.)

W. W. Cleland, in P. D. Boyer (ed.), *The Enzymes,* Vol. II, 3rd Edition. Academic Press, New York, 1970.

M. Dixon and E. C. Webb, *Enzymes* (2nd ed.). Academic Press, New York, 1964.

J. B. Neilands and P. K. Stumpf, *Outlines of Enzyme Chemistry* (2nd ed.). Wiley, New York, 1958.

K. M. Plowman, *Enzyme Kinetics.* McGraw-Hill, New York, 1972.

C. S. Vestling, "Determination of Dissociation Constants for Two-Substrate Enzyme Systems," *Methods of Biochemical Analysis* **10**:137 (1962).

---

11. PROBLEMS A

Answers on page 352

1. If initial velocities were measured in micromoles min^{-1} and concentration of substrate in millimoles liter^{-1} what would be the units of the axes in the three types of rate plots (that is, $1/v$, $1/[S]$, $[S]/v$, and $v/[S]$)?

2. From a Lineweaver-Burk plot of an enzyme reaction, where $1/v = 25$ hr mole^{-1} at $1/[S] = 0$ and $1/[S] = -1.3 \times 10^2$ liters mole^{-1} at $1/v = 0$, calculate V_{max} and K_m.

3. A plot of v versus $v/[S]$ gave a line with slope -5.2×10^{-5} mole liter^{-1} and an intercept on the v-axis of 2.67 mmole min^{-1}. Calculate the intercept of the $v/[S]$-axis giving the correct units.

4. Alanine racemase catalyzes the conversion of L-alanine to the D-isomer. The velocity of one reaction mixture was measured by the rate of formation of D-alanine, which is given in Table 11.10. What is the initial velocity?

Table 11.10

D-*alanine* (*mmole/liter*)	*Time* (*min*)
0.00	0.0
4.06	0.5
5.94	1.0
8.05	2.0
9.10	3.0
9.55	4.0
9.80	equilibrium

5. What information can be obtained from Lineweaver-Burk plots of enzyme kinetics for the reaction S $\rightarrow$ P under the following conditions?
   (a) $1/v$ is zero when $1/[S]$ equals -40 liter mole^{-1}
   (b) $1/v$ is zero when $1/[S] \times 10^{-2}$ equals -2.5 liter mole^{-1}
   (c) $1/v$ is zero when $1/[S] \times 10^{-3}$ equals -3.3 liter mole^{-1} in the presence or absence of ADP
   (d) $1/[S]$ is zero when $1/v$ equals 3.0×10^5 min mole^{-1}
   (e) $1/[S]$ is zero when $1/v \times 10^{-5}$ equals 3.0 min mole^{-1}
   (f) $1/[S]$ is zero when $1/v \times 10^{-3}$ equals 5.0 min mole^{-1} and the slope of the line is 120 min liter^{-1}

6. In plots of v versus $v/[S]$, what is concluded if these conditions hold?
   (a) the slope of the line is -2.3×10^{-3} mole liter^{-1}
   (b) when v is zero, $v/[S]$ equals 60 liter min^{-1}
   (c) when $v/[S]$ is zero v equals 4.6 mmole min^{-1}

7. An enzyme-catalyzed reaction, whereby substrate S is isomerized to product P, using 2.5 mg of enzyme, molecular weight 125,000, was found to have a K_m of 3×10^{-3} *M* and a maximum velocity of 275 μmole per minute. What would be the molecular activity (turnover number) of the enzyme under these conditions? Also calculate the initial velocity at a substrate concentration of 7.5 m*M*.

8. An isomerase reaction was followed by reading the absorbance at given time intervals. The absorbance was proportional to the concentration of the product of the reaction. From the particular reaction summarized in Table 11.11 what is the initial velocity?

Table 11.11

Abs	*Time* (*min*)
0.00	0
0.46	1
0.74	2
0.89	3
0.97	4
1.00	5
1.07	10
1.10	equilibrium

9. The decarboxylation of a β-keto acid proceeded at the initial rates for various substrate concentrations, as determined by the rate of CO_2 formation, given in Table 11.12. Calculate the Michaelis-Menten constant for the enzyme and the maximum velocity by two graphical methods.

Table 11.12

(μmole CO_2/2 min)	Keto acid concentration (mM)
0.588	2.500
0.500	1.000
0.417	0.714
0.370	0.526
0.256	0.250

10. An enzyme was isolated that catalyzed the reaction

$$A + B \rightleftharpoons C + D$$

From the kinetic data summarized in Table 11.13, deduce the type of reaction mechanism and determine the appropriate kinetic constants.

Table 11.13

[A] (mM)	[B] (mM)	v (μmole min^{-1})
1.050	100	5.26
0.524	100	3.91
0.314	100	2.88
0.210	100	2.13
1.050	50	3.74
0.524	50	2.68
0.314	50	1.92
0.210	50	1.43
1.050	30	2.54
0.524	30	1.87
0.314	30	1.32
0.210	30	0.99
1.050	20	2.03
0.524	20	1.39
0.314	20	0.99
0.210	20	0.78

11. The initial rates of acid formation shown in Table 11.14 were obtained for the hydrolysis of an ester at various concentrations, using 1.3 mg esterase protein for each. The acid formation was measured automatically by titration with 0.01 *N* sodium hydroxide at constant pH. Determine K_m and V_{max} by plotting $[S]/v$ versus $[S]$ and, as an alternate solution, v versus $v/[S]$. What is the specific activity of the enzyme?

Table 11.14

NaOH (*μ liter added in first min*)	*Ester concentration* (*mM*)
500	5.00
400	1.43
333	0.83
278	0.59
196	0.31

12. The effect of AMP on the kinetics of the reaction between NADH and acetaldehyde catalyzed by picolinimidylated liver alcohol dehydrogenase was studied.* The reaction mixtures all had a fixed amount of enzyme and NADH (200 *μM*). The data are summarized in Table 11.15. What is the type of inhibition and what are the kinetic constants?

Table 11.15

AMP (*mM*)	*Acetaldehyde* (*mM*)	v ($\Delta Abs\ min^{-1}$)
0	5.0	0.238
0	2.5	0.166
0	1.5	0.126
0	1.0	0.098
0.5	5.0	0.164
0.5	2.5	0.103
0.5	1.5	0.086
0.5	1.0	0.067
1.0	5.0	0.126
1.0	2.5	0.090
1.0	1.5	0.070
1.0	1.0	0.052
2.0	5.0	0.085
2.0	2.5	0.064
2.0	1.5	0.046
2.0	1.0	0.036

*B. V. Plapp, R. L. Brooks, and J. D. Shore, *J. Biol. Chem.* **248**:3470 (1973).

13. The activity of a glycosidase is inhibited by an acid. The initial velocity of the hydrolysis of the glycoside, as determined by the release of reducing sugar, is measured under varying conditions of glycoside and inhibitor concentration. The level of glycosidase is constant. The results are given in Table 11.16. Determine the type of inhibition and the values of K_{ii}, K_{is}, K_m, and the maximum velocity.

Table 11.16

	Glycoside (cleaved in μmole/2 min)		
$[I]$ (mM)	at $[S] = 6.25$ mM	at $[S] = 0.862$ mM	at $[S] = 0.424$ mM
0.0	0.625	0.455	0.340
1.0	0.488	0.352	0.265
2.0	0.400	0.289	0.217
3.0	0.399	0.244	0.184
4.0	0.294	0.213	0.159

14. The reaction kinetics of phosphofructokinase (PFK) from a thermophilic bacterium were studied at 23°C.† The reaction mixture contained, in a final volume of 1.0 ml: 100 mM KCl, 3 mM $MgCl_2$, 1 mM ATP, 0.22 mM NADH, 50 mM Tris HCl, 0.2 mg rabbit muscle aldolase, 0.04 mg triosephosphate isomerase-α-glycerophosphate dehydrogenase mixture, and the indicated concentration of fructose 6-phosphate (F-6-P); pH 8.0 at 23°C.

The reaction was initiated by the addition of an aliquot of PFK and the reaction monitored by following the decrease in absorbance at 340 nm in a Gilford recording spectrophotometer. The rate of NADH reduction is proportional to the rate of formation of fructose 1,6-diphosphate (FDP). From the data in Table 11.17, describe the type of kinetics and

Table 11.17

F-6-P (mM)	*ΔAbs/ 2.5 min*
0.05	0.069
0.10	0.164
0.111	0.232
0.125	0.274
0.142	0.346
0.166	0.401
0.200	0.499
0.250	0.725
0.333	1.020
0.50	1.128
1.00	1.355

*K. Cass, unpublished data.

calculate the appropriate kinetic constants. (Hint: An estimate of V_{max} can be attained from a plot $1/v$ versus 1/F-6-P^2.)

15. The kinetics of the reaction catalyzed by dihydrofolate reductase of a *Streptococcus faecium* mutant have been studied by measuring the initial velocities of the reaction spectrophotometrically at 340 nm. The absorbance changes have been converted to the corresponding changes of the substrate concentrations, which are summarized in Table 11.18.*

$$\text{dihydrofolate} + \text{NADPH} + \text{H}^+ \rightleftharpoons \text{tetrahydrofolate} + \text{NADP}^+$$

The reaction was buffered at pH 6.6. From a double reciprocal plot of $1/v$ versus 1/[DHF] what can be stated concerning the reaction mechanism. What are the values of the kinetic constants? Compare with values calculated by computer in the original paper.

Table 11.18

Initial velocity (*μmole/min*)	[*Dihydrofolate*] (*μmole/liter*)	[*NADPH*] (*μmole/liter*)
0.0634	1.673	49.84
0.0800	1.995	49.84
0.0874	2.489	49.84
0.1164	3.308	49.84
0.1498	4.977	49.84
0.0460	1.673	5.002
0.0536	1.995	5.002
0.0638	2.489	5.002
0.0768	3.308	5.002
0.0998	4.977	5.002
0.0320	1.692	1.690
0.0362	2.018	1.690
0.0422	2.518	1.690
0.0500	3.343	1.690
0.0630	5.035	1.690

16. The initial velocities for the hydrolysis of kestose by β-D-fructofuranosidase is given in Table 11.19. What are the values of K_m and V_{max}?

*R. L. Blakley et al., *Annals N.Y. Acad. Sci.* **186**:119 (1971).

Table 11.19

[*Kestose*] (*M*)	*v* (*units/min*)
0.01	4.1
0.025	9.0
0.05	15.1
0.07	23.9
0.15	38.1
0.20	40.1
0.30	40.0

17. A dehydrogenase enzyme catalyzes the reduction of a substrate S by NADH. The initial velocity of the reaction is determined by the rate of change of absorbance at 340 nm. The following data were collected:

[S] μM	v μmole/min
1.0	0.10
2.0	0.74
3.0	2.13
6.0	6.84
8.0	8.37

$V_{\max}$ is 10.0 μmole min^{-1}. What type of kinetics applies to this reaction? Calculate the appropriate kinetic constants.

11. PROBLEMS B No answers given

1. A derivative of horse liver alcohol dehydrogenase was studied at pH 7.0 by measuring the initial velocity of the reaction

$$NAD^+ + \text{ethanol} \rightleftharpoons NADH + H^+ + \text{acetaldehyde}$$

in the forward and reverse directions under conditions summarized in Tables 11.20 and 11.21.* What type of mechanism fits the data? Compare the appropriate kinetic constants for the two reactions.

*B. V. Plapp, unpublished data.

Table 11.20

$[NAD^+]$ (mM)	[Ethanol] (mM)	v (units/min)
1.076	100	7.24
0.538	100	5.46
0.323	100	3.92
0.215	100	3.20
1.076	50	5.22
0.538	50	3.90
0.323	50	2.75
0.215	50	1.92
1.076	30	3.71
0.538	30	2.62
0.323	30	1.94
0.215	30	1.54
1.076	20	2.78
0.538	20	1.87
0.323	20	1.46
0.215	20	1.00

Table 11.21

NADH (mM)	(Acetaldehyde) (mM)	v (units/min)
0.104	5.0	390.7
0.052	5.0	271.8
0.031	5.0	180.6
0.021	5.0	129.4
0.104	2.5	279.4
0.052	2.5	189.1
0.031	2.5	133.2
0.021	2.5	90.9
0.104	1.5	197.6
0.052	1.5	142.0
0.031	1.5	95.4
0.021	1.5	73.1
0.104	1.0	150.6
0.052	1.0	110.6
0.031	1.0	80.5
0.021	1.0	62.6

2. A study of phosphofructokinase was made at 60°C by measuring the rate of reaction through a coupled enzyme system in which the oxidation of NADH was followed at 340 nm. The effect of ATP was also determined and the results are summarized in Table 11.22.*

The reaction mixture contained in a final volume of 0.5 ml: 100 m*M* KCl, 50 m*M* glycine, 2 m*M* $MgCl_2$, and the indicated concentrations of ATP and fructose 6-phosphate (F-6-P), pH 7.5 at 60°C. The reaction was initiated by the addition of an aliquot of PFK and stopped after exactly two minutes by the addition of EDTA to a final concentration of 60 m*M*. The reaction mixture was then made 0.22 m*M* in NADH, 0.2 mg/ml in rabbit muscle aldolase, 0.04 mg/ml in triosephosphate isomerase: α-glycerophosphate dehydrogenase mixture. The absorbance at 340 nm was read and subtracted from the control, which lacked PFK. What is the type of kinetics and the effect of ATP on the enzyme? Plot $1/v$ versus 1/F-6-P to estimate V_{max}.

*K. Cass, unpublished data.

Table 11.22

	[F-6-P] (mM)	Abs_{340}
	control	1.333
2.0 mM ATP	0.02	1.283
	0.04	1.197
	0.06	1.115
	0.08	1.069
	0.10	1.033
	0.125	0.963
	0.142	0.909
	0.166	0.889
	0.20	0.849
	0.25	0.834
	0.50	0.802
	1.0	0.725
1.0 mM ATP	0.02	1.286
	0.04	1.192
	0.06	1.122
	0.08	1.052
	0.10	1.024
	0.125	0.956
	0.142	0.922
	0.166	0.907
	0.20	0.897
	0.25	0.837
	0.50	0.817
	1.0	0.787
0.5 mM ATP	0.02	1.284
	0.04	1.194
	0.06	1.110
	0.08	1.060
	0.10	1.036
	0.125	0.970
	0.142	0.960
	0.166	0.973
	0.20	0.961
	0.25	0.920
	1.0	0.854

3. Mitochondrial monoamine oxidase.* A reaction mixture of 2.2 ml, containing benzylamine hydrochloride, 0.05 M phosphate buffer, pH 7.6, 3.0 μg enzyme, and either α- or β-naphthol inhibitors, was observed spectrophotometrically at 250 nm. The velocity is expressed as the change in this absorbance per minute.

*V. G. Erwin and L. Hellerman, *J. Biol. Chem.* **242**:4230 (1967).

From a graph of the data given in Table 11.23 determine the K_m value for benzylamine, the different K_{ii} values for the naphthols and the nature of the inhibition.

Table 11.23

$\frac{1}{[S]M} \times 10^{-3}$	$\frac{1}{v} \times 10^{-3}$		
	No inhibitor	2×10^{-4} M α-naphthol	2×10^{-4} M β-naphthol
20	60	103	150
10	39	61	86
8	33	52	70
4	25	34	44

4. Mutarotase.* Weighed amounts of α-D-glucose were dissolved in 12 ml of 5 mM EDTA buffer, pH 7.4, containing 30 units of green pepper mutarotase, 17,000 units mg^{-1} of protein. The change in optical rotation was recorded automatically and converted to micromoles α-D-glucose converted per minute to the β-D-anomer. The values were corrected for the spontaneous uncatalyzed mutarotation and are represented in Table 11.24.

   The enzyme has a mol. wt. of approximately 50,000. Calculate the values of K_m and the molecular activity of the mutarotase.

Table 11.24

v (μmole/min)	$[S]$ (mM)
2.21	27.4
1.97	22.0
1.66	16.5
1.43	12.0

5. Esterase activities of human carbonic anhydrases.† In a kinetic analysis of two carbonic anhydrases acting upon nitrophenylacetate substrates, the modified Lineweaver-Burk plots gave the intercepts at pH 8.0 shown in Table 11.25. What are the values of K_m and V_{max}, in the proper units?

Table 11.25

	$[S]/v \times 10^{+5}$ min^{-1} (at $[S] = 0$)	$[S]$ (mM) (at $[S]/v = 0$)
Carbonic anhydrase B		
Substrate: p-nitrophenyl acetate	4.30	−5.75
Substrate: o-nitrophenyl acetate	1.75	−4.15
Carbonic anhydrase C		
Substrate: o-nitrophenyl acetate	7.25	−7.7

*J. M. Bailey, P. H. Fishman, and P. G. Pentchev, *J. Biol. Chem.* **242**:4263 (1967).
†J. A. Verpoorte, S. Mehta, and J. T. Edsall, *J. Biol. Chem.* **242**:4221 (1967).

6. Dihydrofolate reductase* was isolated from a wild type *Streptococcus faecalis*. From the data given in Table 11.26 what conclusion can be drawn concerning the reaction mechanism and the rate constants? Plot $1/v$ versus $1/[\mathrm{DHF}]$ and $1/v$ versus $1/[\mathrm{NADPH}]$. From the two families of curves, how do the rate constants compare?

Table 11.26

v ($\mu mole/min$)	$[DHF]$ (μM)	$[NADH]$ (μM)
0.034	5.020	1.690
0.027	2.510	1.690
0.031	3.367	1.690
0.033	5.020	1.690
0.041	10.040	1.690
0.043	20.200	1.690
0.031	2.508	2.014
0.035	3.364	2.014
0.040	5.016	2.014
0.044	10.030	2.014
0.049	20.180	2.014
0.032	2.009	2.556
0.037	2.512	2.556
0.041	3.329	2.556
0.046	5.024	2.556
0.054	10.050	2.556
0.038	2.004	3.304
0.043	2.506	3.304
0.049	3.321	3.304
0.057	5.013	3.304
0.067	10.020	3.304
0.048	1.994	4.969
0.055	2.492	4.969
0.063	3.301	4.969
0.075	4.984	4.969
0.088	9.968	4.969
0.073	1.994	49.690
0.089	2.492	49.690
0.108	3.301	49.690
0.147	4.984	49.690
0.196	9.968	49.690

*R. L. Blakley, unpublished data.

7. The effects of ADP and phosphoenolpyruvate (PEP) on phosphofructokinase were studied at 23° in the manner described in Question 2 of this section. Interpret the data summarized in Table 11.27. The reaction mixture contained in a final volume of 1 ml: 100 m*M* KCl, 50 m*M* Tris HCl, 20 m*M* $MgCl_2$, 0.2 m*M* ATP, 0.22 m*M* NADH, 0.2 mg rabbit muscle aldolase, 0.04 mg triosephosphate isomerase-α-glycerophosphate dehydrogenase mixture, and the indicated concentrations of F-6-P and effector (ADP or PEP); pH was 7.1 at 23°C. The reaction was initiated by the addition of an aliquot of PFK and monitored at 340 nm in a Gilford recording spectrophotometer.

Table 11.27

	[*F*-6-*P*] (m*M*)	ΔAbs_{340}/2.5 *min*
No effector	0.10	0.072
	0.111	0.075
	0.125	0.120
	0.142	0.140
	0.166	0.180
	0.20	0.220
	0.25	0.315
	0.333	0.445
	0.50	0.590
	1.0	0.785
0.1 m*M* PEP	0.05	0.015
	0.10	0.039
	0.111	0.053
	0.125	0.048
	0.142	0.067
	0.166	0.069
	0.20	0.090
	0.25	0.140
	0.333	0.200
	0.50	0.311
	1.0	0.502
1 m*M* ADP	0.05	0.172
	0.10	0.364
	0.111	0.400
	0.125	0.436
	0.142	0.440
	0.166	0.512
	0.20	0.550
	0.25	0.606
	0.333	0.700
	0.50	0.732
	1.0	0.778

8. Dextran synthetase:*

$$n \text{ sucrose} \xrightarrow{\text{synthetase}} \text{dextran} + n \text{ fructose}$$

The reaction of sucrose at various concentrations was carried out in 0.01 *M* acetate buffer at pH 5.6 and 23°C with the transglucosylase enzyme. After proper intervals the reaction was stopped by adding five volumes of 0.02 *N* sodium hydroxide and the fructose determined by the reducing sugar procedure of Hagedorn and Jensen. The results are given in Table 11.28. What is the K_m and V_{max} for this system? Graph by v versus $v/[S]$ and $[S]/v$ versus $[S]$.

Table 11.28

Sucrose concentration (mM)	*Initial velocity (mg reducing sugar hour^{-1})*
10	1.58
20	2.33
40	3.08
60	3.48
80	3.60
160	4.03

9. Arginosuccinate cleaving enzyme.† The enzymic activity of arginosuccinate cleaving enzyme was studied by coupling the reaction through the arginase reaction as follows:

$$\text{arginosuccinate} \xrightleftharpoons{\text{cleaving enzyme}} \text{arginine} + \text{fumarate}$$

$$\text{arginine} + H_2O \xrightarrow{\text{arginase } Mn^{++++}} \text{urea} + \text{ornithine}$$

The rate of formation of arginine was therefore measured by reacting the urea formed in the second reaction with 1-phenyl-1,2-propane dione-2-oxime and reading the absorbance of the colored solution at 540 nm.

The kinetic data are collected in Table 11.29. Determine the K_m and V_{max} for the cleaving enzyme, using the method of least squares.

Table 11.29

Arginosuccinate concentration (μM)	*Initial velocity (μmole urea formed hour^{-1})*
20.0	2.22
25.0	2.71
38.5	4.00
77.0	5.89
111.0	7.15

*E. J. Hehre, *J. Biol. Chem.* **163**:221 (1946).
†G. W. Brown and P. P. Cohen, *J. Biol. Chem.* **234**:1769 (1959).

10. α-Chymotrypsin.* The initial velocities for the hydrolysis of N-acetyl phenylalaninamide at various concentrations by α-chymotrypsin are given in Table 11.30. What is K_m? Calculate by two graphical methods and compare to the value obtained by least squares.

Table 11.30

Substrate concentration (mM)	*Initial velocity (mmole alanine min^{-1})*
10.5	3.33×10^{-2}
16.1	4.55×10^{-2}
20.0	5.56×10^{-2}
35.0	7.15×10^{-2}
93.4	10.0×10^{-2}

11. Abscisin inhibition of α-amylase.† Abscisin, a growth-inhibitor substance found in the leaves of sycamore, inhibits α-amylase. With an 0.67% starch concentration in the reaction solutions at pH 4.7 and 30°C, containing various amounts of inhibitor but a constant amount of α-amylase (0.0167 mg), the amount of maltose produced was determined after 30 min of reaction. The results are summarized in Table 11.31. Assuming the inhibition to be noncompetitive in nature, calculate the K_{ii} and the specific activity.

Table 11.31

Abscisin concentration ($M \times 10^5$)	*Reaction velocity (mg maltose formed/30 min)*
2.8	7.69
1.4	9.61
0.7	11.30
0.35	13.69
0	16.58

12. D-Amino acid oxidase. A manometric assay for D-amino acid oxidase was carried out in a Warburg apparatus. The substrate for the oxidation was D-leucine. The main compartment of each flask contained the substrate, buffer, and necessary cofactors in a volume of 2.4 ml; each side arm held 0.4 ml of the enzyme solution and in the center well of each flask 0.2 ml of 10% KOH was placed to absorb any CO_2 that might be produced.

A stock solution of 0.1 *M* D-leucine was diluted as indicated in the data below. The reaction was run at 37°C in air. The heights of the columns of fluid in the manometers were recorded at 5-min intervals and are listed below. Flask number 1 is the thermobarometer control flask.

*H. T. Huang, R. J. Foster and C. Ciemann, *J. Amer. Chem. Soc.* **74**:105 (1952).
†T. Hemberg, *Acta Chem. Scand.* **21**:1665 (1967).

From these data plot $1/v$ versus $1/[S]$. Comment on the results.

Flask	Flask constant*	Substrate dilution	Height (in mm) h_0	h_5	h_{10}	h_{15}	h_{20}
1	1.300	–	193	197	197	198	198
2	1.265	None	198	187	163	138	113
3	1.302	None	185	185	161	137	113
4	1.818	1:12.5	195	189	172	154	135
5	1.836	1:12.5	185	164	164	148	132
6	1.764	1:40	191	172	151	132	131
7	1.302	1:40	220	213	198	182	175
8	1.374	1:100	225	214	204	202	203
9	1.149	1:100	162	145	134	133	133
10	1.265	1:150	155	147	142	142	142
11	1.302	1:150	161	152	147	142	147
12	1.818	1:150	166	165	160	159	160
13	1.374	1:150	162	152	152	153	153
14	1.672	1:200	152	150	147	147	147
15	1.589	1:200	152	150	147	147	146
16	1.612	1:200	152	151	148	149	149
17	1.524	1:200	152	150	148	148	148

Source: From F. Rodden (unpublished).

*The flask constant is the factor by which the change in h (in millimeters) is multiplied to give the change in volume (in microliters).

13. β-D-Galactosidase catalyzes the hydrolysis of *o*-nitrophenyl β-D-galactopyranoside (ONPG), forming *o*-nitrophenol, which is yellow in neutral or basic solution. The data obtained by measuring the rate of change of absorbance at 420 nm in the presence of a constant amount of enzyme at pH 7.30 is given in Table 11.32. If the absorbance at 420 nm of a 2.50×10^{-4} M solution of *o*-nitrophenol at pH 7.3 is 0.785, determine the K_m and V_{max} for the reaction by means of a Lineweaver-Burk plot. How do these values compare with those obtained from v versus $v/[S]$ and $[S]/v$ versus $[S]$ plots?

Table 11.32

Time (min)	*Absorbance at 420 nm*				
0.5	0.008	0.012	0.020	0.025	0.030
1.0	0.016	0.026	0.040	0.050	0.060
1.5	0.023	0.040	0.060	0.075	0.090
2.0	0.030	0.052	0.081	0.100	0.121
2.5	0.038	0.068	0.100	0.127	0.153
3.0	0.045	0.080	0.122	0.150	0.180
3.5	0.052	0.092	0.140	0.173	0.207
4.0	0.059	0.104	0.155	0.193	0.230
4.5	0.065	0.117	0.170	0.214	0.255
5.0	0.071	0.127	0.185	0.230	0.278
Initial concentration ($M \times 10^5$)	5.0	10.0	15.0	20.0	25.0

14. Phosphodiesterase. During the hydrolysis of the dinucleotide, adenyl-3′,5′-adenylic-3′-acid (ApAp) by a phosphodiesterase (mol. wt. 10,000), the U.V. absorbancy decreases somewhat. This forms the basis of a spectrophotometric assay of the enzyme activity. After addition of 0.01 ml of a 0.01 mg ml^{-1} solution of the enzyme to 1 ml of 1×10^{-4} *M* ApAp at pH 6.5, the absorbancies at 2800 Å were recorded (Table 11.33).
    (a) Calculate the initial velocity of the reaction (mole of substrate per mole of enzyme per second). From similar experiments carried out at pH 7.5, the data in Table 11.34 were obtained.
    (b) Calculate the Michaelis constant.
    (c) What is the turnover number of the enzyme?

Table 11.33

Time (sec)	*Absorbance (A_{280} nm)*
0	1.700
40	1.665
80	1.635
120	1.607
160	1.582
200	1.553
Infinite time	1.255

Table 11.34

ApAp (M)	*Initial velocity (moles (mole enzyme)$^{-1}$ sec^{-1})*
4.64×10^{-4}	1023
3.70×10^{-4}	896
2.78×10^{-4}	685
1.85×10^{-4}	525
0.92×10^{-4}	413

15. The hydrolysis of acetyl-L-hexahydroxylphenyl-alaninamide at 25°C by chymotrypsin, at an enzyme protein concentration of 0.208 mg protein *N* per ml, in 0.02 *M* tris buffer, pH 7.9, was followed by formol titration. The corresponding initial velocities and substrate concentrations are recorded in Table 11.35. By plotting v versus $v/[S]$, determine K_m and the specific activity of the enzyme, assuming the enzyme to contain 15.0% *N*.

Table 11.35

Substrate concentration (*mM*)	*Initial velocity* (10^{-5} *mole min*$^{-1}$)
10.0	3.08
15.0	4.30
20.0	4.97
25.0	5.66
30.0	6.50
40.0	7.51

16. Multivalent feedback inhibition of aspartokinase.* Aspartokinase, partially purified from extracts of *Bacillus polymyxa,* is studied kinetically. The kinetics were followed by measuring the release of $^{32}P_i$ from γ-^{32}P-ATP in the coupled reaction

$$\gamma\text{-}^{32}\text{P-ATP} + \text{L-aspartate} \xrightarrow[\text{aspartokinase}]{} {}^{32}\text{P-L-aspartyl-phosphate} \xrightarrow{\text{hydroxylamine}} {}^{32}\text{P}_i$$

The time course of the reaction was not linear (see Table 11.36). Determine the initial velocities and compare them with those obtained by assuming a 30-min reaction time as the initial velocity.

Table 11.36

Time (*min*)	$^{32}P_i$ *released* (*nmole*)	$^{32}P_i$ *released* (*nmole*)
	2 m*M* L-Aspartate	2 m*M* L-Asp + 1 m*M* L-Thr + 1 m*M* L-Lys
5	78	15
10	120	25
20	175	25
30	236	46
	0.5 m*M* L-Aspartate	0.5 m*M* L-Asp + 1 m*M* L-Thr + 1 m*M* L-Lys
30	121	15
60	185	25
90	225	28
120	255	29

17. Hexokinase. Brain hexokinase is an enzyme with a moderately broad substrate specificity and for which the information in Table 11.37 is determined for a constant enzyme concentration. Calculate the initial velocities for each sugar at a concentration of 1 μM.

*H. Paulus and Edith Gray, *J. Biol. Chem.* **242**:4980 (1967).

Table 11.37

Sugar substrate	K_m	*Maximum initial velocity (mole min^{-1})*
D-glucose	8×10^{-6} *M*	1.0
D-fructose	2×10^{-3} *M*	1.4
D-mannose	5×10^{-6} *M*	0.5
D-glucosamine	8×10^{-5} *M*	0.6

18. β-D-Galactosidase. The hydrolysis of phenyl β-D-galactoside by β-D-galactosidase is partially inhibited by beryllium ions. The initial velocities (expressed as moles of substrate per gram of enzyme per second) were measured in the absence and presence of 1×10^{-3} *M* Be^{++}, with the use of the substrate concentrations shown in Table 11.38. The molecular weight of the enzyme is 750,000. From the data determine the following.
    (a) Is the inhibition competitive or noncompetitive?
    (b) Calculate the Michaelis constant. Indicate some possible interpretations concerning its meaning.
    (c) Calculate the turnover number of the enzyme.

Table 11.38

Phenyl β-D-galactoside (M)	*Initial velocity (moles substrate sec^{-1} (g enzyme)$^{-1}$)* without Be^{++}	with Be^{++}
1×10^{-2}	7.70×10^{-5}	3.85×10^{-5}
5×10^{-3}	6.25×10^{-5}	3.00×10^{-5}
2×10^{-3}	4.00×10^{-5}	1.92×10^{-5}
1×10^{-3}	2.50×10^{-5}	1.22×10^{-5}
5×10^{-4}	1.57×10^{-5}	0.69×10^{-5}

19. In studies aimed at identifying the type of sequential mechanism from the nature of product inhibition,* the inhibition of dihydrofolate reductase from *Streptococcus faecalis* (wild type) by $NADP^+$ and THF (tetrahydrofolate) was analyzed. The initial velocities were determined spectrophotometrically and the rates expressed in micromoles min^{-1} for the reaction mixtures given in Tables 11.39 and 11.40.† From the patterns of product inhibition, what are the respective kinetic constants and what sequential mechanism for the enzyme is consistent with the evidence?

*W. W. Cleland, *Biochem. Biophys. Acta* **67**:173 (1963).
†R. L. Blakley et al, *Annals N.Y. Acad. Sci.* **186**:119 (1971).

Table 11.39

Initial velocity (μ*mole/min*)	[*NADPH*] (μM)	[*NADP*] (μM)
0.0958	10.020	0.0
0.0664	5.009	0.0
0.0470	3.312	0.0
0.0406	2.515	0.0
0.0334	2.004	0.0
0.0874	10.000	7.241
0.0548	5.000	7.241
0.0410	3.306	7.241
0.0340	2.510	7.241
0.0274	2.000	7.241
0.0786	10.000	14.480
0.0504	5.000	14.480
0.0334	3.306	14.480
0.0272	2.510	14.480
0.0220	2.000	14.480
0.0974	20.080	28.970
0.0660	10.110	28.970
0.0380	5.117	28.970
0.0308	4.017	28.970
0.0262	3.304	28.970
0.0202	2.509	28.970

Table 11.40

Initial velocity (μ*mole/min*)	[*DHF*] (μM)	[*THF*] (μM)
0.104	10.120	0.00
0.073	5.060	0.00
0.058	3.373	0.00
0.048	2.530	0.00
0.040	2.024	0.00
0.077	10.120	12.300
0.053	5.060	12.300
0.042	3.373	12.300
0.034	2.530	12.300
0.029	2.024	12.300
0.060	10.120	24.710
0.042	5.060	24.710
0.034	3.373	24.710
0.027	2.530	24.710
0.023	2.024	24.710
0.047	10.120	37.07
0.033	5.060	37.07
0.025	3.373	37.07
0.020	2.530	37.07
0.017	2.024	37.07

20. The inhibition by α-hydroxyglutarate of malate thiokinase,* which catalyzes the reaction

$$\text{malate} + \text{ATP} + \text{coenzyme A} \rightleftharpoons \text{malyl coenzyme A} + \text{ADP} + \text{P}_i$$

was determined with respect to malate, ATP and Coenzyme A under the conditions used for the noninhibited reaction. The results are summarized in Tables 11.41 through 11.43.

From this data calculate the rate constants, and interpret the nature of the inhibitions by the distinguishing characteristics of the graphic plots.

---

*L. B. Hersh, *J. Biol. Chem.* **249**:6264 (1974).

Table 11.41
[ATP] and [Coenzyme A] held constant at 0.9 m*M* and 0.08 m*M*, respectively

Specific activity (units)	*[Malate] (mM)*	*[α-hydroxyglutarate] (mM)*
0.726	2.0	0.0
0.598	1.0	0.0
0.555	0.7	0.0
0.492	0.5	0.0
0.449	0.4	0.0
0.535	2.0	2.0
0.396	1.0	2.0
0.328	0.7	2.0
0.265	0.5	2.0
0.224	0.4	2.0
0.396	2.0	5.0
0.268	1.0	5.0
0.207	0.7	5.0
0.156	0.5	5.0

Table 11.42
[Malate] and [Coenzyme A] held constant at 0.5 mM and 0.08 mM, respectively

Specific activity (units)	*[ATP] (mM)*	*[α-hydroxyglutarate] (mM)*
0.477	2.0	0.0
0.436	1.0	0.0
0.321	0.7	0.0
0.282	0.5	0.0
0.246	0.4	0.0
0.263	2.0	2.0
0.250	1.0	2.0
0.207	0.7	2.0
0.191	0.5	2.0
0.173	0.4	2.0
0.175	2.0	5.0
0.169	1.0	5.0
0.149	0.7	5.0
0.140	0.5	5.0
0.130	0.4	5.0

Table 11.43
[Malate] and [ATP] held constant at 0.5 mM and 0.08 mM, respectively

Specific activity (units)	*[Coenzyme A] (mM)*	*[α-hydroxyglutarate] (mM)*
0.555	0.05	0.0
0.448	0.10	0.0
0.386	0.143	0.0
0.305	0.20	0.0
0.261	0.25	0.0
0.386	0.05	2.0
0.302	0.10	2.0
0.246	0.143	2.0
0.206	0.20	2.0
0.183	0.25	2.0
0.240	0.05	5.0
0.199	0.10	5.0
0.168	0.143	5.0
0.145	0.20	5.0
0.122	0.25	5.0

PART THREE

SPECTROPHOTOMETRY AND RADIOCHEMISTRY

12
Spectrophotometry

Spectrophotometry is one of the most widely used and versatile of all biochemical analytical tools. The analysis of systems by measuring light absorption has definite advantages. It is a nondestructive method, unless a photochemical reaction occurs, which is not common, and it offers selectivity in that each compound has a characteristic spectrum. Often a particular component in a solution or group in a molecule can be singled out for observation. It averages the properties of the system over a short time interval, 10^{-14} sec (contrast this with viscosity, which averages the properties of the system over a period of seconds or minutes), and this enables one to follow the details of fast reactions. Finally, probably the most important fact is that measurements of high accuracy can be made in a short time.

The absorption of visible light—that detected by the human eye—is most commonly noted by the color of objects. A solution of vitamin B_{12}, for example, transmits red light because blue, green, and yellow, the other colors also present in white light, are absorbed. The spectrum of vitamin B_{12} is shown in Figure 12.1. Visible light is only a small part of the electromagnetic spectrum, as may be seen in Figure 12.2.

A light wave can be considered to possess oscillating electric and magnetic fields that are in phase but perpendicular to each other and to the direction of propagation. This is illustrated in Figure 12.3. The length of the light wave, λ, in centimeters is related to the velocity of light in a vacuum, c, in centimeters per second, and the frequency, ν, in cycles per second:

$$\lambda = \frac{c}{\nu} \tag{12.1}$$

Figure 12.1

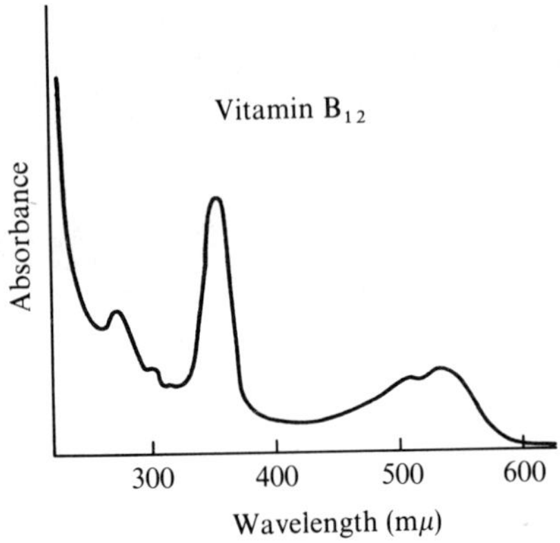

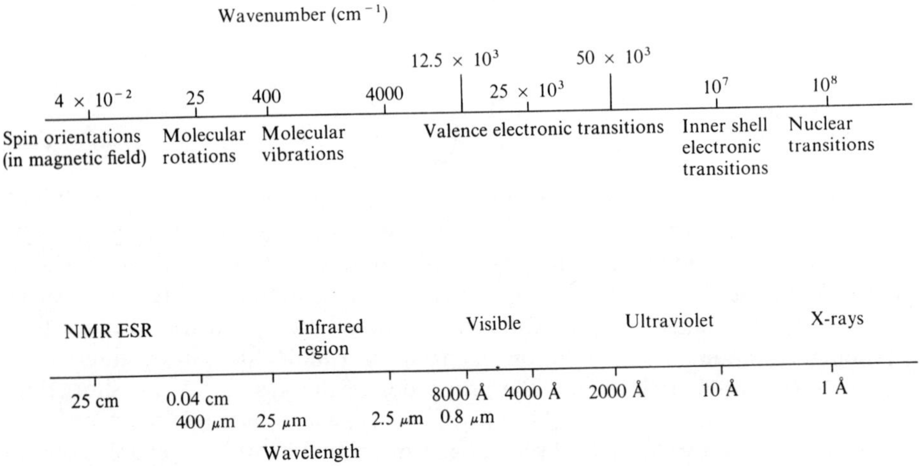

Figure 12.2

Figure 12.3

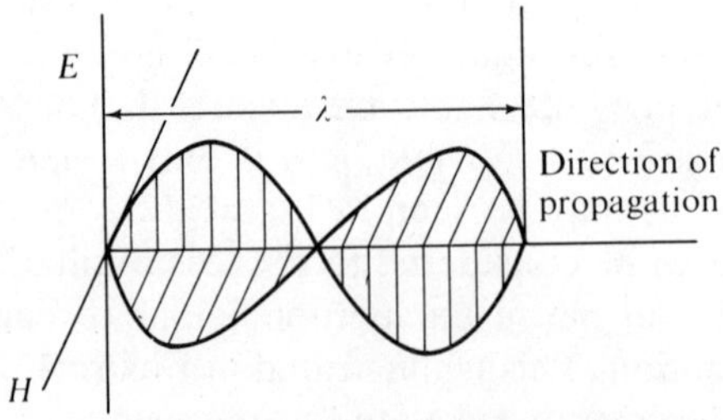

Problem 12.1: Express the wavelength, 3000 Å, as frequency, in cycles per second, and as wave numbers in reciprocal centimeters. Convert also to wavelength in nanometers (nm) and millimicrons (mμ). Note relationship of presently used metric nomenclature to earlier millimicron units of wavelength.

Solution: Equation (12.1) relates frequency and wavelength. Solving for frequency gives

$$\nu = \frac{c}{\lambda} \tag{i}$$

Convert 3000 Å to centimeters:

$$[3000 \text{ Å}][10^{-8} \text{ cm Å}^{-1}] = 3 \times 10^{-5} \text{ cm}$$

Substituting in (i) the velocity of light c and the wavelength, we obtain

$$\nu = \frac{2.997 \times 10^{10} \text{ cm sec}^{-1}}{3.0 \times 10^{-5} \text{ cm}}$$

$$= 9.99 \times 10^{14} \text{ sec}^{-1}$$

To obtain the wave number in reciprocal centimeters we take the reciprocal of Equation (12.1),

$$\frac{1}{\lambda} = \frac{\nu}{c} \tag{ii}$$

and define a new term $\bar{\nu}$, the wave number, which is the reciprocal of wavelength. The reciprocal in our example is

$$\bar{\nu} = \frac{1}{3 \times 10^{-5} \text{ cm}} = 0.333 \times 10^{5} \text{ cm}^{-1} = 3.33 \times 10^{4} \text{ cm}^{-1}$$

$$= 33{,}300 \text{ cm}^{-1}$$

The wavelength in micrometers is

$$[3000 \text{ Å}][10^{-8} \text{ cm Å}^{-1}][10^{-2} \text{ m cm}^{-1}][10^{9} \text{ nm m}^{-1}] = 300 \text{ nm}$$

In millimicrons we obtain

$$[3000 \text{ Å}][10^{-8} \text{ cm Å}^{-1}][10^{4}\ \mu \text{ cm}^{-1}][10^{3} \text{ m}\mu\ \mu^{-1}] = 300 \text{ m}\mu$$

for the wavelength. Note the numerical equivalence of wavelength in nanometers and millimicrons.

Light interacts with matter primarily through the oscillating electric field. For purposes of absorption, light should be thought of as an energy packet or photon, $h\nu$, where h is Planck's constant, 6.627×10^{-27} erg sec. The light energy is directly proportional to frequency and, by Equation (12.1), is inversely proportional to wavelength.

Problem 12.2: Calculate the energy in calories for Avogadro's number of photons with a wavelength of 5000 Å. The velocity of light is 2.997×10^{10} cm sec^{-1}.

Solution: The energy of a single photon is given by

$$E = h\nu \tag{i}$$

Using Equation (12.1), calculate the frequency of 5000 Å light.

$$\nu = \frac{c}{\lambda} = \frac{2.997 \times 10^{10}}{5000\ \text{Å}}\ \text{cm sec}^{-1}$$

$$= \frac{2.997 \times 10^{10}}{5000\ \text{Å}}\ \text{cm sec}^{-1} \cdot 10^{8}\ \text{Å cm}^{-1}$$

$$= \frac{2.997 \times 10^{14}}{0.5}\ \text{sec}^{-1} = 5.994 \times 10^{14}\ \text{sec}^{-1}$$

Substituting the frequency in (i), we obtain

$$E = 6.627 \times 10^{-27}\ \text{erg sec} \times 5.994 \times 10^{14}\ \text{sec}^{-1}$$

$$= 3.972 \times 10^{-12}\ \text{erg per photon}$$

The number of photons equal to Avogadro's number is called a mole of radiation. Therefore

$$E = [6.023 \times 10^{23}\ \text{photon mole}^{-1}] \times 3.972 \times 10^{-12}\ \text{erg photon}^{-1}$$

$$= 23.92 \times 10^{11}\ \text{erg mole}^{-1}$$

Converting to calories, we have

$$E = \frac{23.92 \times 10^{11}\ \text{erg mole}^{-1}}{4.186 \times 10^{7}\ \text{erg cal}^{-1}}$$

$$= 5.7 \times 10^{4}\ \text{cal (mole radiation)}^{-1}$$

One mole of radiation is an einstein, so E is 5.7×10^4 cal einstein^{-1} at this particular wavelength.

Absorption of light occurs when the energy of the photon, $h\nu$, corresponds to the difference between two energy levels in the molecule. These various types of energy levels in order of decreasing energy difference are electronic, vibrational, rotational, and spin orientational (in a magnetic field) (see Figure 12.2). Transitions between electronic levels, promoting to a higher energy electronic configuration, occur with light in the ultraviolet and visible range. This region of the spectrum has been and will continue to be of most general use to biochemists, and we limit our discussion to it. Absorption of light in the infrared, microwave, and radio-wave regions causes the process indicated in Figure 12.2. Of these phenomena, NMR and ESR are becoming increasingly important biochemical research tools as new instruments are developed.

A large number of the molecules of biological interest absorb light in the ultraviolet-visible region of the spectrum. Its range in wavelength units is 2000–8000 Å, or 200–800 nm. Some examples of biochemical molecules that absorb in this region are purines, pyrimidines, and aromatic amino acids, and the macromolecules, nucleic acids, and proteins, which contain residues of them in their structures. Other examples are the heme-containing proteins (the cytochromes, hemoglobin and myoglobin), and the carotenoids and steroids. All of these compounds contain conjugated systems which absorb in the ultraviolet region and some, the cytochromes, hemoglobin and myoglobin, also absorb in the visible range.

ABSORPTION LAWS

Spectrophotometry is primarily used by biochemists to monitor the concentration of various species in reaction mixtures. The absorption of light at a given wavelength is related to the concentration of absorbing species through two laws. The first, which is attributed to Bouguer, states that the fraction of light absorbed is proportional to the thickness of the absorber, x, and independent of the light intensity. Thus for successive layers of absorber of thickness, x, each absorbs the same fraction of the light incident on it. For example, if 20% of the incident light is absorbed by a thickness x, the transmitted light is 80%. For additional thicknesses, x, the transmitted light will be diminished as follows: 64%, 51.2%, and so on, which is the sequence $(0.8)^0$, $(0.8)^1$, $(0.8)^2$, $(0.8)^3$, and so on. It is expressed mathematically as

$$\frac{I}{I_0} = e^{-\alpha x} \tag{12.2}$$

where I is the intensity of the transmitted light, I_0 is the intensity of the incident light, x is the thickness, and α is the linear absorption coefficient, a constant characteristic of the medium. Writing in logarithmic form, we obtain

$$\ln \frac{I_0}{I} = \alpha x \tag{12.3}$$

which in base ten logarithms is

$$\log \frac{I_0}{I} = \frac{\alpha x}{2.303} \tag{12.4}$$

Expression (12.4), which is for a uniform solid absorber, can be modified to include solution concentrations. The path length x is, in a sense, a concentration (uniform solid) as well as a path length. For a solution, which contains an absorber dispersed in a nonabsorbing solvent, these terms need to be separated.

Light is absorbed only when a photon collides with a molecule. Thus the amount of light absorbed or the probability of absorption is proportional to the number of molecules in the light beam. This is a statement of Beer's law, the second law of absorption. In solution the number of molecules of absorber in the light beam is proportional to the product of the concentration c and the path length b. Combining Beer's law with Bouguer's law, we obtain

$$\log \frac{I_0}{I} = abc \tag{12.5}$$

where a is the absortivity or specific absorption coefficient and is characteristic of the absorbing species (analogous to α). The quantities in the log term are as previously defined. When the concentration units are moles per liter, a is called the molar absorptivity or the molar absorption coefficient and is denoted by ϵ. The absorption law is also written in terms of absorbance, A, as in

$$A = \log \frac{I_0}{I} = \epsilon bc \tag{12.6}$$

Absorbance (commonly and less desirably called optical density) is then a linear function of concentration for any given wavelength if Beer's law* is obeyed. The molar absorption coefficient, however, varies with wavelength and solvent.

Beer's law is exact only for parallel, monochromatic light in an isotropic medium (randomly oriented absorbers in solution). One should always check to see that it is obeyed.

SPECTROPHOTOMETERS

A spectrophotometer, which is a device for measuring light absorption as a function of wavelength, has four basic parts; a source of light, the monochromator, the sampling cell, and the detector. Light from the source, which for the region 200–800 nm is either a hydrogen-discharge lamp or a tungsten bulb, is focused on the entrance slit of the monochromator, which disperses the light and focuses on the desired wavelength on the exit slit. This monochromatic light then passes through the sample to the detector.

The quantities I and I_0 for Equation (12.6) are determined for each wavelength as follows. The intensity of the incident light I_0 is obtained from

*It is commonly called Beer's law rather than Beer-Bouguer's or Beer-Lampert's law. The first absorption law has been attributed to both Bouguer (1729) and Lampert (1760); hence both names are used.

a measurement of a suitable blank which contains all the solution components except the absorber. Rather than read I_0 in percent, most spectrophotometers enable the reading of log I_0 directly. The transmitted intensity, I, is obtained from a similar measurement on the solution containing the absorber. If log I is read, then log I_0/I can be obtained by subtraction. Another term in common usage is transmittance, which is the ratio I/I_0.

Problem 12.3: Calculate the molar absorption coefficient, ϵ, at 351 nm for aquocobalamin in 0.1 *M* phosphate buffer, pH 7.0, from the following data, which were obtained in a 1-cm cell.

Solution	$c \times 10^5$ *M*	I_0	I
A	2.23	93.1	27.4
B	1.90	94.2	32.8

[From H. P. C. Hogenkamp, unpublished results.]

Solution: Calculate the I_0/I ratio for each concentration and determine ϵ by substituting Beer's law.

For solution A,

$$\frac{I_0}{I} = \frac{93.1}{27.4} = 3.396$$

$$\log \frac{I_0}{I} = 0.531$$

Substituting in Equation (12.6) gives

$$\epsilon = \frac{1}{bc} \cdot \log \frac{I_0}{I}$$

$$= \frac{1}{1 \cdot (2.23 \times 10^{-5})} \cdot (0.531)$$

$$= 2.38 \times 10^4 \text{ liter cm}^{-1} \text{ mole}^{-1}$$

Similarly, for solution B, I_0/I is 2.870 and

$$\log \frac{I_0}{I} = 0.458$$

Substituting in Equation (12.6) gives

$$\epsilon = \frac{1}{1 \cdot (1.90 \times 10^{-5})} \cdot (0.458)$$

$$= 2.40 \times 10^4 \text{ liter cm}^{-1} \text{ mole}^{-1}$$

The average molar absorption coefficient from these two measurements is 2.39×10^4 liter cm^{-1} mole^{-1}.

Automatic scanning spectrophotometers give the absorbance, log I_0/I, directly as a function of wavelength. This continuous comparison is achieved by using two cells, a sample and a reference, two light beams, and a chopper which enables the detector to look alternately at the light transmitted by each solution.

The spectra are generally presented with either the molar absorption coefficient, ϵ, or absorbance as the ordinate and wavelength or frequency as the abscissa. If absorbance is used, the concentration must be included in addition to other pertinent experimental conditions, such as pH and temperature. For biochemical substances of unknown molecular weight the absorption unit $E_{1\,\mathrm{cm}}^{1\%}$ is convenient. It is the absorptivity for a 1% (w/v) solution in a 1-cm cell. Common abscissa units in the ultraviolet-visible range are angstroms and nanometers.

Problem 12.4: A solution of purified DNA isolated from *Escherichia coli* gives an absorbance of 0.793 at 260 nm in a 1-cm cell at pH 4.5. If $E_{1\,\mathrm{cm}}^{1\%}$ is 197, calculate the concentration of the solution in milligrams per milliliter.

Solution:

$$A = E_{1\,\mathrm{cm}}^{1\%} \cdot c$$

since b is understood to be 1 cm. Solving for c we obtain

$$c = \frac{A}{E_{1\,\mathrm{cm}}^{1\%}}$$

Substitution gives

$$c = \frac{0.793}{197} = 4.03 \times 10^{-3}\%$$

A $4.03 \times 10^{-3}\%$ solution contains 4.03×10^{-3} g per 100 ml. Thus we have 4.03×10^{-5} gm ml^{-1} or 4.03×10^{-2} mg ml^{-1}.

SAMPLING TECHNIQUES

The mathematical statement of Beer's law makes it apparent that accurate values of the absorbance and the molar extinction coefficient depend on the accuracy of the concentration and path-length determinations. Concentrations are generally obtained by the careful weighing of pure sample into suitable volumetric equipment. The path length can be determined by measuring the absorbance of a standard absorber of known concentration such as K_2CrO_4 (see Table 12.1). This procedure assumes that the spectrophotometer faithfully reproduces the values for these standards in a cell with

Table 12.1
Standard values of absorbance, A, for standard potassium chromate solution*

Wavelength (nm)	*A*	*Wavelength (nm)*	*A*
220	0.4559	370	0.9914
230	0.1675	380	0.9281
240	0.2933	390	0.6841
250	0.4962	400	0.3872
260	0.6345	410	0.1972
270	0.7447	420	0.1261
280	0.7235	430	0.0841
290	0.4295	440	0.0535
300	0.1518	450	0.0325
310	0.0458	460	0.0173
320	0.0620	470	0.0083
330	0.1457	480	0.0035
340	0.3143	490	0.0009
350	0.5528	500	0.0000
360	0.8297		

Source: K. S. Gibson, *Spectrophotometry*. Natl. Bur. Standards (U.S.), Circ. 484, 1949.

*0.0400 g liter^{-1} of K_2CrO_4 in 0.05 N KOH solution at ~25°C in a 1.000-cm cell.

a path length of 1 cm. The path lengths of very thin cells can be determined by interferometry.*

A standard solution can also be used to check the accuracy of the absorbance scale of the spectrophotometer (see for example Table 12.1 for standard values for the ultraviolet region). For maximum accuracy the linearity of the absorbance scale of the spectrophotometer should be calibrated at several absorbance values with linearity discs or secondary standard neutral density filters or standard solutions.

The cells for the measurements should be of the proper optical material; for example, quartz for the far ultraviolet. If the cells absorb a significant part of the light, the sensitivity of the instrument will be decreased. Before any measurements are made the cells should be carefully cleaned to remove dust or film that may be present. The presence of dust causes the incident light to be scattered and thereby introduces errors in the absorbance. The intensity of scattered light increases as frequency to the fourth power and thus these errors increase dramatically as the wavelength is

*R. C. Lord, R. S. MacDonald, and F. A. Miller, *J. Opt. Soc. Amer.* **42**:199 (1952).

decreased. Often the solutions themselves will contain suspended dust particles which should be removed by filtration or centrifugation.

Cells should be carefully positioned in the spectrophotometer so that the light beam passes through without reflecting off the sides. This is difficult to achieve when using round cuvettes, so they must be in the same position for each reading. Positioning is extremely important when microcells are used also. For microcells the positioning errors can be eliminated by using the proper mask, which is often supplied by the manufacturer. One may be constructed from blackened metal or paper. The mask should limit the aperture of the light beam so that it doesn't strike the sides of the cell.

DEVIATIONS FROM BEER'S LAW

The linear relationship between absorbance and concentration as predicted by Beer's law is not always obeyed. For the most part the deviations we observe are only apparent. They are due to the failure of the systems observed or the observing system to satisfy requirements of the law as stated earlier. The two types of deviations are then instrumental, which may be optical and/or mechanical, and those related to the sample.

Instrumental Deviation. Beer's law is obeyed only for monochromatic light, which in practice is never realized. The extent to which the light is monochromatic can be varied by changing the slit width in most commercial instruments. Slit widths are varied to provide sufficient light energy to operate the electronic system at an acceptable noise level.

For the visible region of the spectrum a special class of spectrophotometers called colorimeters have long been used. They differ from the usual spectrophotometer in that filters are used to obtain monochromatic light rather than a prism or grating and a slit. The filters pass a broad band of light; for example, a typical 540-nm filter passes light which ranges from 525 to 555 nm.

In spectrophotometers the slit width is given in millimeters; however, it is related to a spectral slit width or spectral band pass which is in units of wavelength, such as nanometers. Let us examine the effect of slit width in the following way. Suppose the true shape of an absorption band for a compound is as shown in Figure 12.4. The dotted vertical lines indicate the spectral band width which is the width of the band in nanometers at half the peak intensity. Now we will observe what the spectrophotometer "sees" as a function of slit width. First scan with a spectral slit width of 1 nm as indicated by the vertical lines. This is a beam of light with a wavelength range of 1 nm, for example, 400–401 nm. As we move across the absorption band the spectrophotometer averages the absorbance values between the

Figure 12.4

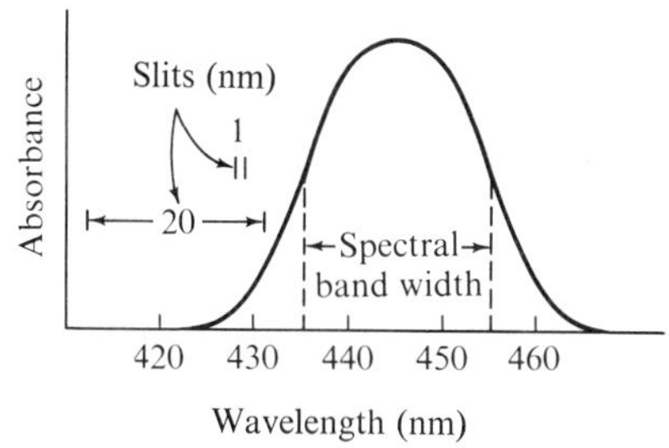

bars. Since the spectral slit width, 1 nm, is small compared to the band width, 15 nm, the spectrophotometer gives an accurate reproduction of the absorption. If the band is now scanned with a spectral slit width of 20 nm, the absorption band will have a lower peak intensity, as the spectrophotometer is averaging the sides of the absorption band with the peak. Deviations from Beer's law are expected whenever absorbance measurements are made on the side of an absorption band.

Slit width effects need to be assessed whenever molar extinction coefficients are being measured or used. One can check for slit effects by observing the absorbance at the band maximum as the slit is varied. Little or no change should be observed. Certain precautions need be taken in this procedure, however.* Fortunately, most absorption bands in aqueous solution in the ultraviolet and visible regions are broad and consequently (in view of our earlier discussion) would not be expected to show a major slit effect. Spectral slit, in nanometers, versus slit width, in millimeters, is plotted in Figure 12.5 for a Beckman DU spectrophotometer. Under normal operating conditions a slit of 0.3 mm is used, for which the spectral

Figure 12.5

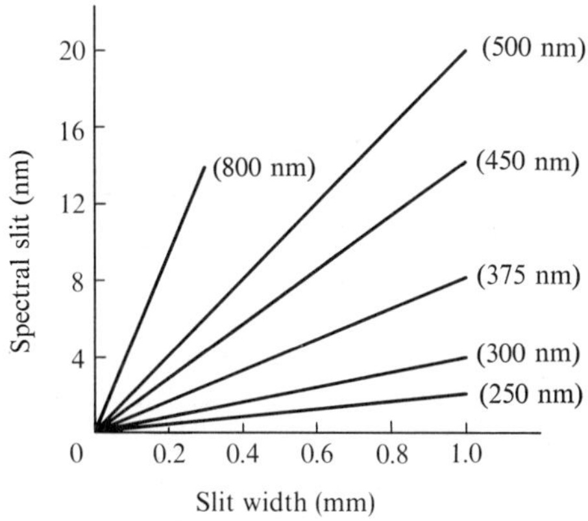

*See C. F. Hiskey, *Physical Techniques in Biological Research,* Vol. 1, G. Oster and A. W. Pollister (eds.). Academic Press, New York, 1955.

slit ranges from 0.3 nm at 250 nm to 6 nm at 500 nm. As a rule of thumb one can assume that the slit effects will be negligible if the ratio of spectral slit width to the spectral band width is 1:10 or less.

Deviations from Beer's law are also expected for nonparallel light. though most instruments do have a converging and/or diverging beam passing through the sample, the errors in absorbance which result are insignificant for nonscattering samples.

Deviations Due to the Nature of the Sample. Most all of these effects result from a change in the effective concentration of the absorbing species or interference by the product of some side reaction.

Many absorbing species contain acidic or basic groups. Since the conjugates in the acid-base pair have different absorption properties, deviations will occur if the pH is not controlled. Similarly, if the absorbing species is involved in an equilibrium which is temperature-dependent, deviations will occur with small temperature changes. Some absorbing molecules may be physically adsorbed on the cell walls during the course of observation, thereby decreasing the effective concentration in solution and thus causing apparent deviations from Beer's law. Solvent effects, such as are caused by changing electrolyte concentration, can also change the absorption characteristics. The absorption coefficient of the absorber may vary with concentration due to association and/or aggregation of the molecules.

Deviations from Beer's law arise from sample fluorescence. The absorbed radiation excites fluorescence radiation at a longer wavelength, which is emitted in all directions. Some is measured by the detector and gives rise to low or even negative absorbance values. Samples can be checked for fluorescence error by measuring the absorbance as a function of distance from the detector. When fluorescence is present the absorbance increases as the sample is moved further from the detector. The fluorescent radiation can often be removed by using the proper filter between the sample and the detector.

It is not always possible to eliminate every factor which causes a deviation from Beer's law. Although this makes it difficult and often impossible to use or to obtain molar absorption coefficients, one can still determine the concentration of an absorber by making a standard curve. Known concentrations of absorber are analyzed and the standard curve constructed. Since the reasons for the deviation from Beer's law are not always known, one should keep all experimental conditions constant. This means that the same instrument, cells, instrumental settings, and sample preparation procedures must be used.

Problem 12.5: The activity of the enzyme tyrosine oxidase from liver can be measured by following the decrease in the concentration of tyrosine spectrophotometrically. The basis of the method is the quantitative reaction of tyrosine with nitrosonaphthol to give a red color. Calculate the μg of tyrosine oxidized per minute from the following data. The initial amount of tyrosine in the standard reaction mixture was 100 μg.

Sample	Klett units ($\propto$ absorbance) Reaction time (5 min)
1	241
2	236

Standard curve for tyrosine

μg tyrosine	Klett units ($\propto$ absorbance)
0	3
10	38
30	138
50	223
75	314
100	368

Solution: Construct standard curve (See Figure 12.6).

Figure 12.6

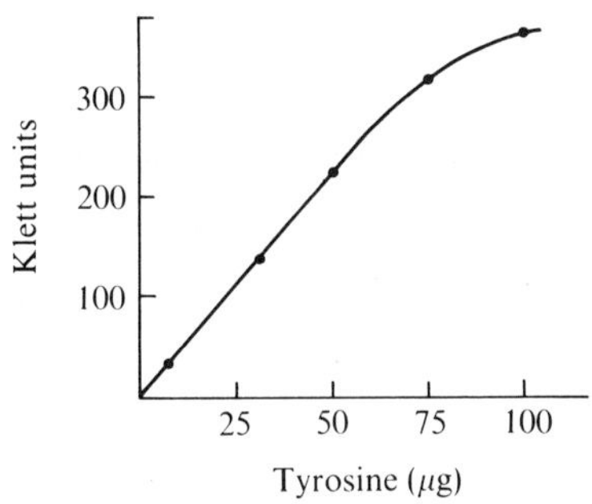

From the standard curve read the μg of tyrosine remaining after the 5-min reaction. Samples 1 and 2 give 54.1 and 52.5 μg, respectively. This means that 45.9 and 47.5 μg had been oxidized in the 5-min reaction. The amount of tyrosine oxidized per minute for the two samples is 9.18 and 9.50 μg, respectively. In the 80–100 μg range the accuracy of the method is decreasing as the number of Klett units per μg is decreasing (that is, as the standard curve is flattening out).

ABSORPTION OF MIXTURES

The absorption of mixtures of noninteracting substances is additive; hence

$$A = (\epsilon_1 c_1 + \epsilon_2 c_2 + \cdots + \epsilon_n c_n) b$$

If the molar absorption coefficients are known for each of n components at n

wavelengths, then the concentrations can be calculated by solving n simultaneous equations. For two components this is

$$A_{\lambda_1} = \epsilon_{1_{\lambda_1}} c_1 + \epsilon_{2_{\lambda_2}} c_2$$

$$A_{\lambda_2} = \epsilon_{1_{\lambda_2}} c_1 + \epsilon_{2_{\lambda_2}} c_2$$

where λ_1 and λ_2 are the two wavelengths of observation.

Problem 12.6: Consider a mixture of compounds B and D. The values of the absorbance for the mixture at 437 nm and 491 nm are 0.732 and 0.361, respectively, in a 1-cm cell. Molar absorption data for B and D are

	437 nm	491 nm
ϵ_B	14,700	3,100
ϵ_D	2,200	9,300

Solution: Set up the simultaneous equations

$$\text{at 437 nm} \qquad A = \epsilon_{B437}[\text{B}] + \epsilon_{D437}[\text{D}]$$

$$\text{at 491 nm} \qquad A = \epsilon_{B491}[\text{B}] + \epsilon_{D491}[\text{D}]$$

When we substitute values

$$0.732 = 14{,}700[\text{B}] + 2200[\text{D}] \qquad \text{(i)}$$

$$0.361 = 3100[\text{B}] + 9300[\text{D}] \qquad \text{(ii)}$$

To obtain the solution, solve for B in terms of D in (i) and substitute in (ii) and solve for D. Substitution of D in (i) gives B.

Solve for B in (i) to give

$$\text{B} = \frac{0.732 - 2200[\text{D}]}{14{,}700} \qquad \text{(iii)}$$

Substituting (iii) in (ii) gives

$$0.361 = \frac{(0.732 - 2200[\text{D}])}{14{,}700}\,3100 + 9300[\text{D}]$$

$$= (0.732 - 2200[\text{D}])0.2109 + 9300[\text{D}]$$

$$= 0.154 - 464.0[\text{D}] + 9300[\text{D}]$$

$$0.207 = 8386[\text{D}]$$

$$[\text{D}] = 2.34 \times 10^{-5}\ M$$

Substituting [D] in (i) gives [B]

$$0.732 = 14{,}700[\text{B}] + 2200(2.34 \times 10^{-5})$$

$$= 14{,}700[\text{B}] + 5.14 \times 10^{-2}$$

$$0.681 = 14{,}700[\text{B}]$$

$$[\text{B}] = 4.63 \times 10^{-5}\ M$$

This principle of additivity of absorption has been used to determine the tyrosine and tryptophan content in proteins. Tyrosine and tryptophan absorb in the region of 280–315 nm where interference from other amino acids is minimal. Thus if values of the molar absorption coefficients can be determined at two wavelengths, their respective concentrations can be calculated from the spectrum of the protein. This determination of the molar absorption coefficient for tyrosine and tryptophan in proteins is beset with difficulties due primarily to environmental effects which change both the wavelength and intensity of their respective absorptions. These and other errors have been thoroughly discussed.*

SPECTROPHOTOMETRIC TITRATIONS

The course of a chemical reaction or the position of an equilibrium can readily be determined spectrophotometrically if the absorption band of one of the components can be measured. Probably the best known examples are the hydrogen ion titration curves of indicators:

$$\mathrm{In^- + H^+ \rightleftharpoons InH}$$

where either the protonated or unprotonated (or both) forms are colored. The absorption maxima for two forms are usually separated so that the concentration of either form can be measured. If an equal quantity of the indicator is dissolved in buffers of various pH's and the absorbance measured at one of the absorption maxima, the pK can be calculated. Spectrophotometric pK determinations have a definite advantage over the methods discussed in Chapter 7 in that the concentrations of the salt and acid are measured directly and do not depend on assumptions made for them in terms of hydrogen ions.

Problem 12.7: The following data were obtained for the indicator phenolphthalein. Solutions for the absorbance measurement at 550 nm contain 1 ml of stock indicator solution plus 2 ml of buffer (0.2 M). Calculate the pK of the indicator.

pH	A_{550} nm
5.0	0.050 (used as blank)
7.0	0.053
8.7	0.076
9.3	0.134
9.9	0.245
10.3	0.315
10.8	0.368
12.0	0.400

*See G. H. Beaven and E. R. Holiday, *Advances in Protein Chem.* 7:319 (1952); J. W. Donovan "Ultraviolet Absorption" in S. Leach (ed.), *Physical Principles and Techniques of Protein Chemistry,* Part A, Academic Press, New York, 1969.

Solution: Plot absorbance versus pH (Figure 12.7). From the data it is apparent that the salt form is the absorbing species in the equilibrium:

$$\text{InH (acid)} \rightleftharpoons \text{In}^- \text{(base)} + \text{H}^+$$

The total absorbance change in going from the acid form to the salt form is 0.350. Thus we can write

$$0.350 = [\text{InH}] + [\text{In}^-] = [\text{In}]_T \tag{i}$$

which is analogous to Equation (7.19) since absorbance is directly proportional to concentration. The concentration of salt at pH 9.3 is 0.134 − 0.050 or 0.084. From Equation (i) the acid concentration is 0.350 − $[\text{In}^-]$ or 0.266 and the salt to acid ratio is 0.315. Substituting the salt to acid ratio and the pH in the Henderson-Hasselbalch equation we obtain

$$9.3 = \text{p}K + \log 0.315$$

$$\text{p}K = 9.3 + \log 3.17 = 9.8$$

A similar calculation can be made at each pH; all should give the same pK.

Figure 12.7

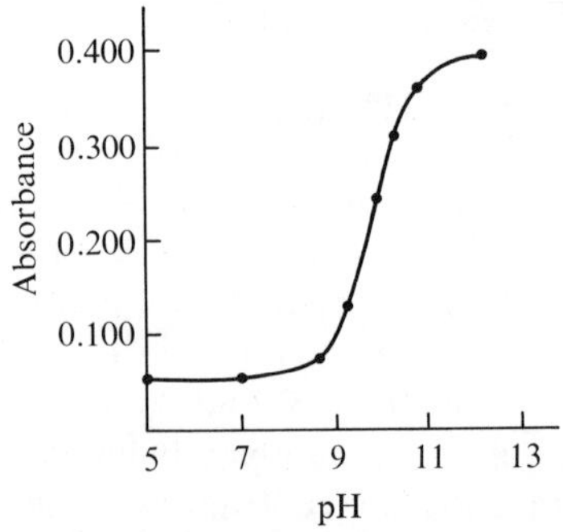

It is apparent that for points which are within one pH unit of the pK the salt to acid ratio has a lower percentage error. Note also that the concentration and molar absorption coefficient need not be used for a spectrophotometric pK determination. The pK should be determined at various concentrations to check for deviations from Beer's law.

Equilibrium between two species in solution, which are responsible for all the absorption in a given region, is indicated by the presence of an isosbestic point in the spectrum, where the absorbance is independent of the concentration ratio of the two components. It can be shown that such a point must exist for two species in equilibrium; however, if the absorption bands of the two species do not overlap, this point will lie in the region of zero absorption. The appearance of an isosbestic point then indicates that an equilibrium exists between two species; however, the converse is not necessarily true. At the isosbestic point the molar absorption coefficients of the two components are equal. The presence of an isosbestic point for methyl red is shown in Figure 12.8.

Figure 12.8

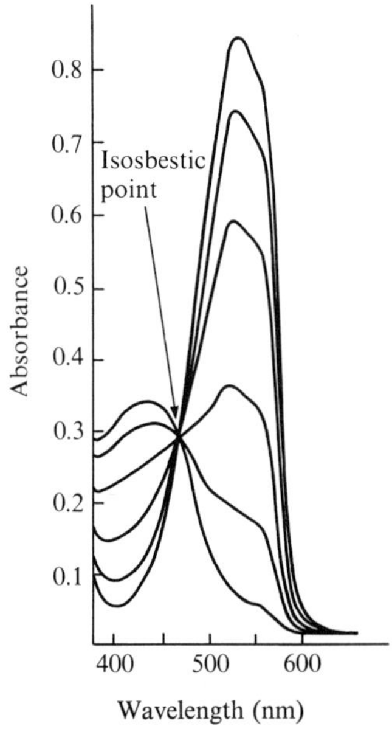

REFERENCES

Robert P. Bauman, *Absorption Spectroscopy*. Wiley, New York, 1962.

John R. Dyer, *Applications of Absorption Spectroscopy of Organic Compounds*. Prentice-Hall, Englewood Cliffs, N.J., 1965.

J. H. Harley and S. E. Wiberley, *Instrumental Analysis*. Wiley, New York, 1954.

12. PROBLEMS A

Answers on page 360

1. Calculate the energy of a mole of photons (an einstein) that have a wavelength of 3700 Å.

2. Calculate the frequency ν in cycles sec^{-1} and the wave numbers $\bar{\nu}$ in reciprocal centimeters for the following wavelengths.
   (a) 200 nm
   (b) 0.7 μ
   (c) 3750 Å
   (d) 2.5 μm

3. A solution of reduced nicotinamide adenine dinucleotide (NADH) absorbs light at wavelengths of 340 nm and 260 nm. What is the spacing of these energy levels in cal $mole^{-1}$?

4. In the preparation of a standard curve for spectrophotometric analysis the following data were obtained.

Concentration (mg $liter^{-1}$)	I_0	I
0	98	98
1	97	77.2
2	100	63.5
3	99.5	50
4	100	41.3
5	100	33.5
6	100	27.9
7	99.0	23.4
8	98.2	20.3
9	100	18.1
10	100	16.4

Calculate the absorbance and plot it versus concentration. Is Beer's law obeyed? What concentration corresponds to an absorbance of 0.42?

5. A particular sample of a solution of a colored substance, which is known to obey Beer's law, shows 70% transmittance when measured in a 1-cm cell.
   (a) Calculate the percent transmittance for a solution of twice the concentration in the same cell.
   (b) What must be the cell length to give 70% transmittance for a solution of twice the original concentration?
   (c) Calculate the percent transmittance of the original solution contained in a 0.5-cm cell.

6. The concentrations of five solutions of cytidine at pH 7.0, along with their respective absorbance values at 271 nm, are given below. Is Beer's law obeyed within experimental error?

A	c (mg liter^{-1})
0.684 ± 0.002	18.30
0.473 ± 0.002	12.65
0.296 ± 0.002	7.92
0.135 ± 0.002	3.61
0.097 ± 0.002	2.59

7. Molar absorption coefficients (ϵ), are usually reported as M^{-1} cm^{-1}. What would be the absorbance at 260 nm measured in a Beckman DU spectrophotometer with a 1-cm light path, of a solution containing 0.01 μmole ml^{-1} if ϵ_{260} is 34.7×10^3 M^{-1} cm^{-1}? What are the units if the molar absorption coefficient of this compound is expressed as 34.7×10^6?

8. The absorption spectra of two concentrations of adenosine triphosphate are given in Figure 12.9. Calculate the concentration of sample two if the concentration of sample one is 39 mg liter^{-1}. What assumptions are necessary for the calculation to be valid?

Figure 12.9

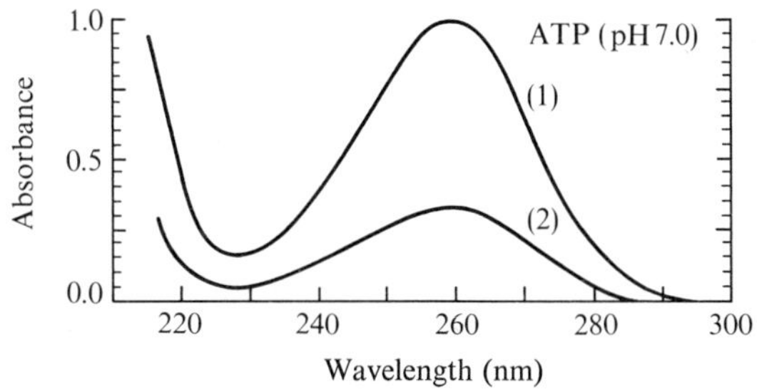

9. Part of setting up a spectrophotometric method is choosing the proper wavelength of observation. For a phosphate determination the standard solutions were run using two different filters with the following results.

Standard (μmole PO_4)	A (Filter A)	A (Filter B)
0	0	0
0.2	0.09	0.05
0.4	0.18	0.10
0.7	0.29	0.16
1.0	0.42	0.23

Which would be the better filter for the determination? Explain.

10. The equilibrium constant for the reaction

$$AB \rightleftharpoons A + B$$

is 5×10^{-3} mole liter^{-1}. The complex, AB, absorbs at 370 nm, with a molar absorption coefficient of 300. Calculate the absorbance at 370 nm in a 1-cm cell of a solution prepared by mixing 10 ml of 0.01 M solution of A with 10 ml of 0.01 M solution of B.

11. A compound A has a molar absorption coefficient at 260 nm of 34.7×10^3 M^{-1} cm^{-1}. A second compound B has absorbance maxima at 261, 340, 375, and 524 nm. Both compounds are quantitatively decomposed to compound C. A solution of compound A with $A_{260} = 0.805$, yields a solution of compound C, with $A_{351} = 0.324$. A solution of B gives the following absorbance readings: $A_{261} = 0.653$, $A_{340} = 0.227$, $A_{375} = 0.195$, and $A_{524} = 0.147$. Decomposition of this solution of B yields a solution of C with $A_{351} = 0.265$. Calculate the molar absorption coefficients of compound B at 261, 340, 375, and 524 nm.

---

12. PROBLEMS B

No answers given

1. The fundamental event in photosynthesis is the absorption of light. Two different wavelengths of light, 440 nm and 660 nm, are absorbed by the chlorophylls *a* and *b*, respectively, which are present in the leaves of plants. Calculate the light energy absorbed by a mole of chlorophyll *a* in calories. What is the reduction potential in volts which corresponds to this energy?

2. The fact that coenzyme A is required for the acetylation of sulfanilamide is the basis of a quantitative assay for the coenzyme. Construct a standard curve for the assay from the following data. Is Beer's law obeyed? Give possible reasons for deviations.

Sample	Transmittance (%)	CoA (μg)
1 (blank)	64.0	0
2	68.0	0.73
3	71.5	1.46
4	75.0	2.23
5	78.2	2.91
6	79.0	3.64
7	82.5	4.62

3. Ribonucleic acid can be hydrolyzed by base to yield a mixture of the four component nucleotides. Column chromatography achieved a separation of cytidylic and uridylic acids but adenylic and guanylic acids remained mixed. The absorption spectra of the mixture at pH 7.0 gave the following results.

$$A = 0.731 \text{ at } 260 \text{ nm}$$

$$A = 0.341 \text{ at } 280 \text{ nm}$$

Adenylic acid pH 7.0

$$\epsilon_{260} = 15{,}400$$

$$\epsilon_{280} = 2{,}500$$

Guanylic acid pH 7.0

$$\epsilon_{260} = 11{,}700$$

$$\epsilon_{280} = 7{,}700$$

Calculate the ratio of the concentrations in the mixture.

4. The following data were obtained for the absorbance of methylene tetrahydrofolic acid at 400 nm as a function of pH at 25°C. Determine the pK_a for the group titrated by fitting the data. Compare with the results of Robinson and Jencks.*

pH	A_{400} nm
6.95	0.459
7.30	0.463
7.68	0.476
7.72	0.479
8.03	0.491
8.41	0.532
8.56	0.555
8.65	0.579
8.84	0.608
8.92	0.627
9.03	0.647
9.10	0.661
9.31	0.695
9.41	0.707

5. The specifications of a spectrophotometer state that the transmittance values are reproduced to +0.5% full scale. Find the percentage uncertainty in the concentration at transmittances of 37% and 10%.

6. The velocity of light v varies with the refractive index, n, of the medium. Refractive index is defined as the ratio

$$n = \frac{c}{v}$$

where c is the velocity of light in cm sec^{-1} in vacuum. Calculate the velocity and the wavelength of 5200 Å light in water ($n^{25°} = 1.333$).

7. Certain morphological changes in plants, such as the inception of flowering and dormancy, seem to be controlled by the light-sensitive pigment phytochrome, ϵ_m 76,000 at 664 nm. What fraction of the incident light at this wavelength would be absorbed by a 6.1×10^{-7} M solution in a 0.2-mm cell?

*D. R. Robinson and W. P. Jencks, *J. Amer. Chem. Soc.* **89**:7098 (1967).

8. The pK_a of an organic dye changes when it is bound to a macromolecule. This property is often used to assess alterations in protein conformation. The azo mercurial dye, 4-(*p*-hydroxybenzeneazo) phenylmercuriacetate forms a complex with the protein myosin.* The amount of dye bound per gram of protein can be determined from the absorption coefficient of the bound dye ($\epsilon = 2.73 \times 10^4$ at 434 nm) and the Lowry method for protein.
   (a) The absorbance of a solution of the complex in a 1-cm cell was 0.19. The same solution contained 7 mg of protein ml^{-1}. Calculate the moles of dye bound per 10^5 g of protein.
   (b) Calculate the pK_a of the dye in the free and bound states from the following data.

Free dye (in 0.5 *M* KCl)		Bound dye (in 0.5 *M* KCl)	
pH	A_{434} nm	pH	A_{434} nm
6.1	0.017	7.2	0.020
6.9	0.042	7.8	0.040
7.3	0.074	8.1	0.059
7.9	0.179	9.3	0.283
8.7	0.353	10.0	0.401
9.3	0.437	11.2	0.450

9. β-Lactoglobulin has a specific extinction, $E_{1\%}^{1\,cm}$, of 9.3 at pH 4.8. Calculate the range of the molar absorption coefficient if the molecular weight is 38,500 ± 10%.

10. Vitamin D_2 (calciferol) measured at 264 nm, its absorption maximum, in the solvent alcohol, follows Beer's law over a wide concentration range, with an ϵ of 18,200. What range of concentrations, expressed in w/v percent, can be used for analysis if it is desirable to keep the absorbance between 0.4 and 0.9?

11. The spectrum of the colored product formed in the orcinol reaction for glyceraldehyde, shows two maxima at 432 nm and 665 nm with the following molar absorption coefficients: $\epsilon_{432} = 5.8 \times 10^3\ M^{-1}\ cm^{-1}$, $\epsilon_{665} = 2.3 \times 10^3\ M^{-1}\ cm^{-1}$. What would be the absorbances at 432 and 665 nm of a solution containing 0.03 μmole ml^{-1}?

---

12. PROBLEMS IN CLINICAL BIOCHEMISTRY

1. A 24-hour specimen of urine (950 ml), obtained by catheter, was brought to the laboratory. A 1-ml aliquot, pH 5.4, was diluted to 200 ml with distilled water. A standard containing 0.005 mg ml^{-1} of creatinine was prepared. To each of three test tubes, containing 4 ml of diluted urine, creatinine standard, and water, respectively, was added 2 ml of alkaline picrate solution. After 10 min, the absorbance readings at 530 nm were as

*P. W. Mattocks, Jr., G. B. Keswani, and R. M. Dowben, *Biochemistry* **6**:3751 (1967).

follows: diluted urine 0.256, standard 0.485, and water 0. What was the total urinary creatinine excretion (mg day^{-1})?

2. In the Folin method for the determination of blood sugar, 0.1 ml of blood is added to 10 ml of tungstic acid in order to remove the proteins. A 4-ml aliquot of the protein-free filtrate that has been thus obtained is oxidized with ferricyanide. The ferrocyanide formed in this reaction is treated with Fe^{+++} to produce a colored (Prussian blue) solution. The final solution has a volume of 25 ml. The intensity of the color is compared with that produced when 4 ml of a standard glucose solution is treated with ferricyanide in a manner similar to the protein-free filtrate. A standard glucose solution (1 mg/100 ml) yields a color 1.05 times as intense as that produced by a blood filtrate. What was the concentration of sugar in the final solution? in the protein-free filtrate? in the whole blood?

3. The flame photometer is often used for the determination of serum potassium. The intensity of monochromatic light emitted when a solution of potassium is atomized in a flame is approximately proportional to the concentration of potassium. Four standards containing 2.5, 5.0, 7.5, and 10 meq $liter^{-1}$ of potassium gave respective intensities of 26, 54, 77, and 100 units of deflection. What is the concentration of potassium in solutions which give 22, 45, and 68 units of deflection?

4. A patient excreted 1130 ml of urine in 24 hours. Its pH was 6.2. A 2-ml aliquot of the urine was placed in a 100-ml volumetric flask, treated with 10 ml of molybdate reagent and 4 ml of aminonaphtholsulfonic acid reagent and diluted to the mark. A series of five color standards was similarly prepared in which 0, 1, 2, 3, and 4 ml of a phosphate standard (containing 0.4 mg of P in 5 ml) were used, respectively. These were also diluted to 100 ml after addition of the molybdate and aminonaphtholsulfonic acid reagents. The colors were measured in a colorimeter at 660 nm.

Standards (0.08 mg P per ml)	Absorbance
0 ml	0.000
1 ml	0.175
2 ml	0.350
3 ml	0.525
4 ml	0.700
Urine unknown	0.545

(a) How many grams of phosphorus were excreted per day?
(b) Calculate the phosphate concentration of the urine as millimoles per liter.
(c) Calculate the ratio of HPO_4^{--} to $H_2PO_4^-$ in this sample. The pK values of H_3PO_4 are 2.1, 6.7, and 12.3.

5. An aliquot of whole blood (0.2 ml) was treated with 0.5 ml of glycerol-urease solution in a stoppered bottle. After 15 min, 1.0 ml of saturated potassium carbonate solution was added and the liberated ammonia was

allowed to diffuse onto a filter paper roll impregnated with saturated boric acid solution. When the diffusion was complete, the filter paper was transferred to 10 ml of distilled water. Blanks and standards (containing 0.001 mmole NH_3 per ml) were also diluted to 10 ml with distilled water. Nessler's solution (1 ml) was added to each tube, and the yellow color was measured in the colorimeter at 480 nm.

	A
Blank	0.012
Standard (1 ml)	0.082
Standard (2 ml)	0.152
Standard (3 ml)	0.222
Unknown	0.186

What was the total urea plus ammonia nitrogen content of the blood sample, expressed as molarity of NH_3? as mg% of N? as mg% of urea?

6. A 2-ml portion of an oxalated blood specimen was mixed with 16 ml of $N/12$ H_2SO_4, then treated with 2 ml of 10% sodium tungstate solution and filtered. Three test tubes were numbered 1, 2, and 3. Into tube 1 was placed 4 ml of the above filtrate. Into tube 2 was placed 4 ml of distilled water. Into tube 3 was placed 4 ml of a solution containing 0.005 mg of creatinine/ml. To each of the test tubes was added 2 ml of a freshly prepared mixture of half-saturated picric acid in 5% sodium hydroxide. After the solutions were mixed and allowed to stand for 15 min, the intensity of the yellow-orange color of each was measured in a colorimeter at 530 nm. The room temperature was 27°C and the barometric pressure was 744 mm. The colorimeter readings were as follows.

Tube No.	A
1	0.087
2	0.000
3	0.106

From the information given, calculate the concentration of creatinine in the blood specimen in milligram percent.

13

Radiochemistry

Analytical procedures for chemical compounds, reaction pathways, or the visualization of cellular components, require ideally that the system under study be disturbed as little as possible. An alternate to the spectrophotometric methods described in the previous chapter is the use of isotopic labeling, in particular with those isotopes that are radioactive.

RADIOACTIVITY

A number of isotopes spontaneously and continuously emit a characteristic radiation. These are called radionuclides. Each emits at a rate that is governed by the first-order rate equation

$$N = N_0 e^{-\lambda t} \tag{13.1}$$

where N_0 is the number of atoms present at a reference time taken as zero, N is the number of atoms after time t, and λ is the disintegration constant, often expressed in the units per second. The negative exponent is consistent with the fact that the number of the atoms of the isotope in question is decreasing with time. Equation (13.1) may be expressed in logarithmic form,

$$\lambda = -\frac{1}{t} \ln \frac{N}{N_0} \tag{13.2}$$

or, to base 10,

$$\lambda = -\frac{2.303}{t} \log \frac{N}{N_0} \tag{13.3}$$

Since it is more usual to measure the radioactivity in terms of the disintegrations and not to count the ratio of atoms, N/N_0, (13.3) can be restated as

$$\lambda = -\frac{2.303}{t} \log \frac{A}{A_0} \tag{13.4}$$

where A_0 and A are the absolute activities expressed as disintegrations per second (dps). The unit of absolute activity is a curie, C_i, which is now defined as 3.700×10^{10} dps for any radionuclide.

Problem 13.1: A sample of ^{32}P disintegrates at a rate of 30,120 dpm. What is the activity in microcuries?

Solution:

$$30{,}120 \text{ dpm} = \frac{30{,}120}{60} \text{ dps}$$

$$= 502 \text{ dps}$$

$$1 \text{ microcurie} = \frac{3.70 \times 10^{10}}{10^6} \text{ dps}$$

$$= 3.70 \times 10^4 \text{ dps}$$

Therefore the activity of the ^{32}P sample is

$$\frac{502}{3.70 \times 10^4} \mu C_i = 1.36 \times 10^{-2} \mu C_i$$

Depending upon the method of counting and the instrument used, the counting rate, R, will be less than the absolute disintegration rate by a factor which is the counting yield, Y. This takes into account those disintegrations that are not recorded in each measurement because of poor geometry, scattering, self-absorption in the sample, counter losses, and so on:

$$R = YA \tag{13.5}$$

It follows from (13.5) that

$$\frac{R_0}{R} = \frac{A_0}{A}$$

and Equation (13.4) becomes

$$\lambda = -\frac{2.303}{t} \log \frac{R}{R_0} \tag{13.6}$$

Problem 13.2: A primary standard radium disk with 500 dps recorded 3000 cpm above the background count. What is the counting efficiency, Y?

Solution:

$$3000 \text{ cpm} = 50 \text{ cps}$$

$$Y = \frac{50}{500} \times 100\% = 10\%$$

HALF-LIFE, MEAN LIFE, AND EFFECTIVE HALF-LIFE

The activity of each radionuclide is characterized by a constant called the half-life, which is the time for the disintegration rate to decrease by 50%.

At the half-life, T, $A = A_0/2$ or

$$\frac{A}{A_0} = \frac{\frac{A_0}{2}}{A_0} = \frac{1}{2} \tag{13.7}$$

and from (13.4)

$$2.303 \log \tfrac{1}{2} = -\lambda T$$

$$\lambda = \frac{0.693}{T} \tag{13.8}$$

The average life of a radionuclide atom is called the mean life ($\bar{T}$) and is the reciprocal of the disintegration constant:

$$\bar{T} = \frac{1}{\lambda}$$

From Equation (13.8)

$$\bar{T} = \frac{T}{0.693} = 1.443\ T \tag{13.9}$$

Problem 13.3: The half-life of ^{64}Cu is 12.8 hours; calculate λ and $\bar{T}$.

Solution:

$$\lambda = \frac{0.693}{12.8 \times 3600} \text{ sec}^{-1} = 1.5 \times 10^{-5} \text{ sec}^{-1}$$

$$\bar{T} = 1.443 \times 12.8 \text{ hours} = 18.5 \text{ hours}$$

Problem 13.4: ^{42}K has a half-life of 12.4 hours. What fraction of the initial activity remains after 4 hours, 12.4 hours, and 1 week?

Solution: If the initial activity is represented by A_0, then at any later time, A, the relationship is given by Equations (13.4) and (13.8):

$$-\frac{2.303}{t}\log\frac{A}{A_0} = \lambda = \frac{0.693}{T}$$

$$= \frac{0.693}{12.4}\text{ hour}^{-1}$$

$$\log\frac{A}{A_0} = -\frac{0.693}{12.4 \times 2.303}t$$

$$= -(2.425 \times 10^{-2})t$$

After 4 hours,

$$\log\frac{A}{A_0} = -2.425 \times 10^{-2} \times 4$$

$$= -9.7 \times 10^{-2} = -(0.097) = \bar{1}.903$$

$$\frac{A}{A_0} = \text{antilog } \bar{1}.903$$

$$= 0.80$$

After 12.4 hours,

$$\frac{A}{A_0} = 0.50$$

since 12.4 hours is the half-life. After 1 week or 7 × 24 hours,

$$\frac{A}{A_0} = \text{antilog } (-2.425 \times 10^{-2} \times 7 \times 24)$$

$$= \text{antilog } (-4.07) = \text{antilog } \bar{5}.93$$

$$= 8.51 \times 10^{-5}$$

Problem 13.5: What is the half-life of a radionuclide in a sample with 5000 cpm and then 3500 cpm 5 hours later?

Solution: From Equation (13.6),

$$\lambda = -\frac{2.303}{5}\log\frac{3500}{5000}\text{ hour}^{-1}$$

$$= -0.4606 \log 0.700 = -0.4606 \times \bar{1}.8451$$

$$= -0.4606 \times (-0.1549) = +0.0714 \text{ hour}^{-1}$$

$$T = \frac{0.693}{0.0714}\text{ hour} = 9.7 \text{ hours}$$

The amount of a radionuclide required to provide a level of activity is given by

$$N = \frac{\text{number of atoms}}{\text{curie}} = \frac{3.700 \times 10^{10}}{\lambda}$$

Since

$$N = \frac{\text{Avogadro's number} \times \text{grams/curie}}{\text{atomic weight}}$$

then the grams of a radionuclide giving a curie of activity is

$$\frac{3.700 \times 10^{10}}{\lambda} \times \frac{\text{at. wt.}}{6.02 \times 10^{23}}$$

Problem 13.6: How many atoms of ^{59}Fe, half-life 46 days, would be required to give 1 μC_i of activity? What is the amount in grams?

Solution:

$$\text{Half-life} = 46 \text{ days} = 46 \times 24 \times 3600 \text{ sec} = 3.98 \times 10^6 \text{ sec}$$

$$1\ \mu C_i = 3.70 \times 10^4 \text{ dps} = \frac{0.693}{3.98 \times 10^6} N$$

where N = number of atoms of ^{59}Fe,

$$N = \frac{3.70 \times 3.98 \times 10^{10}}{0.693}$$

$$= 2.12 \times 10^{11} \text{ atoms}$$

or

$$N = \frac{2.12 \times 10^{11}}{6.02 \times 10^{23}} \times 59 \text{ g}$$

$$= 2.08 \times 10^{-11} \text{ g}$$

The radionuclide in a biochemical compound is subject to elimination or turnover from the organism or cell according to the rate reflected in the biological half-life of that compound. Assuming this rate to follow first-order kinetics, then the biological rate constant, λ_b, is related to the half-life, T_b, by

$$T_b = \frac{0.693}{\lambda_b} \tag{13.10}$$

The overall or effective half-life of the radionuclide, T_{eff}, is

$$\frac{1}{T_{\text{eff}}} = \frac{1}{T_a} + \frac{1}{T_b} \tag{13.11}$$

where T_a is the physical half-life of decay of the radionuclide. If the time span of the experiment is short compared to T_a or T_b, then this correction is not significant.

Problem 13.7: If the biological half-life of Fe in erythrocytes is 60 days, what is the effective half-life of ^{59}Fe in this application?

Solution: T_a for ^{59}Fe is 46 days. Then

$$\frac{1}{T_{\text{eff}}} = \frac{1}{46} + \frac{1}{60}$$

$$T_{\text{eff}} = \frac{46 \times 60}{60 + 46} \text{ days} = 26 \text{ days}$$

Problem 13.8: ^{3}H has a half-life of 12.3 years. The biological half-life (T_b) of H_2O (as water) in humans is 12 days. What is the effective half-life (T_{eff}) of $^{3}H_2O$?

Solution:

$$T_a = 12.3 \times 365.25 \text{ days}$$

$$T_{\text{eff}} = \frac{12 \times (12.3 \times 365.25)}{12 + (12.3 \times 365.25)} \text{ days} = 12 \text{ days}$$

SOME ERRORS IN RADIOTRACER ASSAYS

Radiotracer assays must be done with a realization that several inherent experimental errors exist. These are summarized.

Background Radiation. With no apparent source of radiation in the laboratory, the detector will measure a background count due to cosmic radiations, radioactive contaminants in the phototube, contamination from prior spills in the laboratory, lead shielding, and the like. The background will vary with the instrument and the laboratory. Thus a background count must be made and subtracted from each experimental reading.

Random Decay. The rate of decay of a radionuclide at any instant is variable. During a small number of counts this variation is large so that by taking more counts the error is reduced. The desired accuracy in any experiment can be obtained within certain limits by calculating the uncertainty due to randomness from the statistical considerations discussed in Chapter 4. It will be recalled that the counts N, will for 68% of the time lie within the range $N \pm \sigma$ where, for all practical purposes (since N is usually large), $\sigma = \pm\sqrt{N}$. From the t-tables, any other confidence level can be calculated, but, because of the square root relationship, errors below $\pm 0.1\%$ are rarely obtained (counts of over a million would be required).

Dead Time. After every count is recorded there is a certain time (a fraction of a microsecond to over a hundred microseconds) before the instrumentation can record the next count. Dead time, as this is called, is most important for very active samples. For example, if an instrument with a 100-microsecond dead time recorded 100,000 counts in one minute, then the total dead time was $(100{,}000 \times 100 \times 10^{-6})$, or 10 sec min^{-1}. The total counts that would have been detected using a zero dead-time instrument are

$$100{,}000 \times \left(\frac{60}{60 - 10}\right) = 120{,}000 \text{ cpm}$$

Coincidence Losses. When more than one disintegration occurs at the same instant the instrument records only one count and thus loses the coincident events. This coincidence loss is particularly relevant at high counting rates. The error is sometimes included under dead time.

Quenching. In liquid scintillation counting, the presence of inert material or color may reduce the recorded scintillation. The problem can be solved in several ways, such as by adding a small amount of a standard solution of the isotope being counted (internal standard), but details of these procedures are outside the scope of this book.

Problem 13.9: Several fractions, some of them colored, were obtained from the ion exchange chromatography of the hydrolysate of a ^{14}C-labeled protein. Each was treated identically in its preparation for liquid scintillation counting, which proceeded as indicated below. To check for quenching, 50 μl of ^{14}C-L-alanine solution was added to each vial and the counting repeated.

	Fraction alone (cpm)	Fraction and internal standard (cpm)
Blank	25	305
Fraction *a*	50	335
Fraction *b*	1030	1220
Fraction *c*	1180	1360
Fraction *d*	705	955

What are the relative activities of the fractions?

Solution: Correcting the above values for background gives

	Fraction alone (cpm)	Fraction and internal standard (cpm)
Fraction *a*	25	310
Fraction *b*	1005	1995
Fraction *c*	1155	1335
Fraction *d*	680	930

The internal standard should have added $(305 - 25) = 280$ cpm to each fraction. Actually the added activity was

Fraction *a*	$335 - 50 = 285$
Fraction *b*	$1220 - 1030 = 190$
Fraction *c*	$1360 - 1180 = 180$
Fraction *d*	$955 - 705 = 250$

It would not be unrealistic for fraction *a* to be a little higher than expected, especially if the counting times were short. Apparently there is no quench in fraction *a*. Fractions *b*, *c*, and *d* were "quenched" to differing degrees and should be proportionately corrected:

Fraction *b* $\quad 1005 \times \dfrac{280}{190} = 1480$ cpm

Fraction *c* $\quad 1155 \times \dfrac{280}{180} = 1797$ cpm

Fraction *d* $\quad 680 \times \dfrac{280}{250} = 761$ cpm

DECAY SCHEMES

Each radionuclide undergoes some nuclear change as a result of the decay processes, which may be simple, as in the case of ^{32}P, ^{60}Co and ^{45}Ca:

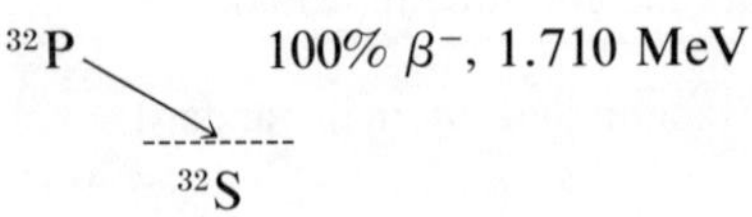

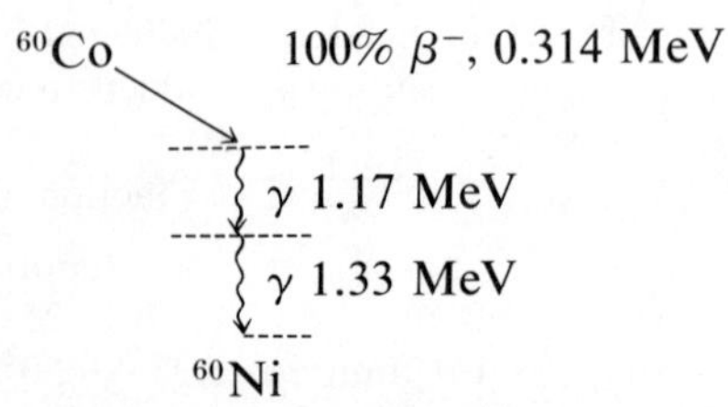

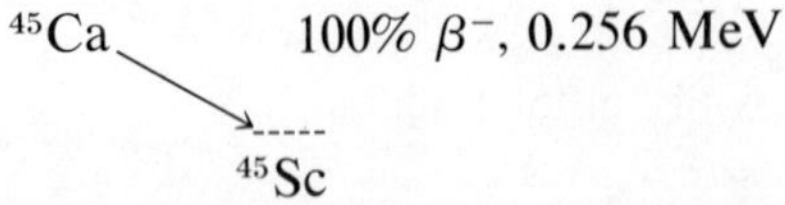

or more complex, such as for ^{64}Cu where branches are present,

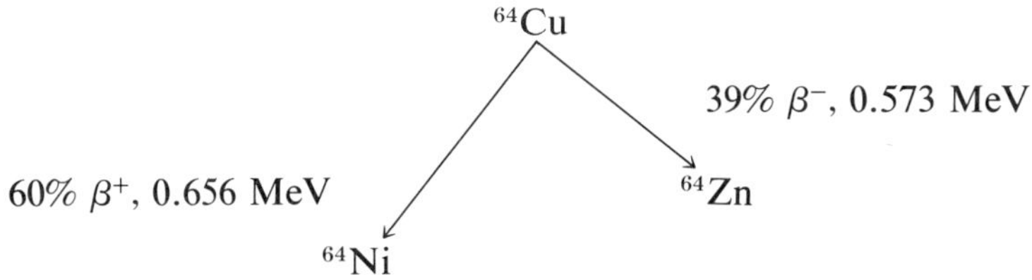

Some of the decay products (daughter isotopes) are radioactive. The radioactive daughter, now as a parent atom, produces a new daughter. The series of decays, each with its decay constant, may be represented as

$$A \xrightarrow{\lambda_1} B \xrightarrow{\lambda_2} C$$

Each decay rate is independently governed by the relationship in Equation (13.1) as would be any mixture of radionuclides, except of course in a decay series, where the daughter atoms are being replenished by the decay of the respective parents. For the simplest case given above, the number of atoms of B at any time t is

$$N_B = \frac{\lambda_1}{\lambda_2 - \lambda_1} N_A^0 (e^{-\lambda_1 t} - e^{-\lambda_2 t}) + N_B^0 e^{-\lambda_2 t} \tag{13.12}$$

Equation (13.12) could well be written in terms of activities as in the case of Equation (13.4). Different kinds of radioactive equilibria are established depending upon the relative values of the component decay constants. The reader is referred to more advanced texts for such considerations.*

In a mixture of unrelated radionuclides the total activity, A_{tot}, is the sum of the individual activities:

$$A_{tot} = A_A^0 e^{-\lambda_A t} + A_B^0 e^{-\lambda_B t} + \cdots \tag{13.13}$$

Problem 13.10: The danger of ^{90}Sr in radioactive fallout may be due more to the daughter ^{90}Y, as indicated by the decay scheme

^{90}Sr	β^-, 0.544 MeV,	T 28.0 years
^{90}Y	β^-, 2.23 MeV,	T 64 hours
^{90}Zr stable		

What are the relative proportions of ^{90}Sr and ^{90}Y?

Solution: For ^{90}Sr,

$$\lambda_{Sr} = \frac{0.693}{28 \times 365.25 \times 24} \text{ hour}^{-1}$$

---

*See R. T. Overman and H. M. Clark, *Radioisotope Techniques*. McGraw-Hill, New York, 1960.

For ^{90}Y,

$$\lambda_Y = \frac{0.693}{64} \text{ hour}^{-1}$$

Since λ_{Sr} is much smaller than λ_Y, Equation (13.12) reduces to

$$N_Y = \frac{\lambda_{Sr}}{\lambda_Y} N_{Sr}^0 e^{-\lambda_{Sr} t} \tag{i}$$

The ^{90}Y is thus essentially decaying as fast as it is being formed, giving a type of radiochemical equilibrium in which the dps is twice the number of ^{90}Sr disintegrations.

Since from Equation (13.1)

$$N_{Sr} = N_{Sr}^0 e^{-\lambda_{Sr} t}$$

then from (i)

$$N_Y = \frac{\lambda_{Sr}}{\lambda_Y} N_{Sr}$$

$$\frac{N_Y}{N_{Sr}} = \frac{\lambda_{Sr}}{\lambda_Y} = \frac{T_Y}{T_{Sr}} = \frac{64}{28 \times 365.25 \times 24} = 2.61 \times 10^{-4}$$

Each radionuclide produces one or more kinds of radiation according to its decay scheme. The forms of radioactivity, which are commonly encountered in counting and analysis, are α- and β-particles and γ- or X-radiations. Each radiation has a characteristic energy. All the α-particles, ^4_2He, from a given isotope have the same energy, between 5 and 9 MeV, in contrast to β-particles, which have a continuous spectrum from zero to a characteristic kinetic energy maximum, E_{max}, usually 0 – 4 MeV. The γ- and X-radiations are of discrete energies and up to 3 MeV. Modern instrumentation permits the analysis of such spectra in much the same way as for ultraviolet or visible spectra. For example, the scintillation spectrometer may be adjusted to count only radiation above a certain energy, or the energy window ("slit width") may be set to count over a selected energy range. This ability of the spectrometer to resolve radiations of differing energy allows two isotopes to be counted in the same sample if the two spectra are not seriously overlapping. For example, the monochromatic γ-peak at 0.320 MeV of ^{51}Cr can be separated from the spectrum of ^{59}Fe, which has two peaks at 1.100 and 1.290 MeV.

ISOTOPIC DILUTION

By the use of radioactive tracer techniques it is possible to analyze very small amounts of compounds. The method is termed isotope dilution and can involve either the determination of an inactive compound by dilution with the radioactive compound or vice versa. Both methods are based on the same principle.

If a weight, W_a, of radioactive compound with a known specific activity, A_{sp} (the specific activity is the activity per unit weight or mole of a substance), is mixed with a weight, W_i, of the inert compound, then the specific activity, A'_{sp}, of the reisolated compound is

$$A'_{sp} = A_{sp} \frac{W_a}{W_a + W_i} \tag{13.14}$$

From this relationship either W_a or W_i can be calculated if the other weight is known. It will be realized that by either procedure it is necessary to isolate radiochemically pure diluted material and to weigh a sample for the determination of its specific activity.

Problem 13.11: The amount of sucrose in an actively photosynthesizing canna leaf was determined by extracting the leaf with boiling 80% ethanol. To the extract was added 5 mg of ^{14}C-labeled sucrose, specific activity 80 μC_i/mg. Sucrose was obtained chromatographically pure from a portion of the extract, and a 3-mg sample showed an activity of 16 μC_i. How much sucrose was extracted from the leaf?

Solution: From Equation (13.14),

$$A'_{sp} = \frac{16}{3}\ \mu C_i/\text{mg}$$

$$A_{sp} = 80\ \mu C_i/\text{mg}$$

$$W_a = 5\ \text{mg}$$

$$\frac{16}{3} = \frac{80 \times 5}{5 + W_i}$$

Rearranging, we have

$$80 + 16W_i = 1200$$

$$W_i = \frac{1200 - 80}{16}$$

$$= 70\ \text{mg}$$

Problem 13.12: A protein which contained 10% (w/w) glycine was labeled with ^{3}H by the Wilzbach method. The product was equilibrated several times with H_2O to remove all exchangeable ^{3}H. A 0.50-mg sample, specific activity 0.1 μC_i/mg, was hydrolyzed in the presence of 1.00 mg of unlabeled glycine. The glycine, isolated quantitatively from the hydrolyzate, had an activity of 0.1 μC_i/mg. What was the specific activity of the glycine in the protein?

Solution: From Equation (13.14),

$$A'_{\text{sp}} = 0.1 \text{ nC}_i/\text{mg}$$

$$W_i = 1.00 \text{ mg}$$

$$W_a = 10\% \text{ of } 0.50 \text{ mg} = 0.05 \text{ mg}$$

$$0.1 \text{ n}C_i = \frac{0.05 \times A_{\text{sp}}}{1.00 + 0.05}$$

$$A_{\text{sp}} = \frac{1.05 \times 0.1 \text{ nC}_i}{0.05 \text{ mg}} = 2.1 \text{ nC}_i/\text{mg}$$

REFERENCES

G. D. Chase and J. L. Rabinowitz, *Principles of Radioisotope Methodology.* Burgess, Minneapolis, 1969.

M. D. Kamen, *Isotopic Tracers in Biology.* Academic Press, New York, 1957.

E. R. King and T. G. Mitchell, *A Manual for Nuclear Medicine.* Thomas, Springfield, Ill., 1961.

D. A. Lambie, *Techniques for Use of Radioactive Isotopes in Analysis.* Van Nostrand, Princeton, N.J., 1964.

R. T. Overman and H. M. Clark, *Radioisotope Techniques.* McGraw-Hill, New York, 1960.

G. K. Schweitzer and I. B. Whitney, *Radioactive Tracer Techniques.* Van Nostrand, Princeton, N.J., 1949.

D. Steinberg and S. Udenfriend, "The Measurement of Radioisotopes," in S. P. Colowick and N. O. Kaplan (eds.), *Methods in Enzymology,* Vol. IV. Academic Press, New York, 1957.

G. Wolf, *Isotopes in Biology.* Academic Press, New York, 1964.

---

13. PROBLEMS A

Answers on page 361

1. 12 μC_i of ^3H_2O was injected into the tail vein of a 1.6-kg rabbit. Assuming 65% of the rabbit is water, what would be the specific activity of the deproteinized serum if the ^3H were evenly distributed?

2. ^{35}S has a half-life of 87.2 days. What is the loss of activity of $^{35}SO_4$ after 1 year?

3. What is the effective half-life of ^{24}Na (T_a 15 hours), given as NaCl, in man, where the biologic half-life, T_b, is 11 days?

4. Histamine was extracted from 0.1 g of tissue. The extract was treated with ^{131}I-labeled *p*-iodobenzene sulfonyl chloride (pipsyl chloride), which had a specific activity of 10^6 cpm per μmole. After removing excess reagent, the reaction mixture was diluted with 4.0 mmole of unlabeled

pipsylhistamine. The derivative was recrystallized to constant activity, which was 200 cpm for a 0.4-mmole sample. How much histamine was extracted from the tissue?

5. Calculate the percentage of a radionuclide, half-life T, remaining after time t for the ratios t/T 0.10, 0.50, 0.80, 1.00, 2.00, 3.00, and 10.00.

6. A solution of $^{24}NaCl$ was 1.0% (w/v). 1 ml of this solution contained 0.20 μC_i of activity and when plated out gave 47,000 cpm above background. What is the specific activity of the $^{24}NaCl$ in μC_i/mg? Calculate the counting efficiency.

7. The extracellular fluid space in a 70-kg man was determined by injecting 300 μC_i $^{35}SO_4$ in 6 ml of sterile saline. Twenty minutes later 10 ml of blood was withdrawn and the specific activity of the serum was found to be 27 nC_i/ml. If we ignore various correction factors for losses into the urine and some diffusion into the cells, what is the fluid space expressed as a percentage of the total body weight?

8. ^{32}P has a half-life of 14.3 days. What fraction of the original activity of a sample of glucose 1-phosphate (^{32}P-labeled) remains after 3 weeks?

9. The effective half-life (T_{eff}) of ^{35}S in the mucopolysaccharides of skin is 82.5 days. The half-life (T_a) of ^{35}S is 87.2 days. Calculate its biologic half-life.

10. A mixture of $Ba^{14}CO_3$ and $Ba^{35}SO_4$ in a planchet had a count above background of 10,500 cpm. Concentrated HCl was added until all effervescence had just ceased, and, after drying, the planchet then gave 2,500 cpm. The planchet was stored for 3 months and recounted. Assuming 8% counting efficiency in all cases, what were the activities in μC_i of the $Ba^{14}CO_3$ and $Ba^{35}SO_4$ originally and what was the final count after 3 months? Half-life of ^{14}C and ^{35}S: 5568 years and 87.2 days, respectively.

11. Six counts of $Ba^{14}CO_3$ were made successively, followed by three background counts. An accumulation of 5000 counts each time from the sample required 18.16, 17.82, 18.12, 18.00, 18.51, and 17.70 min. Background counts of 1000 were registered in 8.10, 9.00, and 8.66 min. What is the average count and its reliability?

12. Ovalbumin was iodinated, with the use of lactoperoxidase and $^{131}I_2$, which had a specific activity of 0.5 $\mu C_i/\mu$mole. The ^{131}I-labeled ovalbumin was isolated from the reaction mixture by gel chromatography and had a specific activity of 0.45 μC_i/g. What was the extent of iodination, expressed as moles I_2 per mole ovalbumin, molecular weight 45000? How long will the sample remain above 1000 cpm/g, assuming 29% counting efficiency? $^{131}I_2$ has a half-life of 8.1 days.

13. Libby has shown that the biological specific activity of carbon is 15.3 dpm/g, half-life 5568 ± 30 years, and therefore the specific activity of a sample of carbon t years old is given by

$$A_{^{14}C} = 15.3\, e^{-0.693t/5568}$$

The CO_2 of some crustacean shells was liberated by acidification with HCl, purified, and reduced to carbon. This carbon had a specific activity of 6.12 dpm/g. What is the probable age of the shells?

14. A radioactive solution was found in the laboratory, an aliquot of which gave 20,000 cpm when tested. The next day, 20 hours later, the radioactivity of the sample was 19,910 cpm. From a calculation of the half-life, what might the radionuclide be?

15. A 1 mg-sample of ^{3}H-labeled lactic acid (510 cpm per μmole) was equiliberated with H_2O and then oxidized to pyruvate. Unlabeled pyruvic acid (5.0 mg) was added to the reaction mixture and the pyruvate was isolated chromatographically. 1 mg of the isolated mixture had an activity of 710 cpm above background. What was the distribution of the ^{3}H label in the lactate?

16. A 25-g mouse was injected with 0.4 μC_i of ^{3}H-thymidine per gram body weight. Two hours after injection blood and tissue samples were obtained and the following data collected.

Tissue	Sample weight	cpm
Intestine	13.9 mg	21,400
Spleen	26.8 mg	35,500
Skin	49.8 mg	4,730
Blood	100.0 mg	4,922

Counting efficiency was 35%. Sp. Act. of ^{3}H-thymidine was 5000 mC_i/mmole. Calculate the moles of ^{3}H-thymidine uptake per gram of tissue.

---

13. PROBLEMS B No answers given

1. A slice of normal liver tissue was incubated in Ringer's solution containing ^{14}C-glucose (specific activity 0.55 $\mu C_i/\mu$mole). After one hour, the glycogen was extracted into boiling water, cold glycogen (10 mg) was added and precipitated from solution with alcohol. After further purification the glycogen was dissolved in 5 ml water. An aliquot (0.1 ml) was plated out and showed 15045 counts in 5 min. The background count of a similar amount of cold glycogen was 900 counts in 30 min. Assuming that the plated sample is infinitely thin and the counting efficiency is 15%, calculate the amount of glucose incorporated into glycogen.

2. A sample of sodium borotritiide was received, labeled "10^{-2}% enriched ^{3}H, total weight 3.000 g." Assuming all radiochemical reactions to yield 50% of theoretical, how much NaB^3H_4 would be required to synthesize 10 mC_i of lactate-2-^{3}H from pyruvate? Remember that any ^{3}H linked to oxygen will rapidly exchange with ^{1}H in water. Use 12.3 years as the half-life of ^{3}H.

3. Trans-dichlorobisethylene diamine cobalt III chloride (0.100 g), dissolved in 10 ml of 0.03 M Na_2CO_3, is diluted to 25 ml with water and a 20-ml

aliquot is transferred to a reaction vessel. At a constant temperature of 51.2°C, 1 ml of $Na_2^{14}CO_3$ (0.5 μC_i) is added. Periodically thereafter a 1-ml sample of the mixture is withdrawn and excess $BaCl_2$ added immediately. The $BaCO_3$ so formed is carefully freed from contaminating radioactivity and plated out. The dried sample is weighed and counted. The results are as follows.

Time (min)	0	22	51	82	107
Wt. sample (mg)	1.42	1.19	1.01	1.10	1.06
Cpm per mg (above background)	1042	931	808	724	606

From the reaction

$$\text{Co en}_2\,CO_3^+ + {}^{14}CO_3^{--} \rightleftharpoons \text{Co en}_2^{14}CO_3^+ + CO_3^{--}$$

calculate the rate of exchange.*

4. Pernicious anemia patients, even in remission, do not absorb vitamin B_{12} in the absence of intrinsic factor. Absorbed B_{12} is stored normally in the liver and undergoes a slow renal excretion. If $^{58}CoB_{12}$ is given orally and about 2 hours later a large saturating dose of unlabeled B_{12} is injected, it will increase the rate of renal excretion and significant amounts of $^{58}CoB_{12}$ will appear in the urine. Normally a 24-hour urine will contain 13–15% of the $^{58}CoB_{12}$ given initially. The above procedure is known as the Schilling test.

   In a patient suspected of pernicious anemia, the Schilling test proceeded as follows.

   (i) 0.5 μc of $^{58}CoB_{12}$ in a test capsule was given orally to the fasting patient.

   (ii) 2 hours later 1 mg of crystalline B_{12} was given and the bladder was emptied.

   (iii) 1200 ml of urine was collected over the next 24 hours.

   (iv) A 3-ml aliquot of this urine gave 83 cpm.

   (v) Counting 3 ml of a standard, containing 0.4 nC_i/ml, gave 910 cpm.

   (vi) The test was repeated 7 days later; 30 mg of intrinsic factor was given orally with the 0.5 μC_i dose of $^{58}CoB_{12}$.

   (vii) The 3-ml urine sample corresponding to (iv) above was 120 cpm compared to the standard in (v) of 900 cpm.

   (a) Calculate the percentage of ^{58}Co excreted with and without the intrinsic factor, assuming no residual excretion of ^{58}Co from the first dose into the second test.

   (b) The half-life of ^{58}Co is 72 days. What would be the activity of $^{58}CoB_{12}$ in a "0.5 μC_i-capsule" after 7 days?

   (c) If the biological half-life of B_{12} is 180 days, what is the effective half-life of $^{58}CoB_{12}$?

   (d) If the normal rate of renal excretion of B_{12} is 0.1% per day after the first 24 hours from the large saturating dose, what fraction of the $^{58}CoB_{12}$ counted in step (vii) would be due to the original dose given in step (i)? Take into account the decay of the ^{58}Co.

*See D. Barton and K. Winter, *J. Chem. Ed.* **43**:93 (1966).

(e) ^{58}Co decays to ^{59}Fe (half-life 46 days.) If all the ^{59}Fe so produced was incorporated into hemoglobin, what percentage of the total iron would this radionuclide represent at step (iii)? The total iron in the patient was 2.5 g. Ignore the decay of ^{59}Fe.

5. A patient, with a uric acid pool of 1253 mg, which was replaced at a rate of 942 mg/day, was injected with 100 mg ^{15}N-uric acid-U-^{14}C, 0.1 μC_i/mg. A 24-hour urine collection of 1200 ml gave uric acid from 500 ml of the urine with a specific activity of 10 nC_i/mg. To a second 500-ml portion of the urine was added 100 mg of unlabeled uric acid, and a crystalline sample of the isolated material had a specific activity of 7.8 nC_i/mg. What was the total amount of uric acid in the 24-hour urine specimen? What was the percent recovery of the radioactive label?*

6. Castor bean seeds, germinated for 5 days, were sliced. 1 g of the slices was incubated in a Warburg vessel with 2 ml of phosphate buffer, pH 5.0, containing 5.0 μC_i sodium acetate-^{14}C (7.4 mC_i/mmole) for 1 hour. The mixture was then boiled with 80% ethanol and extracted several times in this way. Sucrose, among other products, was isolated by chromatography and found by colorimetric analysis to amount to 7.0 μmole. The specific activity was 1584 cpm/μmole. The sucrose was hydrolyzed, the glucose and fructose separated, and the distribution of ^{14}C in the glucose residue determined.

Carbon	%
1	21.9
2	23.8
3	6.0
4	8.8
5	20.8
6	18.8

Assuming all counting to be corrected for background and self-absorption in the continuous gas flow counter, counting efficiency 5%, and that the glucose and fructose were equally labeled, calculate the activity in each carbon of glucose, expressed in $\mu C_i/\mu$mole.

7. 20 ml of blood is withdrawn from an animal, weight 3.2 kg. To 5 ml of blood are added 100 μC_i ^{32}P and after incubation for about 60 min the red cells are isolated, washed three times, and reconstituted to the original 5 ml. A portion of the active cells are diluted serially with inert whole blood to give a 1:256 dilution. 1 ml of the active undiluted cells is injected into the animal and after 20 min a sample of blood is withdrawn and a 1-ml portion is pipetted into a planchet. A similar 1-ml portion of the 1:256 diluted tagged cells, as well as the inert blood, is prepared for counting, which is continued for a time interval corresponding to several thousand counts. The counts per minute for each sample are

*For discussion, see J. Buzard et al., *J. Biol. Chem.* **196**:179 (1952).

Diluted tagged cells	955.4 cpm
Experimental sample	1417.0 cpm
Background (from inert blood sample)	24.2 cpm
Hemacrit	42.4%

Calculate the whole blood volume, the red cell volume, and the plasma volume per kilogram animal weight.

8. The reversible binding of 2′-cytidylic acid to ribonuclease can be studied by dialysis equilibrium using ^{14}C-labeled nucleotide. From the data given below calculate the moles of 2′-cytidylic acid bound per mole of ribonuclease as a function of pH.

cpm at various pH values (control)

pH	5.22	5.80	6.22	6.60	7.00	7.42
	1628	1728	1799	1732	1740	1730
	1646	1663	1720	1737	1741	1750
	1650	1724	1726	1720	1755	1757

cpm at equilibrium at various pH values

pH	5.22	5.80	6.22	6.60	7.00	7.42
	1020	1094	1162	1248	1432	1618
	1066	1092	1145	1252	1421	1622
	1064	1100	1171	1262	1422	1614
	1025	1114	1158	1254	1414	1655

Each experiment is the average of four samples.

The volume of the solution bathing each dialysis bag was 2 ml, and the volume of the solution inside, which contained the ribonuclease at a concentration of 1.68×10^{-4} *M*, was 1 ml. A standard aliquot (0.2 ml) from the solution bathing the dialysis bag was used for each counting. The specific activity of the cytidylic acid was 3624 cpm per standard aliquot for a concentration of 3.25×10^{-4} *M*. The control experiments excluded the protein.

Compare the results with the literature values.*

9. The following data were obtained from a double label experiment. Sample 1 contains only ^{3}H, sample 2 contains only ^{14}C and sample 3 contains both isotopes. All counts are *net cpm*

Sample	Window	
	^{3}H	^{14}C above ^{3}H
1	26,700	0
2	3,148	4,523
3	10,150	6,981

Sample 1 was a standard containing 6.50×10^{4} dpm ^{3}H. Sample 2 was a standard containing 1.12×10^{4} cpm ^{14}C. Determine the ^{3}H and ^{14}C in disintegrations per minute in sample 3.

*See C. A. Nelson, J. P. Hummel, C. A. Swenson, and L. Friedman, *J. Biol. Chem.* **237**:1575 (1962).

APPENDIXES

APPENDIX 1 TABLE OF FOUR-PLACE LOGARITHMS

N	0	1	2	3	4	5	6	7	8	9	Proportional parts 1	2	3	4	5	6	7	8	9
10	0000	0043	0086	0128	0170	0212	0253	0294	0334	0374	4	8	12	17	21	25	29	33	37
11	0414	0453	0492	0531	0569	0607	0645	0682	0719	0755	4	8	11	15	19	23	26	30	34
12	0792	0828	0864	0899	0934	0969	1004	1038	1072	1106	3	7	10	14	17	21	24	28	31
13	1139	1173	1206	1239	1271	1303	1335	1367	1399	1430	3	6	10	13	16	19	23	26	29
14	1461	1492	1523	1553	1584	1614	1644	1673	1703	1732	3	6	9	12	15	18	21	24	27
15	1761	1790	1818	1847	1875	1903	1931	1959	1987	2014	3	6	8	11	14	17	20	22	25
16	2041	2068	2095	2122	2148	2175	2201	2227	2253	2279	3	5	8	11	13	16	18	21	24
17	2304	2330	2355	2380	2405	2430	2455	2480	2504	2529	2	5	7	10	12	15	17	20	22
18	2533	2577	2601	2625	2648	2672	2695	2718	2742	2765	2	5	7	9	12	14	16	19	21
19	2788	2810	2833	2856	2878	2900	2923	2945	2967	2989	2	4	7	9	11	13	16	18	20
20	3010	3032	3054	3075	3096	3118	3139	3160	3181	3201	2	4	6	8	11	13	15	17	19
21	3222	3243	3263	3284	3304	3324	3345	3365	3385	3404	2	4	6	8	10	12	14	16	18
22	3424	3444	3464	3483	3502	3522	3541	3560	3579	3598	2	4	6	8	10	12	14	15	17
23	3617	3636	3655	3674	3692	3711	3729	3747	3766	3784	2	4	6	7	9	11	13	15	17
24	3802	3820	3838	3856	3874	3892	3909	3927	3945	3962	2	4	5	7	9	11	12	14	16
25	3979	3997	4014	4031	4048	4065	4082	4099	4116	4133	2	3	5	7	9	10	12	14	15
26	4150	4166	4183	4200	4216	4232	4249	4265	4281	4298	2	3	5	7	8	10	11	13	15
27	4314	4330	4346	4362	4378	4393	4409	4425	4440	4456	2	3	5	6	8	9	11	13	14
28	4472	4487	4502	4518	4533	4548	4564	4579	4594	4609	2	3	5	6	8	9	11	12	14
29	4624	4639	4654	4669	4683	4698	4713	4728	4742	4757	1	3	4	6	7	9	10	12	13
30	4771	4786	4800	4814	4829	4843	4857	4871	4886	4900	1	3	4	6	7	9	10	11	13
31	4914	4928	4942	4955	4969	4983	4997	5011	5024	5038	1	3	4	6	7	8	10	11	12
32	5051	5065	5079	5092	5105	5119	5132	5145	5159	5172	1	3	4	5	7	8	9	11	12
33	5185	5198	5211	5224	5237	5250	5263	5276	5289	5302	1	3	4	5	6	8	9	10	12
34	5315	5328	5340	5353	5366	5378	5391	5403	5416	5428	1	3	4	5	6	8	9	10	11
35	5441	5453	5465	5478	5490	5502	5514	5527	5539	5551	1	2	4	5	6	7	9	10	11
36	5563	5575	5587	5599	5611	5623	5635	5647	5658	5670	1	2	4	5	6	7	8	10	11
37	5682	5694	5705	5717	5729	5740	5752	5763	5775	5786	1	2	3	5	6	7	8	9	10
38	5798	5809	5821	5832	5843	5855	5866	5877	5888	5899	1	2	3	5	6	7	8	9	10
39	5911	5922	5933	5944	5955	5966	5977	5988	5999	6010	1	2	3	4	5	7	8	9	10
40	6021	6031	6042	6053	6064	6075	6085	6096	6107	6117	1	2	3	4	5	6	8	9	10
41	6128	6138	6149	6160	6170	6180	6191	6201	6212	6222	1	2	3	4	5	6	7	8	9
42	6232	6243	6253	6263	6274	6284	6294	6304	6314	6325	1	2	3	4	5	6	7	8	9
43	6335	6345	6355	6365	6375	6385	6395	6405	6415	6425	1	2	3	4	5	6	7	8	9
44	6435	6444	6454	6464	6474	6484	6493	6503	6513	6522	1	2	3	4	5	6	7	8	9
45	6532	6542	6551	6561	6571	6580	6590	6599	6609	6618	1	2	3	4	5	6	7	8	9
46	6628	6637	6646	6656	6665	6675	6684	6693	6702	6712	1	2	3	4	5	6	7	7	8
47	6721	6730	6739	6749	6758	6767	6776	6785	6794	6803	1	2	3	4	5	5	6	7	8
48	6812	6821	6830	6839	6848	6857	6866	6875	6884	6893	1	2	3	4	4	5	6	7	8
49	6902	6911	6920	6928	6937	6946	6955	6964	6972	6981	1	2	3	4	4	5	6	7	8
50	6990	6998	7007	7016	7024	7033	7042	7050	7059	7067	1	2	3	3	4	5	6	7	8
51	7076	7084	7093	7101	7110	7118	7126	7135	7143	7152	1	2	3	3	4	5	6	7	8
52	7160	7168	7177	7185	7193	7202	7210	7218	7226	7235	1	2	3	3	4	5	6	7	7
53	7243	7251	7259	7267	7275	7284	7292	7300	7308	7316	1	2	2	3	4	5	6	6	7
54	7324	7332	7340	7348	7356	7364	7372	7380	7388	7396	1	2	2	3	4	5	6	6	7
N	0	1	2	3	4	5	6	7	8	9	1	2	3	4	5	6	7	8	9

											Proportional parts								
N	0	1	2	3	4	5	6	7	8	9	1	2	3	4	5	6	7	8	9
55	7404	7412	7419	7427	7435	7443	7451	7459	7466	7474	1	2	2	3	4	5	5	6	7
56	7482	7490	7497	7505	7513	7520	7528	7536	7543	7551	1	2	2	3	4	5	5	6	7
57	7559	7566	7574	7582	7589	7597	7604	7612	7619	7627	1	2	2	3	4	5	5	6	7
58	7634	7642	7649	7657	7664	7672	7679	7686	7694	7701	1	1	2	3	4	4	5	6	7
59	7709	7716	7723	7731	7738	7745	7752	7760	7767	7774	1	1	2	3	4	4	5	6	7
60	7782	7789	7796	7803	7810	7818	7825	7832	7839	7846	1	1	2	3	4	4	5	6	6
61	7853	7860	7868	7875	7882	7889	7896	7903	7910	7917	1	1	2	3	4	4	5	6	6
62	7924	7931	7938	7945	7952	7959	7966	7973	7980	7987	1	1	2	3	3	4	5	6	6
63	7993	8000	8007	8014	8021	8028	8035	8041	8048	8055	1	1	2	3	3	4	5	5	6
64	8062	8069	8075	8082	8089	8096	8102	8109	8116	8122	1	1	2	3	3	4	5	5	6
65	8129	8136	8142	8149	8156	8162	8169	8176	8182	8189	1	1	2	3	3	4	5	5	6
66	8195	8202	8209	8215	8222	8228	8235	8241	8248	8254	1	1	2	3	3	4	5	5	6
67	8261	8267	8274	8280	8287	8293	8299	8306	8312	8319	1	1	2	3	3	4	5	5	6
68	8325	8331	8338	8344	8351	8357	8363	8370	8376	8382	1	1	2	3	3	4	4	5	6
69	8388	8395	8401	8407	8414	8420	8426	8432	8439	8445	1	1	2	2	3	4	4	5	6
70	8451	8457	8463	8470	8476	8482	8488	8494	8500	8506	1	1	2	2	3	4	4	5	6
71	8513	8519	8525	8531	8537	8543	8549	8555	8561	8567	1	1	2	2	3	4	4	5	5
72	8573	8579	8585	8591	8597	8603	8609	8615	8621	8627	1	1	2	2	3	4	4	5	5
73	8633	8639	8645	8651	8657	8663	8669	8675	8681	8686	1	1	2	2	3	4	4	5	5
74	8692	8698	8704	8710	8716	8722	8727	8733	8739	8745	1	1	2	2	3	4	4	5	5
75	8751	8756	8762	8768	8774	8779	8785	8791	8797	8802	1	1	2	2	3	4	4	5	5
76	8808	8814	8820	8825	8831	8837	8842	8848	8854	8859	1	1	2	2	3	4	4	5	5
77	8865	8871	8876	8882	8887	8893	8899	8904	8910	8915	1	1	2	2	3	3	4	4	5
78	8921	8927	8932	8938	8943	8949	8954	8960	8965	8971	1	1	2	2	3	3	4	4	5
79	8976	8982	8987	8993	8998	9004	9009	9015	9020	9025	1	1	2	2	3	3	4	4	5
80	9031	9036	9042	9047	9053	9058	9063	9069	9074	9079	1	1	2	2	3	3	4	4	5
81	9085	9090	9096	9101	9106	9112	9117	9122	9128	9133	1	1	2	2	3	3	4	4	5
82	9138	9143	9149	9154	9159	9165	9170	9175	9180	9186	1	1	2	2	3	3	4	4	5
83	9191	9196	9201	9206	9212	9217	9222	9227	9232	9238	1	1	2	2	3	3	4	4	5
84	9243	9248	9253	9258	9263	9269	9274	9279	9284	9289	1	1	2	2	3	3	4	4	5
85	9294	9299	9304	9309	9315	9320	9325	9330	9335	9340	1	1	2	2	3	3	4	4	5
86	9345	9350	9355	9360	9365	9370	9375	9380	9385	9390	1	1	2	2	3	3	4	4	5
87	9395	9400	9405	9410	9415	9420	9425	9430	9435	9440	0	1	1	2	2	3	3	4	4
88	9445	9450	9455	9460	9465	9469	9474	9479	9484	9489	0	1	1	2	2	3	3	4	4
89	9494	9499	9504	9509	9513	9518	9523	9528	9533	9538	0	1	1	2	2	3	3	4	4
90	9542	9547	9552	9557	9562	9566	9571	9576	9581	9586	0	1	1	2	2	3	3	4	4
91	9590	9595	9600	9605	9609	9614	9619	9624	9628	9633	0	1	1	2	2	3	3	4	4
92	9638	9643	9647	9652	9657	9661	9666	9671	9675	9680	0	1	1	2	2	3	3	4	4
93	9685	9689	9694	9699	9703	9708	9713	9717	9722	9727	0	1	1	2	2	3	3	4	4
94	9731	9736	9741	9745	9750	9754	9759	9763	9768	9773	0	1	1	2	2	3	3	4	4
95	9777	9782	9786	9791	9795	9800	9805	9809	9814	9818	0	1	1	2	2	3	3	4	4
96	9823	9827	9832	9836	9841	9845	9850	9854	9859	9863	0	1	1	2	2	3	3	4	4
97	9868	9872	9877	9881	9886	9890	9894	9899	9903	9908	0	1	1	2	2	3	3	4	4
98	9912	9917	9921	9926	9930	9934	9939	9943	9948	9952	0	1	1	2	2	3	3	4	4
99	9956	9961	9965	9969	9974	9978	9983	9987	9991	9996	0	1	1	2	2	3	3	3	4
N	0	1	2	3	4	5	6	7	8	9	1	2	3	4	5	6	7	8	9

APPENDIX 2

Abbreviations, Symbols, and Constants

A	absorbance (less desirably called optical density)
	radioactivity of a radionuclide
a	activity
atm	atmosphere
Å	angstrom (= 10^{-8} cm)
C	temperature in Celsius
c	concentration
	velocity of light (= 2.997×10^{10} cm sec^{-1} in vacuum)
C_i	curie (= 3.700×10^{10} dps)
cal	calorie
Cal	kilocalorie (kcal)
cps	counts per second
cpm	counts per minute
NAD	nicotinamide adenine dinucleotide
dps	disintegrations per second
e	natural log base (= 2.71828)
E	energy (in general)
	half-reaction electrode potential (volt)
E_a	activation energy
E_0	E(volt) in standard state and [H$^+$] 1 M
E_0'	E(volt) in standard state and [H$^+$] at some other defined value, usually 10^{-7} M
$E_{1\,\mathrm{cm}}^{1\%}$	absorbance of light by a 1% (w/v) solution of 1-cm light path
ΔE	change of internal energy of a system following some process

ΔE_0	redox potential (volt) of coupled half-reactions in their standard states, $[H^+]$ 1 M if applicable
$\Delta E_0'$	*as for* ΔE_0 but $[H^+]$ at some other defined concentration
ΔE_{obs}	redox potential (volt) of coupled half-reactions, with reactants and products not in their standard states
eq	equivalent
F	temperature in Fahrenheit
F	Faraday constant (= 96,494 coulombs)
FAD	flavin adenine dinucleotide
g	gram
g mole	gram mole
ΔG	free energy change, calories per amount of material changed
$\Delta G°$	standard free energy change, with all components of the system in their standard states
$\Delta G°'$	ΔG with all but $[H^+]$ in the standard states; $[H^+]$ is defined at some value, usually 10^{-7} M
h	Planck's constant ($= 6.6238 \times 10^{-27}$ erg sec^{-1})
ΔH	change in enthalpy, calories per amount of material changed
$\Delta H°$	standard enthalpy change
I	an integration constant
	intensity of transmitted light
I_0	incident light intensity
J	joule
k	Boltzmann's constant ($= 1.38026 \times 10^{-16}$ erg degree^{-1})
k_1, k_2	reaction rate constants
K	Henry's Law constant
K_a	ionization constant of a conjugate acid
K_b	ionization constant of a conjugate base
K_{eq}	equilibrium constant
K_{inst}	instability constant of a complex ion
K_r	apparent equilibrium constant, at a given pH, but not including $[H^+]$
K_{st}	stability constant of a complex ion
K_w	dissociation constant of water, $[H^+][OH^-]$
ln	natural logarithm of a number (to base e)
log	logarithm of a number to base 10
M	molar
m	molal or Hill coefficient
m	milli or meter
min	minute
mole	mole
mmole	millimole
mM	millimolar
mol. wt.	molecular weight
n	nano
n	number of binding sites
	number of electrons in redox reaction
N	normal
nm	wavelength of light in nanometers (= 10^{-7} cm)

P	pressures or partial pressures of gases, for example P_{O_2}, P_{CO_2}
p	as used in the general context of pX it corresponds to $-\log x$, for example, pH, pK_a; also pico
pI	isoelectric point of amphoteric molecule
R	gas constant (see Table 4.1)
	counting rate of radioactivity
ΔS	change in entropy, as cal degree^{-1} mole^{-1}, or entropy units, e.u.
SC	standard conditions of temperature (273°K) and pressure (760 mm Hg)
sec	second
t	confidence level factor (see Table 4.2)
	time
T	temperature in degrees Kelvin
T or T_a	physical half-life of a radionuclide
T_b	biological half-life
$\overline{T}$	mean half-life of a radionuclide
NADP	nicotinamide adenine dinucleotide phosphate
v	velocity of a reaction
	initial velocity of a reaction in enzyme kinetics
V	volume
V_{max}	maximum initial velocity
w	work
w'	useful work
X	mole fraction
Y	counting yield efficiency
Z_i	charge on an ion of the ith species
[ ]	concentration of (material within the brackets)
α	fraction of dissociation
	linear absorption coefficient in spectrophotometry
γ	gamma radiation
$\Gamma/2$	ionic strength
ϵ	molar extinction coefficient (liter cm^{-1} mole^{-1})
λ	disintegration constant of a radionuclide
	wavelength of light
μ	ionic strength
ν	frequency of light in cycles sec^{-1}
$\bar{\nu}$	moles ligand bound per mole macromolecule
σ	standard deviation
Σ	sum of
θ	fraction of binding sites occupied

APPENDIX **3**

Answers to Problems A

Chapter 2

1. (a) 7.231×10^{3}
   (b) 2.89×10^{-3}
   (c) 4.7×10^{-4}
   (d) 4.34×10^{2}
   (e) 1.456×10^{-2}
   (f) 5.368000×10^{6}
   (g) 2.27×10^{2}
   (h) 1.7450×10^{4}
   (i) 1.27×10^{-3}
   (j) 3.824434×10^{6}
   (k) 1.89×10^{-1}
   (l) 4.387×10^{3}
   (m) 8.21×10^{-2}
   (n) 7.3×10^{-5}
   (o) 6.20×10^{2}
   (p) 7.13000×10^{5}

2. (a) 1.4×10^{2}
   (b) 5.2×10^{3}
   (c) 3.5×10^{4}
   (d) 9.5×10^{3}
   (e) 5.9×10^{7}
   (f) 1.8×10^{-16}
   (g) 1.2×10^{-3}
   (h) 3.1×10^{-9}
   (i) 4.1
   (j) 3.0×10^{5}
   (k) 1.0
   (l) 1.1×10^{-2}

3. (a) 2.4330
   (b) −2.4461
   (c) −4.2924
   (d) 1.7708
   (e) −2.1427
   (f) −1.3820
   (g) 3.9499
   (h) −5.7959
   (i) −1.0862
   (j) 0.0934
   (k) 1.7619
   (l) 2.5353
   (m) −3.3768
   (n) 2.5172
   (o) −2.4935
   (p) −0.2381
   (q) 1.6798
   (r) −3.3757
   (s) −2.4685
   (t) 2.7482
   (u) −4.6778
   (v) 2.8162
   (w) −4.5686
   (x) 1.8921

4. (a) 2.6×10^{-6}
(b) 4.0
(c) 3.7×10^{7}
(d) 8.63×10^{4}
(e) 3.4×10^{-2}
(f) 2.72×10^{-1}
(g) 2.79×10^{-1}
(h) 2.6
(i) 2.0×10^{-1}
(j) 4.4×10^{5}
(k) 1.90
(l) 3.0×10^{4}
(m) 1.6×10^{-2}
(n) 1.6×10^{-7}
(o) 4.9
(p) 9700
(q) 3.3×10^{-1}
(r) 6.3×10^{-6}
(s) 1.53×10^{-1}
(t) 4.3×10^{-1}
(u) 1.4×10^{1}
(v) 1.82×10^{5}
(w) 1.18×10^{5}
(x) 5.94
(y) 0.854
(z) 4.93
(a′) 1.12
(b′) 1.16×10^{-5}
(c′) 4.05
(d′) 5.7×10^{-10}
(e′) 5.3×10^{-5}
(f′) 2.5×10^{1}
(g′) 3.36
(h′) 8.9×10^{4}
(i′) 2.2×10^{-2}
(j′) 1.39×10^{-5}
(k′) 9.0×10^{-10}
(l′) 1.71×10^{2}
(m′) 5.4×10^{6}
(n′) 6.42×10^{-5}
(o′) 1.7×10^{-1}
(p′) 8.38
(q′) 8220
(r′) 25.5
(s′) 4.1×10^{-2}
(t′) 8.50×10^{13}
(u′) 4.15×10^{2}
(v′) 0.000361

Chapter 3

1. (a) $\alpha = \dfrac{f}{m}$
(b) $x = \pm 1.73$
(c) $x = \frac{3}{4}, -3$
(d) $t = \pm\sqrt{\dfrac{25}{g}}$
(e) $H = \dfrac{2A}{B}$
(f′) $M_1 = \dfrac{M_2\mu}{M_2 - \mu}$
(g) $h = \dfrac{E\lambda}{C}$
(h) $z = \dfrac{3a^2 + 1}{4}$
(i) $0, 2\frac{1}{3}$
(j) $x = 7$
(k) $x = \dfrac{3 + c}{b}$
(l) $x + \pm \frac{1}{2}\sqrt{c}$
(m) $x = -\frac{1}{4}, -\frac{1}{2}$
(n) $V = \dfrac{RTn}{P}$
(o) $T_2 = \dfrac{P_2V_2T_1}{P_1V_1}$
(p) $A = 4.58$
(q) $d = \dfrac{S}{\pi r^2}$
(r) $x = \pm 2.45$
(s) $x = \dfrac{e \pm \sqrt{e^2 + 40}}{2}$
(t) $r = \sqrt[3]{\dfrac{3V}{4\pi}}$
(u) $n = \pm\sqrt{\dfrac{3\lambda - 2M}{M}}$
(v) $r = \pm\sqrt{\dfrac{s - 1}{7}}$
(w) $E_m = \dfrac{A}{bc}$
(x) $y = -\frac{1}{4}$
(y) $V = \pm\sqrt{\dfrac{fr}{m}}$
(z) $x = \pm 2\sqrt{\dfrac{1 - 2a}{2b}}$
(a′) $x = 1, -\frac{3}{4}$
(b′) $x = 1.22$
(c′) $x = 0.324$
(d′) $K_m = \dfrac{S(V_{\max} - v)}{v}$
(e′) $x = -\frac{1}{2}, -1\frac{1}{2}$
(f′) $x = \pm 0.707$
(g′) $r = \dfrac{I}{tP}$
(h′) $x = \pm 0.775$
(i′) $x = \frac{2}{5}, \frac{2}{5}$
(j′) $R = \dfrac{PV}{nT}$
(k′) $x = 1.78, -0.281$
(l′) $x = \dfrac{4c}{c - 5}$
(m′) $x = \dfrac{8}{4 - a}$
(n′) $x = 1\frac{1}{4}$
(o′) $x = \pm 10$
(p′) $x = -1, \frac{4}{7}$
(q′) $x = \pm 2$
(r′) $x = 3.83, -1.83$
(s′) $x = 1, -\frac{4}{9}$
(t′) $v = \dfrac{2S + gt^2}{2t}$
(u′) $x = \pm\sqrt{\dfrac{h}{3a}}$
(v′) $x = 1.09, -1.84$

2. (a) $m = 4,\ b = 0$
   (b) $m = \frac{1}{4},\ b = 4$
   (c) $m = \frac{1}{8},\ b = \frac{3}{8}$
   (d) $m = \frac{4}{3},\ b = \frac{2}{3}$
   (e) $m = \frac{1}{6},\ b = \frac{1}{6}$
   (f) $m = -3,\ b = 0$
   (g) $m = 2\frac{1}{2},\ b = 2$
   (h) $m = \dfrac{7}{a},\ b = \dfrac{1}{a}$
   (i) $m = 6,\ b = 3$
   (j) $m = \frac{1}{4},\ b = -\frac{1}{4}$
   (k) $m = 6,\ b = 14$
   (l) $m = -\frac{3}{7},\ b = 0$
   (m) $m = \frac{4}{3},\ b = \frac{1}{3}$
   (n) $m = 1,\ b = 0$
   (o) $m = -3,\ b = \frac{1}{2}$
   (p) $m = 2,\ b = 4$

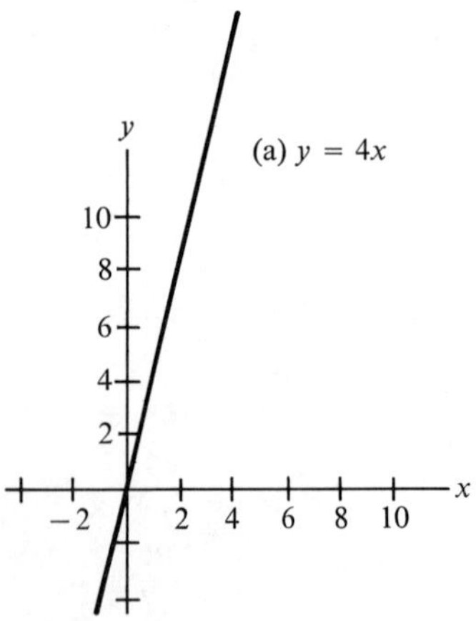

(a) $y = 4x$

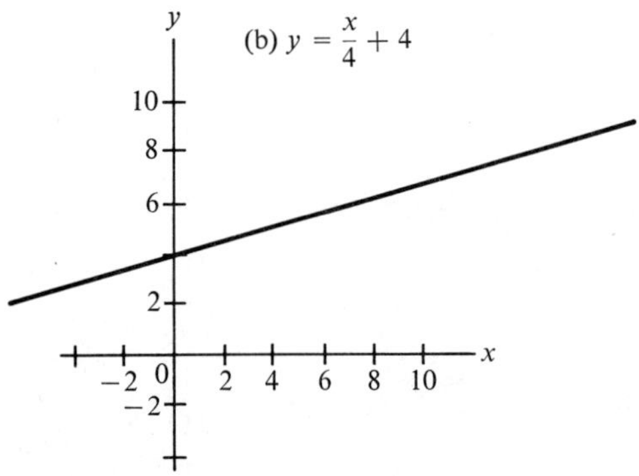

(b) $y = \dfrac{x}{4} + 4$

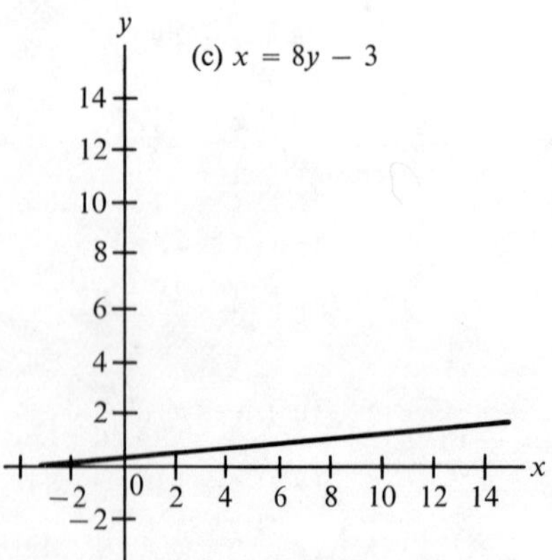

(c) $x = 8y - 3$

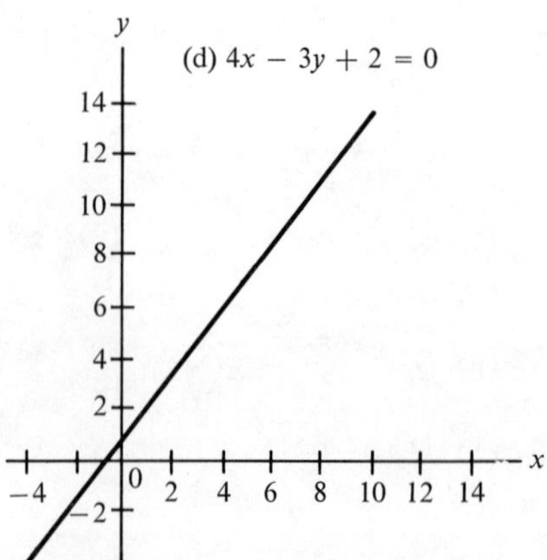

(d) $4x - 3y + 2 = 0$

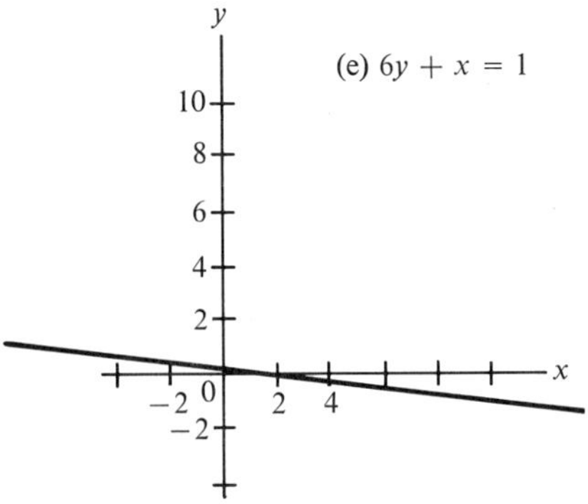
(e) $6y + x = 1$

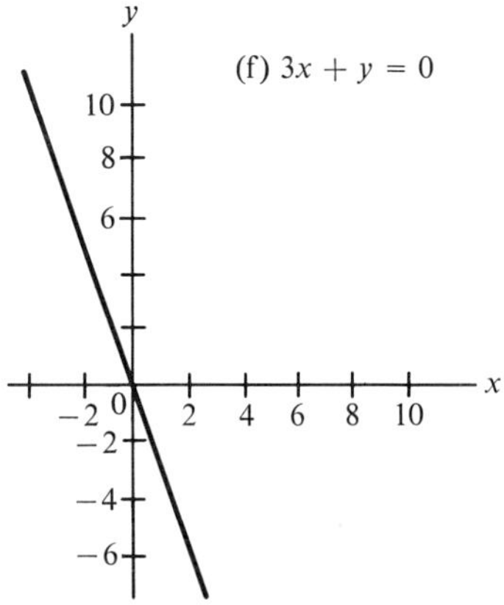
(f) $3x + y = 0$

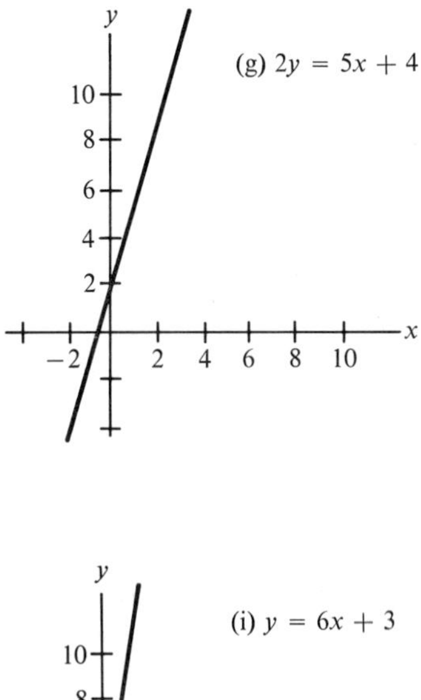
(g) $2y = 5x + 4$

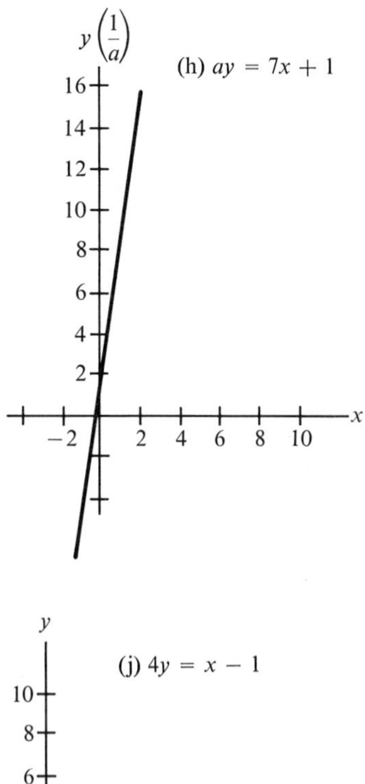
(h) $ay = 7x + 1$

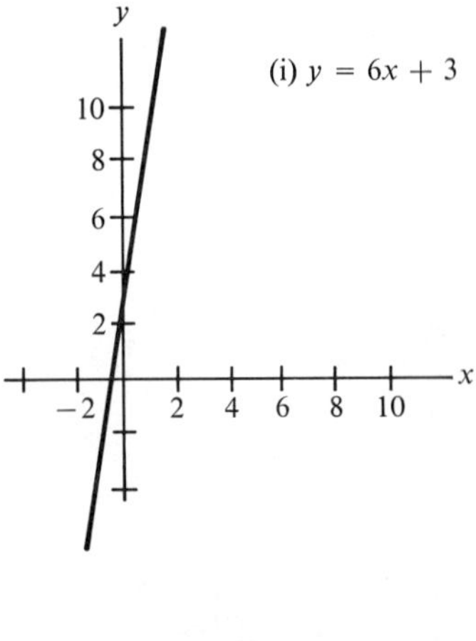
(i) $y = 6x + 3$

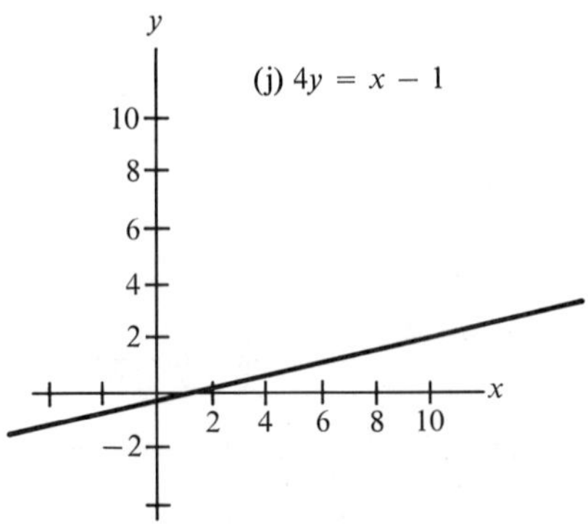
(j) $4y = x - 1$

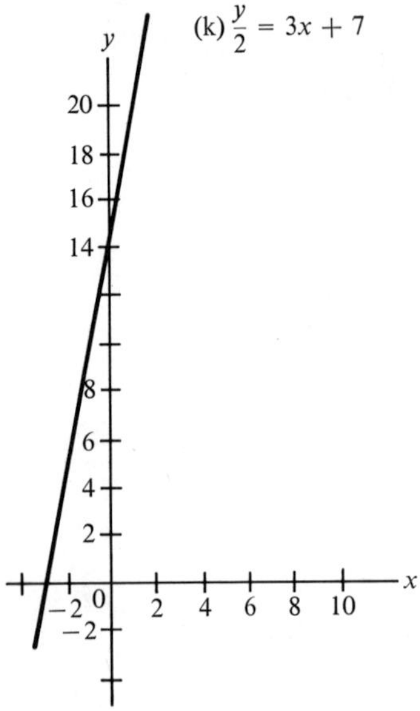

(k) $\frac{y}{2} = 3x + 7$

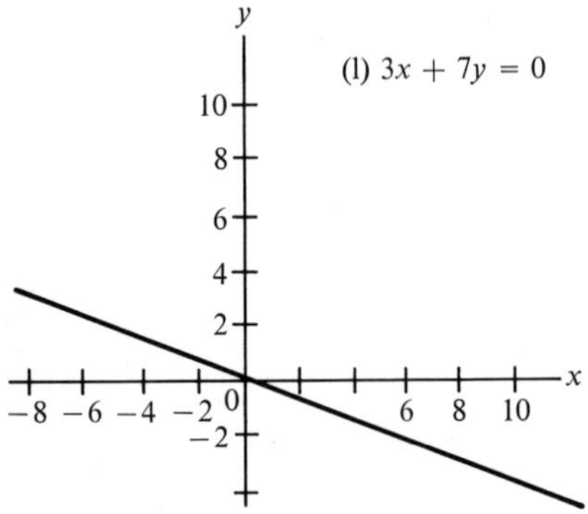

(l) $3x + 7y = 0$

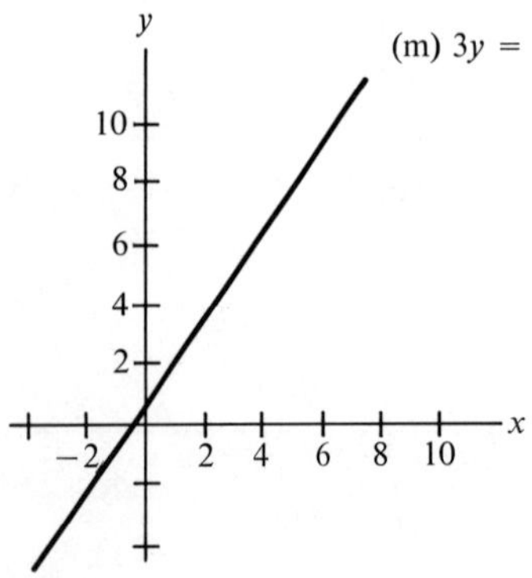

(m) $3y = 4x + 1$

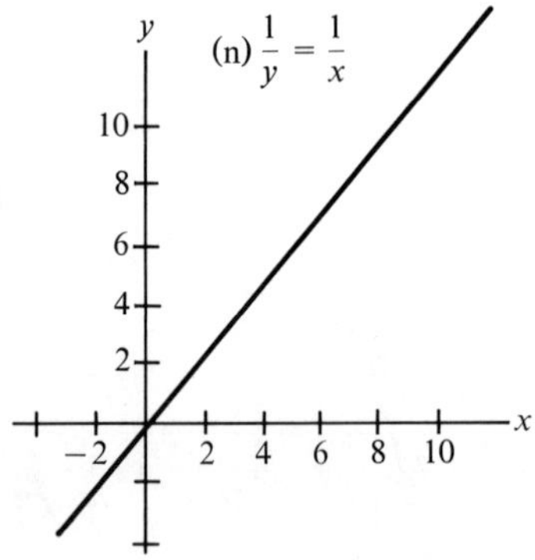

(n) $\frac{1}{y} = \frac{1}{x}$

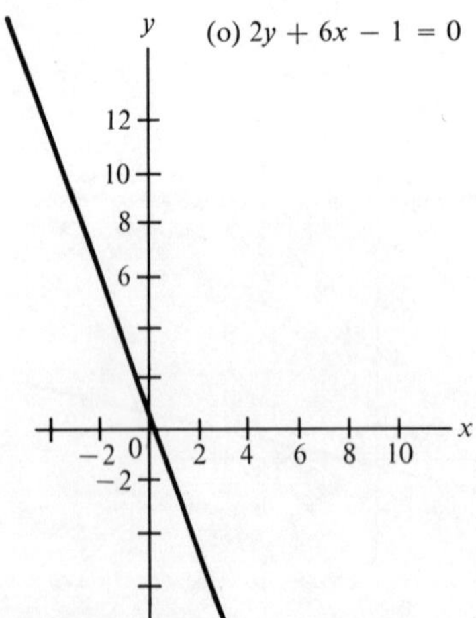

(o) $2y + 6x - 1 = 0$

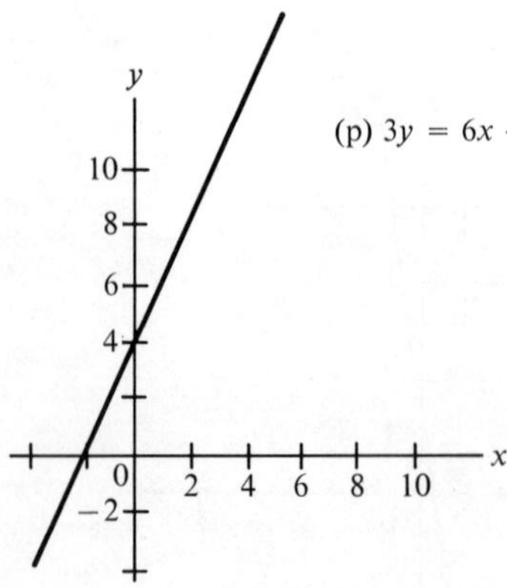

(p) $3y = 6x + 12$

3. (a) y versus x^2
   (b) y versus $\dfrac{1}{x^2}$
   (c) y versus x^2
   (d) y versus x^2
   (e) y versus x^2
   (f) y versus $(x^2)^{-1}$
   (g) S versus r^2
   (h) y versus x^3

4. (a) $4.16 \times 10^{-3}, -4.18 \times 10^{-3}$
   (b) -3.7, two imaginary roots
   (c) 4 imaginary roots
   (d) $-3.6, 0.3, 3.3$
   (e) $-1.4, 1.4, 4.0$

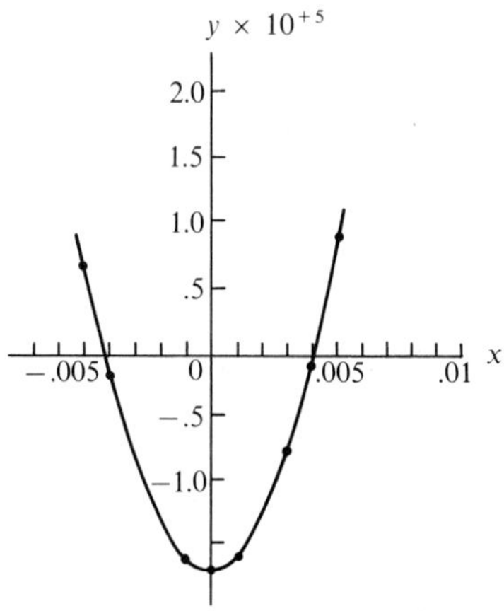

(a) $\dfrac{x^2}{(0.1 - x)} = 1.74 \times 10^{-4}$

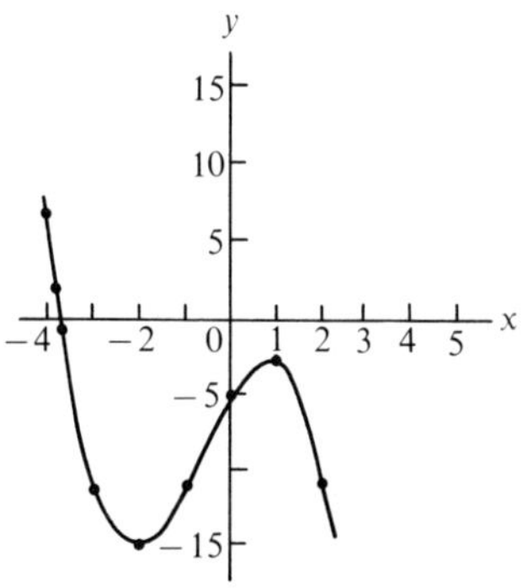

(b) $-x^3 - 2x^2 + 5x - 5 = 0$

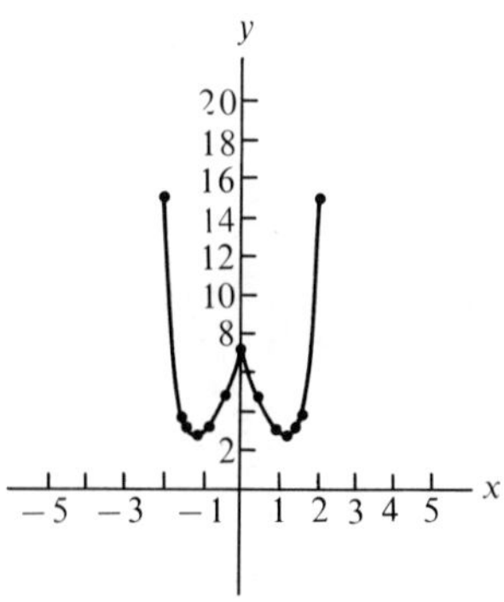

(c) $2x^4 - 6x^2 + 7 = 0$

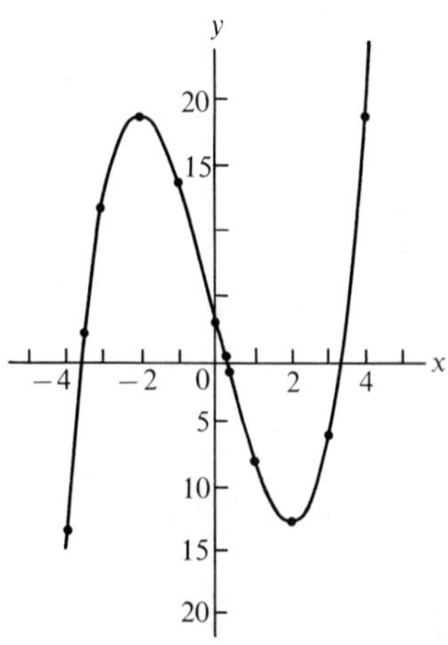

(d) $x^3 - 12x + 3 = 0$

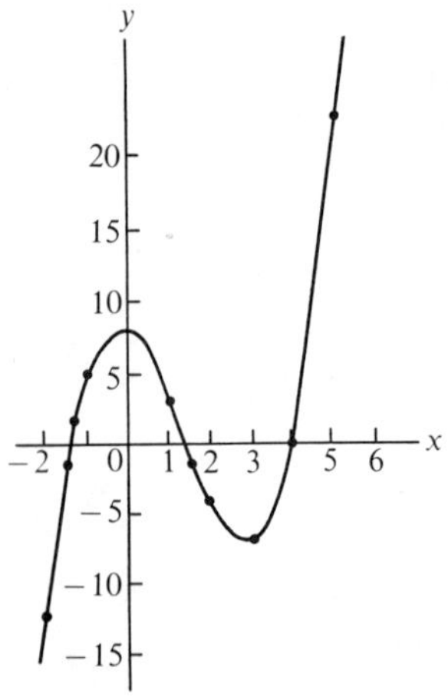

(e) $x^3 - 4x^2 - 2x + 8 = 0$

Chapter 4

1. (a) $27.8 \frac{\text{cm}}{\text{sec}} \Big/ \frac{\text{km}}{\text{hr}}$
   (b) $3484 \frac{\text{ml}}{\text{sec}} \Big/ \frac{\text{gal}}{\text{sec}}$
   (c) $2.45 \times 10^6 \frac{\text{liters}}{\text{day}} \Big/ \frac{\text{ft}^3}{\text{sec}}$
   (d) 1.65×10^{-6} week/sec
   (e) 2.54×10^4 μ/in.
   (f) $6.2 \times 10^{-6} \frac{\text{mile}}{\text{sec}} \Big/ \frac{\text{cm}}{\text{sec}}$

2. (a) 4.5×10^{14} molecules
   (b) 1.82×10^{14} molecules
   (c) 4.1×10^{14} molecules
   (d) 5.2×10^{14} molecules

3. 5.023 liters 4. 253 Å^3 5. 46.19 cm^3 6. 1.053 g 7. 13 min

8. (a) 796.5
   (b) 0.0062
   (c) 2.1×10^2
   (d) 7.46×10^1
   (e) 0.12
   (f) 18.3
   (g) 6.2×10^1
   (h) 739.85
   (i) 0.068
   (j) 13.9
   (k) 7681
   (l) 0.00867
   (m) 0.0335
   (n) 2.68×10^3
   (o) 57.73
   (p) 6.50
   (q) 4.7×10^2
   (r) 0.0119
   (s) 3785
   (t) 0.0111

9. 4.18 10. 0.314268

11. Average deviation 0.1018 ± 0.00030
    Standard deviation $\pm 3.7 \times 10^{-1}$
    90% confidence interval 0.1018 ± 0.00075

12. Average deviation 22400 ± 630.0
Standard deviation ±782.3
80% confidence interval 22400 ± 1081.9

Chapter 5

1. Dissolve the following amounts of materials in distilled water and dilute to the volume requested in the problem. In some problems, such as (e), the procedure of dilution would be followed only in special situations, for example, kinetic studies in reaction vessels of specific volume.

(a) 6.607 g	(h) 4.040 g	(o) 1.00 ml	(v) 62.50 ml
(b) 69.02 g	(i) 0.050 g	(p) 0.375 ml	(w) 147.67 g
(c) 0.60 g	(j) 0.011 g	(q) 3.00 ml	(x) 6.00 ml
(d) 0.832 ml	(k) 0.05 ml	(r) 0.856 ml	(y) 0.586 ml
(e) 0.065 ml	(l) 2.50 ml	(s) 178.0 g	(z) 0.67 ml
(f) 12.50 ml	(m) 1.0 ml	(t) 1.344 ml	
(g) 2.25 ml	(n) 2.19 ml	(u) 0.818 ml	

2. (a) 9.7 g (c) 40,020 g (e) 0.102 mole
(b) 20.73 g (d) 0.201 mole

3. (a) 8.557 g (b) 1.000 g (c) 2.381 g

4.

	Molarity	Molality
(a)	0.346	0.349
(b)	0.105	0.105
(c)	2.34×10^{-2}	2.35×10^{-2}
(d)	0.250	0.240
(e)	0.250	0.253
(f)	0.154	0.154

5. Units are milliequivalents.

(a) 32	(d) 0.3	(g) 0.2	(j) 75
(b) 5	(e) 5	(h) 30	
(c) 1.8	(f) 12	(i) 10.8	

6. Units are pH.

(a) 1.76	(c) 2.78	(e) 3.07	(g) 1.40
(b) 1.12	(d) 11.8	(f) 1.076	

7. Units are molarity of hydrogen ion.

(a) 0.5	(c) 3.98×10^{-8}	(e) 1.82×10^{-5}
(b) 0.5	(d) 2.19×10^{-11}	(f) 3.98×10^{-7}

8. Concentration in milliequivalents per liter.

(a) 151 (b) 4.8 (c) 2.5 (d) 1.0

9. 49.4

10. (a) 29.45 g (b) 9.75 ml (c) 204 mmole (d) 16.50 liters

11. 8.144 mmole O_2; 0.026 mmole CO_2; 30.734 mmole N_2; 0.257 mmole H_2O; total pressure is 1 atm

12. 5.972 mmole O_2; 1.467 mmole CO_2; 29.240 mmole N_2

13. 0.218 mmole O_2

14. 1.507 mM CO_2

15. 32.72 g

16. 4.80 mmole O_2; 1076 ml at SC

17. 4.52 mmole H_2O

18. 198.6

Chapter 6

1. Pyruvate + CO_2 + ATP $\rightleftharpoons$ oxalacetate + ADP + P_i

2. $K_{eq_1}, \frac{38}{18} = 2.1$; $K_{eq_2} = \frac{2}{38} = 0.0526$

   $K_{eq_3} = \frac{18}{2} = 9$

3. $K_{eq} = \frac{7.3}{92.7} = 7.87 \times 10^{-2}$

   $$pK_{eq} = -\log 7.87 \times 10^{-2} = -(0.896 - 2.000) = 1.1$$

4. (a) $pK_{eq} = -\log 7.8 \times 10^3 = -3.89$

   (b) D-fructose 6-P = 2 mM
   D-fructose 1,6-diphosphate = 4 mM

5. $K_{eq} = 2$; $pK_{eq} = -0.30$; α-D-glucose $\rightleftharpoons$ β-D-glucose

6. 38.4 mg

7. 0.444 mmole of uracil; 0.444 mmole of D-ribose 1-PO_4

8. (a) D-glucose + creatine P $\rightleftharpoons$ D-glucose 6-P + creatine

$$K_{eq} = 4 \times 10^4$$

(b) $\dfrac{[\text{ATP}]}{[\text{ADP}]} = 3.33$; $\dfrac{[\text{G-6-P}]}{[\text{G}]} = 1.33 \times 10^4$

9. (a) No
   (b) 2.0×10^{-1}
   (c) 1,3-diphospho-D-glycerate + ADP $\rightleftharpoons$ 3-phospho-D-glycerate + ATP
   (d) 6.4×10^2

10. 2.0×10^{-4} M

11. (a) pyruvate + aspartate $\rightleftharpoons$ oxalacetate + alanine; K_{eq} 2.2×10^{-1}

(b) $\dfrac{[\text{alanine}]}{[\text{aspartate}]} = 4.4 \times 10^{-1}$

12. oxalacetate + acetate $\rightleftharpoons$ citrate K_{eq} 0.5
citrate $\rightleftharpoons$ *cis*-aconitate + H_2O K_{eq} 0.07
iso-aconitate + H_2O $\rightleftharpoons$ *iso*-citrate K_{eq} 2.0

---

oxalacetate + acetate $\rightleftharpoons$ *iso*-citrate

$$K_{eq} = 0.5 \times 0.07 \times 2.0 = 0.07$$

$$-\log 0.07 = -(0.85 - 2.00)$$

$$pK_{eq} = 1.15$$

13. (a) ATP + 2-phosphoglycerate $\rightleftharpoons$ ADP + 1,3-diphosphoglycerate

$$K_{eq} \ 1.4 \times 10^{-3}$$

(b) 0.66 mmole

Chapter 7

1. 1176

2. 8.60

3. 0.307 g/liter by Equation (7.24)

4. 6.57×10^{-3} g/liter by Equation (7.24)

5. 11.5

6. Buffer capacity 0.24 (by calculation)

7. 0.05% lactic acid

8. pH of 3 is 100 times stronger, pH of 3 is 10^{-3} M in $[H^+]$, and pH of 5 is 10^{-5} M in $[H^+]$

9. 1/20 N H_2SO_4 solution; the 1/10 N acetic acid is only 1.3% ionized whereas the 1/20 N H_2SO_4 is 100% ionized

10. Aqueous pyridine

11. (a) 2.22×10^{-4}
    (b) 3.77

12. 0.31 g sodium acetate; 0.36 ml glacial acetic acid

13. (a) 10.5 by Equation (7.27)
    (b) 5.30 by Equation (7.24)
    (c) 12.12 by Equation (7.27)
    (d) 5.78 by Equation (7.24)
    (e) 2.60 by Equation (7.27)

14. 8.87

15. 3.68; 0.35 is the buffer capacity (by calculation)

16. $[H^+]$ 2.69×10^{-5} M; pH 4.43

17. 2.24×10^{-11}

18. (a) 3.4
    (b) 4.8×10^{-5} M

19. 0.04 is the buffer capacity

20. 6.91 g; 2 g NaOH

21. 3.72

22. 28.50 g acetic acid (27.17 ml glacial acetic acid); 2.0 g NaOH

23. 10.55

24. 5.69

25. 3.025 g Tris; 21.55 ml N HCl

26. 393 ml 0.1 N HCl; 199 ml 0.1 N NaOH

Chapter 8

Buffers, Complex Ions, and Titration Curves

1. The phosphate buffer is superior for two reasons:
   (a) the concentration is greater
   (b) a drop of one pH unit in the phosphate buffer will put the pH near its pK_2 (6.7) whereas a drop of one pH unit still leaves the acetate buffer considerably above its pK (4.73)

2. (a) 7.0
   (b) decrease by 0.06 pH units
   (c) decrease by 5 pH units

3. 4.76 ml 0.1 N HCl
   100 ml 0.1 N $NaHCO_3$
   295.24 ml H_2O

4. 93% as monoprotonated acid
   7% as salt of this acid (neutral)

5. 16.4 g Na_3PO_4
   10.6 ml conc. HCl
   H_2O to 1000 ml

6. 22.80 g histidine dihydrochloride
   378 ml 0.5 N NaOH
   H_2O to 1 liter

7. 37.2 ml

8. 41.5 ml 1 M Na_3PO_4; 58.0 ml 1 N NaH_2PO_4; 400.5 ml H_2O

9. (a) 1.486 g $Na_2HPO_4 \cdot 12H_2O$; 0.113 g KH_2PO_4
   (b) pH 6.46; $[H^+]$ 3.47×10^{-7} M

10. 15.7 ml of 0.1 N HCl; 0.152 g xanthine, distilled H_2O to 1000 ml

11. 100%

12. See graph on the top of the next page

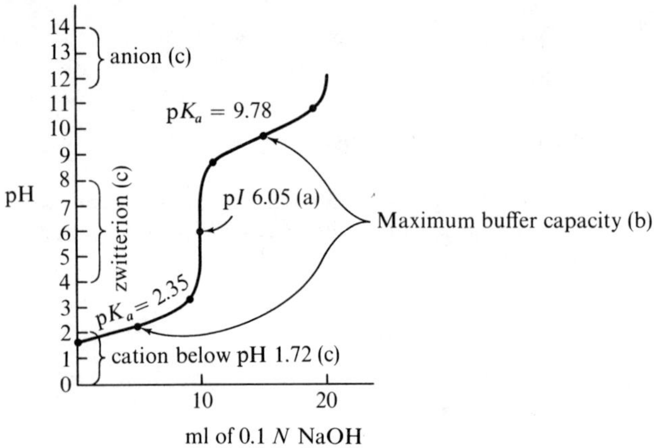

13. 20 ml

14. 22.28 g citric acid; 8 g NaOH

15. 1.13 g NaH_2PO_4; 0.082 g Na_2HPO_4

16. 50 ml

17. 4.68; acetic acid

18. p*I* 5.5–6.0

19. See graph

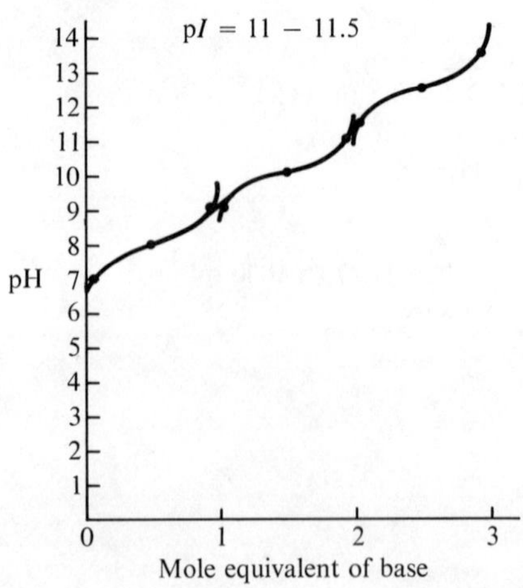

20. 2.5 ml

21. (a) 51.83
(b) histidine
(c) p*I* 7.6

22. 16.87 g glycylglycine hydrochloride; 6.49 g KOH, water to 1 liter.

23. 0.1 mole

24. See graph

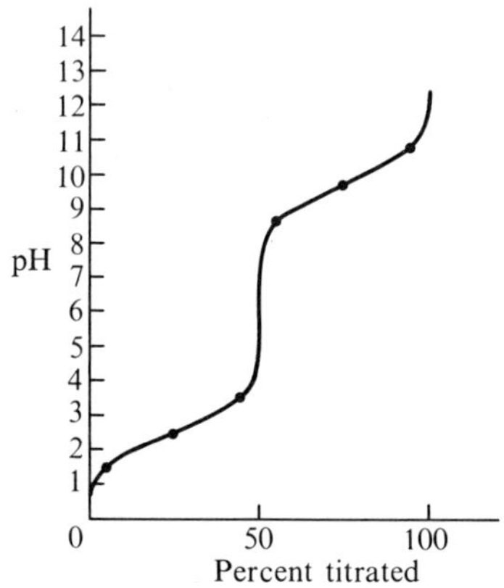

25. p*I* = 5–6

26. (a)

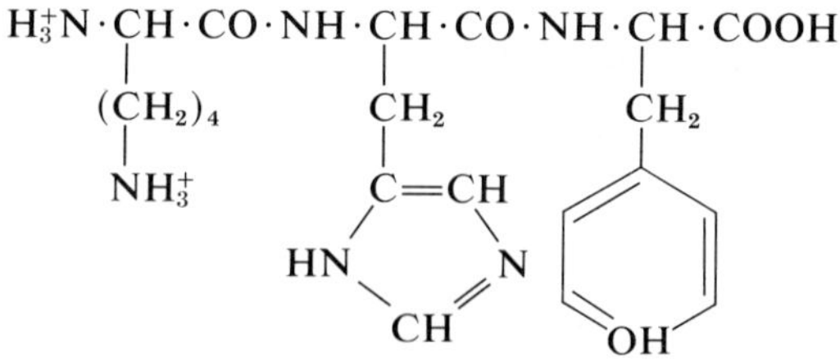

(b) p*I* = 9
(c) 1.2 – 3.3; 5.3 – 11.5

27. 6.66×10^{-3} *M*

28. 33.85 g $Na_2HPO_4 \cdot 7H_2O$; 2.61 g of $NaH_2PO_4 \cdot H_2O$

29. 8 ml HCl, 500 ml Na_3citrate, 492 ml H_2O

30. p*I* is 9–10 (see graph on the top of the next page)

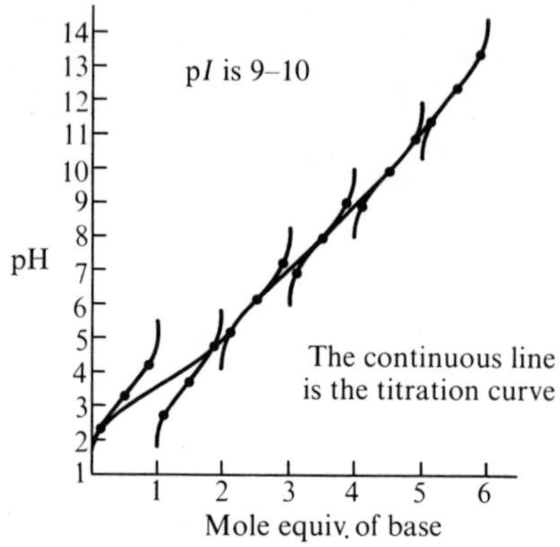

31. See graph

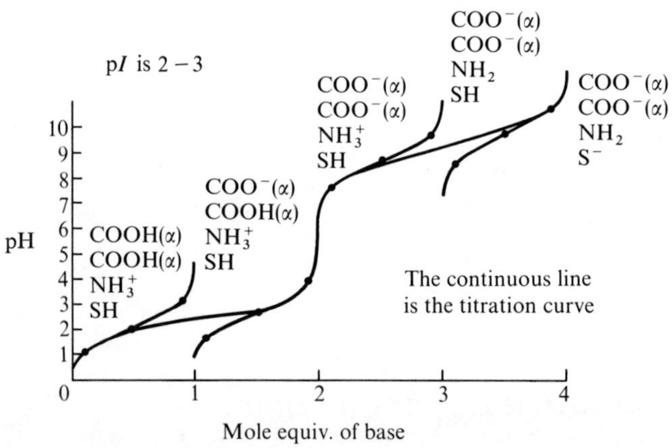

32. 50% with α-amino group protonated and 50% not protonated. The carboxyl group is as the carboxylate ion and the guanidino group is protonated in both cases

33. See graph

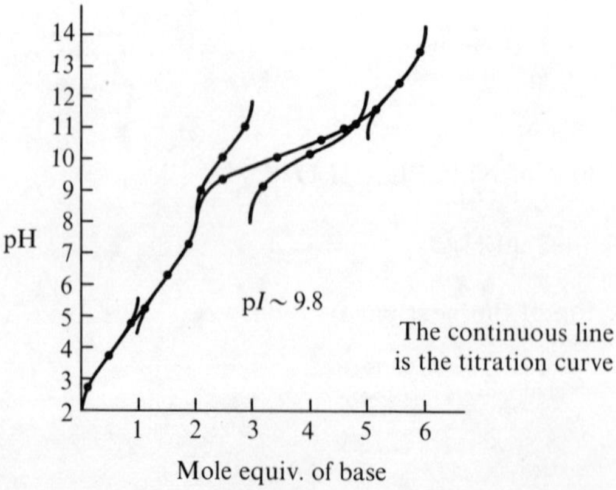

34. 18.36 g glutamine hydrochloride; 532 ml 0.1 *N* NaOH; H_2O to 1000 ml total volume

35. 5.86

36. 2.5 ml

37. 112.5 ml

38. 2.6 g malonic acid; 191 ml 0.25 *N* NaOH

39. 6.65

40. 11.2 g glycine hydrochloride; 5.18 g NaOH; H_2O to 1 liter

41. (a)

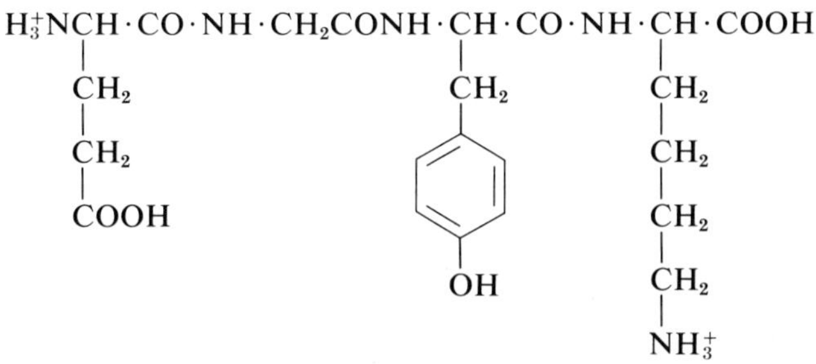

(b) 7

(c) 1.2–3.2; 3.3–5.3; 8.7–11.5

42. See graph.

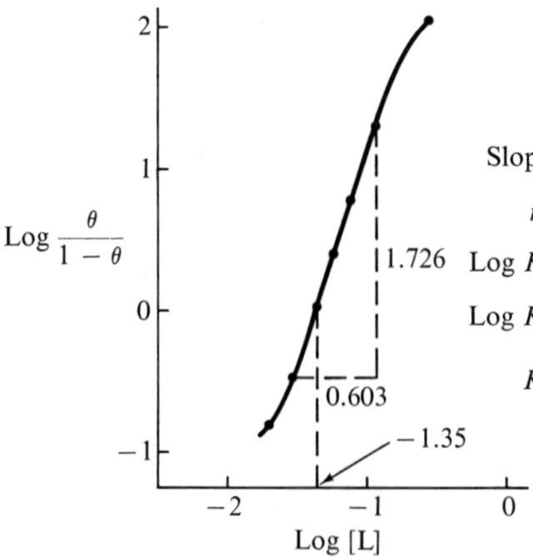

$$\text{Slope} = \frac{1.726}{0.603}$$

$$m = 2.86$$

$$\text{Log } K' = -n \text{ Log [L] at Log } \frac{\theta}{1-\theta} = 0$$

$$\text{Log } K' = -2.86 \times (-1.35)$$

$$= 3.86$$

$$K' = 9.35 \times 10^3$$

43. See graph.

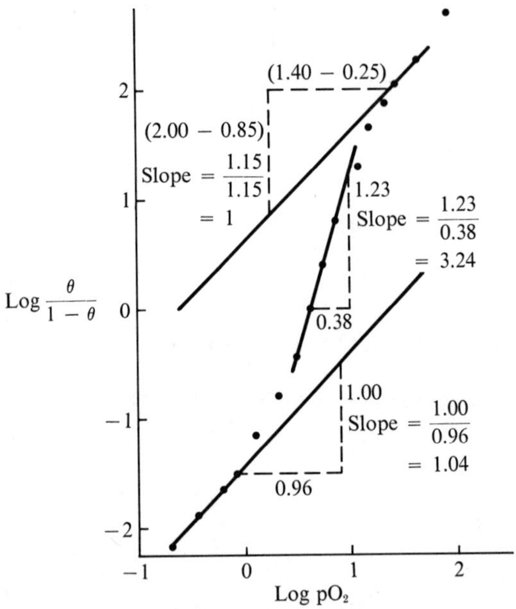

H_2CO_3 and Respiration

1. 0.024 M

2. 54.8 volume %; 18.12 mM

3. 7.35; $\frac{[HCO_3^-]}{[CO_2]} = 17.86$; yes.

4. 7.38

5. (a) 16.84 meq
   (b) 2.13 mmole liter^{-1}

6. 7.1

7. P_{CO_2} 47 mm Hg; $[HCO_3^-]$ 26.89 mM

8. 7.04

9. 7.4

10. 7.29

11. P_{CO_2} 92.6 mm Hg

12. 90%

13. 7.58

14. P_{CO_2} 50 mm Hg; respiratory acidosis

15. 14.4 meq of acid

16. $[HCO_3^-] = 24$ mM; [total CO_2] = 56.1 ml/100 ml

17. 7.35; $\frac{[HCO_3^-]}{[CO_2]} = 17.8$; very slightly acidic

18.

	P_{CO_2}	$\frac{[HCO_3^-]}{[H_2CO_3]}$	$[CO_2]$ (mM)	$[HCO_3]$ (mM)	[Total CO_2] (vol %)
(a)	36.7	18.2	1.10	20.0	47.0
(b)	20.7	31.6	0.62	20.0	47.3
(c)	84.3	15.8	2.53	40.0	94.5
(d)	30.0	19.5	0.90	17.5	40.9
(e)	40.0	27.5	1.90	32.8	75.5

Chapter 9

1. (a) $\frac{1}{2}$
   (b) $\frac{1}{2}$
   (c) $\frac{1}{2}$
   (d) $\frac{1}{2}$
   (e) $\frac{1}{6}$
   (f) $\frac{1}{10}$

$$R{-}S{-}S{-}R \xrightarrow[H_2O]{+2e} 2R{-}SH + O_2$$

$$2R{-}SH + 2O_2 \xrightarrow{-8e} 2R{-}\overset{\overset{\displaystyle O}{\|}}{S}{-}OH$$

$$2R{-}\overset{\overset{\displaystyle O}{\|}}{S}{-}OH + O_2 \xrightarrow{-4e} 2R{-}\underset{\underset{\displaystyle O}{\|}}{\overset{\overset{\displaystyle O}{\|}}{S}}{-}OH$$

Total electron change 10

2. (a) L-β-hydroxybutyrate + $NAD^+ \rightleftharpoons NADH + H^+$ + acetoacetate
   (b) $\frac{1}{2}O_2 + 2H^+ + 2$ cytochrome-a-$Fe^{++} \rightleftharpoons H_2O + 2$ cytochrome-a-Fe^{+++}

3. At pH 4, −0.24 volt; pH 5, −0.30 volt; pH 6, −0.36 volt; pH 7, −0.42 volt; pH 8, −0.48 volt

4. At pH 6, 0.869 volt; pH 5, 0.929 volt; pH 4.5, 0.959 volt

5. (a) Yes
   (b) K_r 1.0×10^{22} at pH 7.0
   (c) K_r 1.0×10^{22} at pH 7.0

6. No pH dependence

7. (a) Yes
   (b) K_r 2.14×10^{7} at pH 7.0

8. (a) FAD 2H + 2 cytochrome-c-Fe^{+++} $\rightleftharpoons$ FAD + $2H^+$ + 2 cytochrome-c-Fe^{++}
   FAD 2H $\rightleftharpoons$ FAD + $2H^+$ + $2e^-$
   cytochrome-c-Fe^{+++} + e^- $\rightleftharpoons$ cytochrome-c-Fe^{++}
   (b) $K_r = 2.14 \times 10^{10}$ at pH 7
   (c) $K_r = 2.14 \times 10^{7}$ at pH 5.5

9. (a) Pyruvate + NADH + H^+ $\rightleftharpoons$ NAD^+ + lactate
   $K_r = 2.14 \times 10^{4}$ at pH 7
   $\Delta E_0' = +0.130$ volt
   (b) $\dfrac{[NAD^+]}{[NADH]} = 2.14 \times 10^{4}$ at pH 7
   (c) $\Delta E_0' = 0.189$ volt

10.

D-glucose + H_2O $\rightleftharpoons$ D-gluconic acid + $2H^+$ + $2e^-$	E_0' + 0.45 volt
O_2 + $2H^+$ + $2e^-$ $\rightleftharpoons$ H_2O_2	E_0' + 0.30 volt
D-glucose + O_2 + H_2O $\rightleftharpoons$ D-gluconic acid + H_2O_2	$\Delta E_0'$ + 0.75 volt
D-glucose + H_2O $\rightleftharpoons$ D-gluconic acid + $2H^+$ + $2e^-$	E_0' + 0.45 volt
Br_2 + $2e^-$ $\rightleftharpoons$ $2Br^-$	E_0' + 0.65 volt
D-glucose + Br_2 + H_2O $\rightleftharpoons$ D-gluconic acid + $2H^+$ + $2Br^-$	$\Delta E_0'$ + 1.10 volt

11.

%	k (volt)
99	0.0298
90	0.0285
80	0.018
70	0.010
60	0.005
50	0.0
40	−0.005
30	−0.010
20	−0.018
10	−0.0285
1	−0.0298

12. $E_{obs} = 0.209$ volt

13.

$$\text{Lipoic acid}\begin{matrix}\diagup \text{SH}\\ \diagdown \text{SH}\end{matrix} \rightleftharpoons \text{lipoic acid}\begin{matrix}\diagup \text{S}\\ \vert\\ \diagdown \text{S}\end{matrix} + 2\text{H}^+ + 2e^- \qquad E_0' + 0.29 \text{ volt}$$

$$\text{dithiothreitol}\begin{matrix}\diagup \text{S}\\ \vert\\ \diagdown \text{S}\end{matrix} + 2\text{H}^+ + 2e^- \rightleftharpoons \text{dithiothreitol}\begin{matrix}\diagup \text{SH}\\ \diagdown \text{SH}\end{matrix} \qquad E_0' - 0.34 \text{ volt}$$

$$\text{lipoic acid}\begin{matrix}\diagup \text{SH}\\ \diagdown \text{SH}\end{matrix} + \text{dithiothreitol}\begin{matrix}\diagup \text{S}\\ \vert\\ \diagdown \text{S}\end{matrix} \rightleftharpoons \text{lipoic acid}\begin{matrix}\diagup \text{S}\\ \vert\\ \diagdown \text{S}\end{matrix} + \text{dithiothreitol}\begin{matrix}\diagup \text{SH}\\ \diagdown \text{SH}\end{matrix} \qquad \Delta E_0' - 0.05 \text{ volt}$$

K_{eq} 2.16×10^{-2} at pH 7.0 and 5.0

14. (a) $(\text{cyst-S})_2 + 2C_6H_5SH \rightleftharpoons (C_6H_5S)_2 + 2$ cysteine
    $\Delta E_0' = 0.08$ volt
    (b) 1.94

15. (a) K_{eq} 1.0×10^{14}
    (b) $\dfrac{[\text{cytochrome f Fe}^{++}]}{[\text{cytochrome f Fe}^{+++}]} > 1.0 \times 10^{14}$

Chapter 10

1. $\Delta G^{\circ\prime} = -1640$ cal mole^{-1}

2. (a) $\Delta G^{\circ\prime} = 4102$ cal mole^{-1}
   (b) $\dfrac{[\text{ATP}]}{[\text{ADP}]} = 3.87 \times 10^2$

3. $\Delta G = +1105$ cal mole^{-1}

4. $\Delta G^{\circ} = -89.4$ kcal mole^{-1}

5. $K_{eq} = 1.50$

6. (a) $\Delta G^{\circ\prime} = 58.2$ cal mole^{-1}
   (b) $\dfrac{[\text{Prod}]}{[\text{React}]} = 0.007$

7. $\Delta G^{\circ\prime} = +12{,}000$ cal mole^{-1}

8. $\Delta G^{\circ\prime} = 16.2$ kcal mole^{-1}

9. $\Delta G^{\circ\prime} = -607$ cal mole^{-1}

10. $\Delta G^{\circ\prime} = -852$ cal mole^{-1}

11. $k_{\text{urease}} = 1.6 \times 10^4$ sec^{-1}

12. $\Delta G^\circ = -51.77$ kcal mole^{-1}
    $\Delta S^\circ = -48.92$ e.u.

13. (a) 20,200 cal
    (b) -322 kcal mole^{-1} versus -306.82 kcal mole^{-1}

Chapter 11

1. $\dfrac{1}{v}$ min μmole^{-1} $\qquad \dfrac{1}{[\text{S}]}$ liter mmole^{-1}

   $\dfrac{[\text{S}]}{v}$ min liter $\times 10^3$ $\qquad \dfrac{v}{[\text{S}]}$ liter min$^{-1} \times 10^{-3}$

2. $V_{\max}$ 0.04 mole hr^{-1}
   K_m 7.7×10^{-3} mole liter^{-1}

3. $\dfrac{V_{\max}}{K_m}$ 51.4 liter min^{-1}

4. See graph.

$$2.303 \log_{10} \frac{[P_{\text{eq}}] - [P_t]}{[P_{\text{eq}}]} = -\frac{vt}{[P_{\text{eq}}]}$$

$$\therefore \text{Slope of line} = -\frac{v}{2.303[P_{\text{eq}}]}$$

$$\text{From graph} = -\frac{1.00}{2.51}$$

$$\therefore v = \frac{2.303 \times [P_{\text{eq}}] \times 1.00}{2.51} = \frac{2.303 \times 9.80}{2.51} = 8.99 \text{ mmole liter}^{-1}\text{ min}^{-1}$$

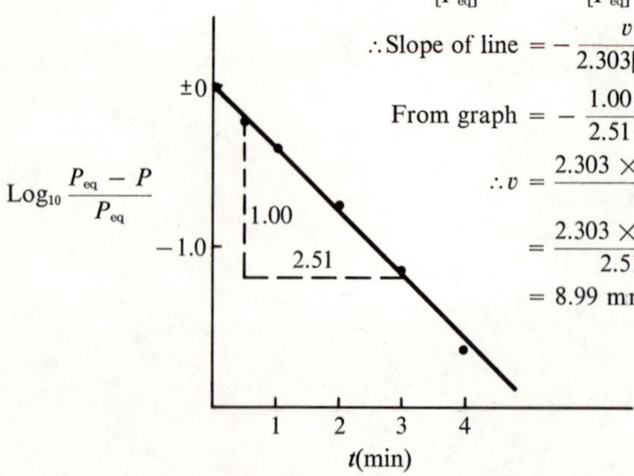

5. (a) K_m 2.5×10^{-2} M
   (b) K_m 4×10^{-3} M
   (c) K_m 3×10^{-4} M; ADP is not a competitive inhibitor
   (d) $V_{\max}$ 3.3×10^{-6} mole min^{-1}
   (e) $V_{\max}$ 3.3×10^{-6} mole min^{-1}
   (f) $V_{\max}$ 2×10^{-4} mole min^{-1}; K_m 2.4×10^{-2} M

6. (a) K_m 2.3 × 10^{-3} M

(b) $\dfrac{V_{\max}}{K_m}$ 60 liter min^{-1}

(c) $V_{\max}$ 4.6 mmole min^{-1}

7. Molecular activity 1.38 × 10^4 min^{-1} mole^{-1}; v 0.196 mmole min^{-1}

8. See graph.

$$\text{Slope} = -\frac{v}{2.303\,[P_{eq}]}$$

$$\text{From graph} = -\frac{0.85}{3.70}$$

$$\therefore v = \frac{0.85 \times 2.303 \times 200}{3.70}\ \mu\text{mole liter}^{-1}\ \text{min}^{-1}$$

$$= 105.8\ \mu\text{mole liter}^{-1}\ \text{min}^{-1}$$

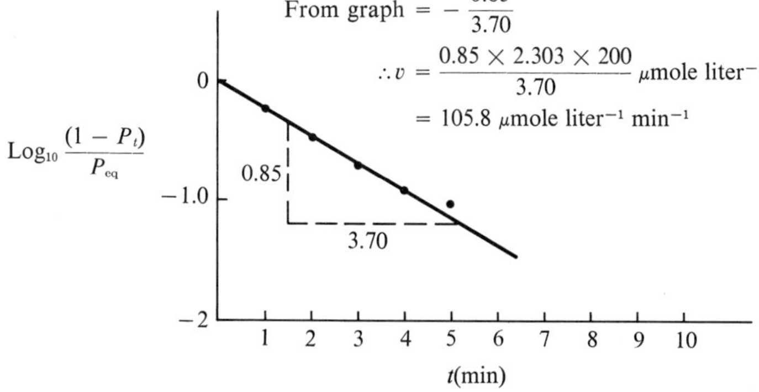

9.

	$V_{\max}$ μmole min^{-1}	K_m $M \times 10^4$
$\dfrac{1}{v}$ versus $\dfrac{1}{[S]}$	0.35	4.3
$\dfrac{[S]}{v}$ versus [S]	0.41	4.2
v versus $\dfrac{v}{[S]}$	0.35	4.6

See the three graphs, below and on the next page.

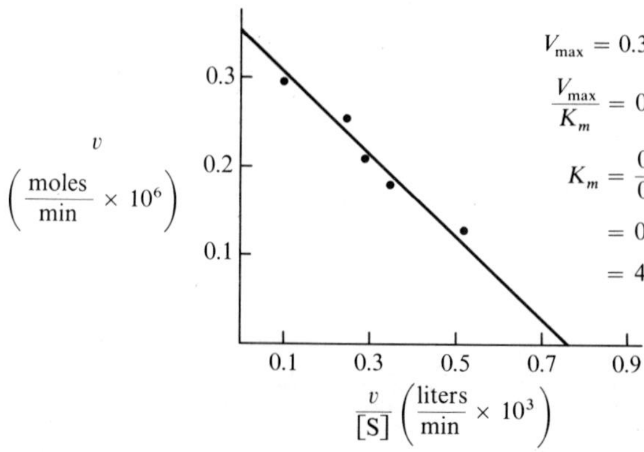

$$V_{\max} = 0.35 \times 10^{-6}\ \text{moles}\ CO_2/\text{min}$$

$$\frac{V_{\max}}{K_m} = 0.759 \times 10^{-3}\ \text{liters/min}$$

$$K_m = \frac{0.351}{0.759} \times 10^{-3}\ \text{moles/liter}$$

$$= 0.463 \times 10^{-3}\ M$$

$$= 4.6 \times 10^{-4}\ M$$

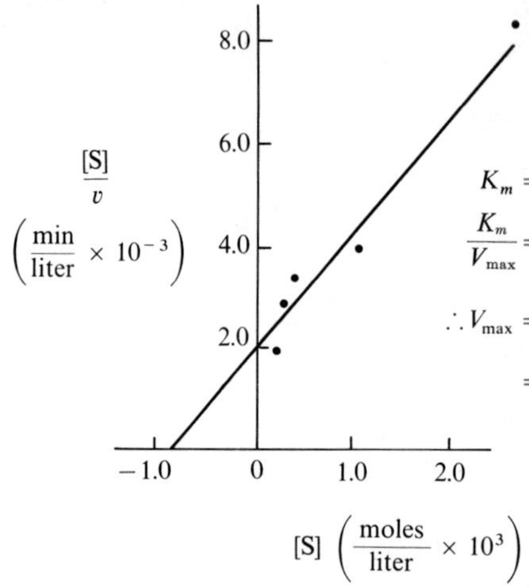

$$K_m = 0.82 \times 10^{-3} \text{ moles/liter}$$

$$\frac{K_m}{V_{\max}} = 2.01 \, \frac{\text{min}}{\text{liter}} \times 10^3$$

$$\therefore V_{\max} = \frac{0.82}{2.01} \times 10^{-6} \text{ moles/min}$$

$$= 0.41 \times 10^{-6} \text{ moles } CO_2\text{/min}$$

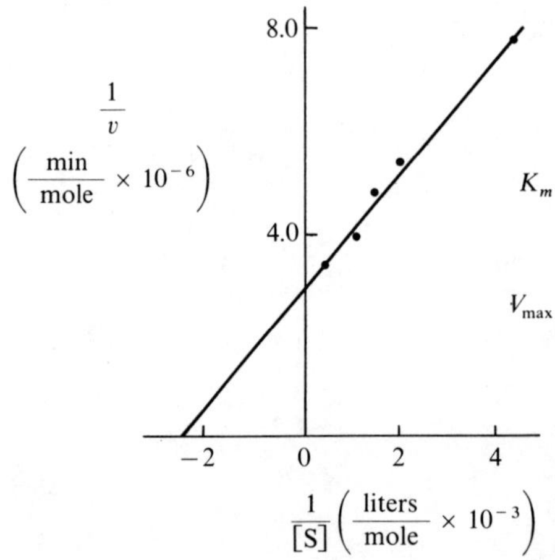

$$K_m = \frac{1}{2.35} \, \frac{\text{moles}}{\text{liter}} \times 10^{-3}$$

$$= 0.426 \times 10^{-3} \text{ moles/liter}$$

$$V_{\max} = \frac{1}{2.86} \, \frac{\text{moles}}{\text{min}} \times 10^{-6}$$

$$= 0.35 \times 10^{-6} \text{ moles } CO_2\text{/min}$$

10. See the three graphs, below and on the top of the next page.

Computer fits:

V_{max}	13.76 μmole min^{-1}
K_a	0.49 mM
K_b	63.7 mM
K_{ia}	0.82 mM

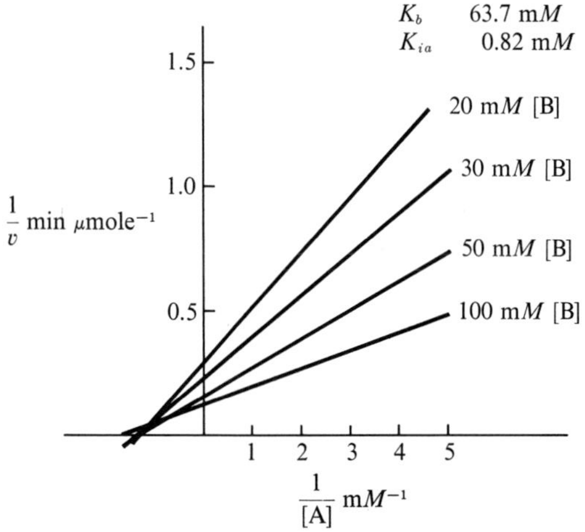

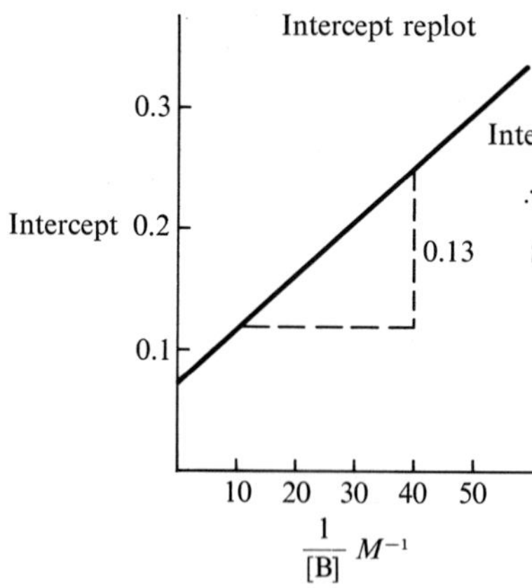

$$\text{Intercept} = 0.0725 = \frac{1}{V_{max}}$$

$$\therefore V_{max} = 13.8\ \mu\text{mole min}^{-1}$$

$$\text{Slope} = \frac{0.13}{29} = \frac{K_b}{V_{max}}$$

$$\therefore K_b = \frac{0.13\ V_{max}}{29}$$

$$= 61.8\ \text{m}M$$

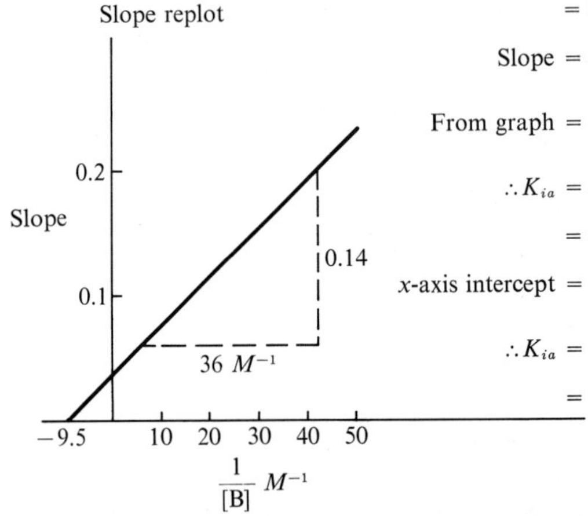

$$y\text{-axis intercept} = \frac{K_a}{V_{max}}$$
$$\text{From graph} = 0.036$$
$$\therefore K_a = 0.036\ V_{max}$$
$$= 0.036 \times 13.8$$
$$= 0.50\ \text{m}M$$
$$\text{Slope} = \frac{K_{ia}K_b}{V_{max}}$$
$$\text{From graph} = \frac{0.14}{36}$$
$$\therefore K_{ia} = \frac{0.14}{36} \times \frac{13.8}{61.8}$$
$$= 0.87\ \text{m}M$$
$$x\text{-axis intercept} = -\frac{K_a}{K_{ia}K_b}$$
$$\therefore K_{ia} = \frac{0.50}{61.8 \times 0.0095}$$
$$= 0.85\ \text{m}M$$

	by graphical calculation	by computer fit
V_{max}	13.8	13.76 μmole min^{-1}
K_a	0.50	0.49 mM
K_b	61.8	63.7 mM
K_{ia}	0.86 (average)	0.82 mM

11.

	V_{max} (μmoles min^{-1})	K_m (mM)
$\frac{[S]}{v}$ versus [S]	5.6	0.57
v versus $\frac{V}{[S]}$	5.6	0.59

See the two graphs on the next page.

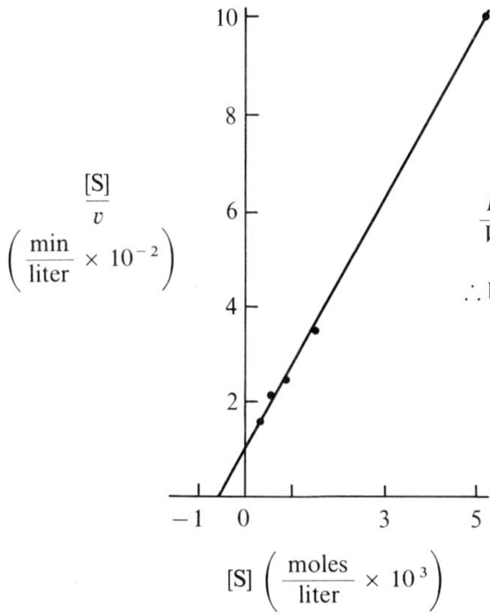

$$K_m = 0.57 \times 10^{-3} \text{ moles/liter}$$

$$\frac{K_m}{V_{max}} = 1.02 \times 10^{2} \text{ min/liter}$$

$$\therefore V_{max} = \frac{0.57}{1.02} \times 10^{-5} \text{ mole/min}$$

$$= 5.6 \times 10^{-6} \text{ mole/min}$$

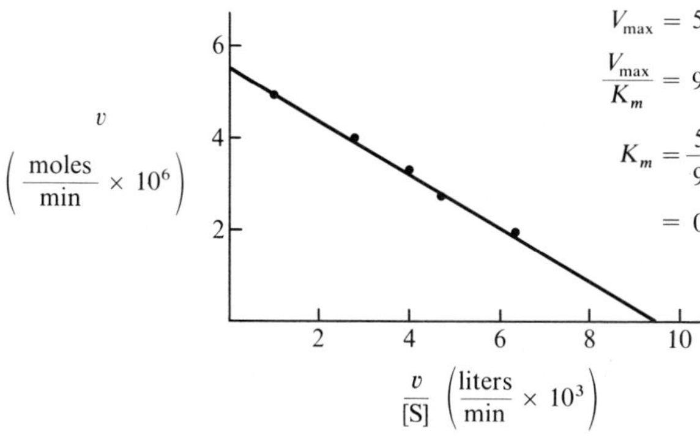

$$V_{max} = 5.6 \times 10^{-6} \text{ moles/min}$$

$$\frac{V_{max}}{K_m} = 9.5 \times 10^{-3} \text{ liters/min}$$

$$K_m = \frac{5.6 \times 10^{-6}}{9.5 \times 10^{-3}} \text{ moles/liter}$$

$$= 0.59 \times 10^{-3} \text{ moles/liter}$$

12. Noncompetitive inhibition

computer fit of data

$K_{acetaldehyde}$	3.06 mM
K_{is}	1.27 mM
K_{ii}	1.03 mM
V_{max}	0.39 ΔAbs unit min^{-1}

13. Noncompetitive inhibition (see the three graphs on the next page)

V_{max}	0.31 μmole min^{-1}
K_m	0.22 mM
K_{is}	1.50 mM
K_{ii}	4.55 mM

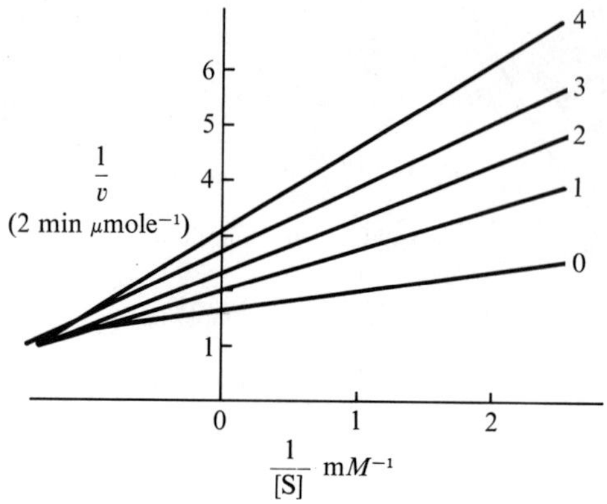

4
3
2
1
0
6
5
4
1
$\frac{1}{v}$
(2 min μmole⁻¹)
0
1
2
$\frac{1}{[S]}$ mM^{-1}

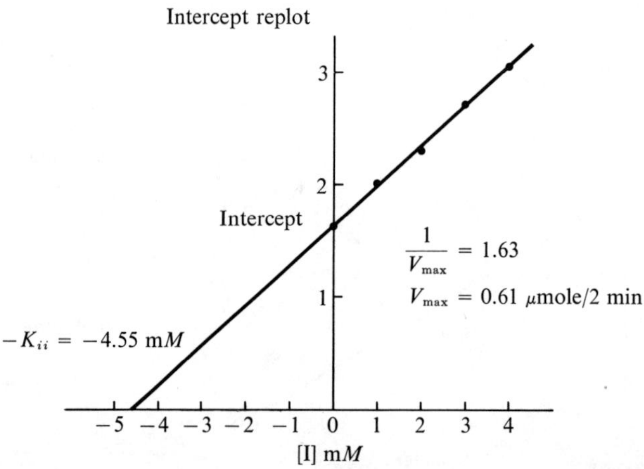

Intercept replot
3
2
1
Intercept
$\frac{1}{V_{max}} = 1.63$
$V_{max} = 0.61$ μmole/2 min
$-K_{ii} = -4.55$ mM
−5 −4 −3 −2 −1 0 1 2 3 4
[I] mM

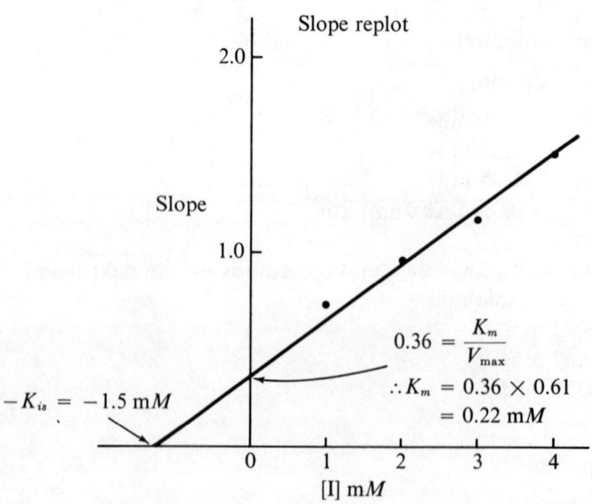

Slope replot
2.0
Slope
1.0
$0.36 = \frac{K_m}{V_{max}}$
$\therefore K_m = 0.36 \times 0.61$
$= 0.22$ mM
$-K_{is} = -1.5$ mM
0 1 2 3 4
[I] mM

14. Sigmoidal kinetics
    m 2.0
    K' 7.1 $^{-}$ 10^{-5} M (see graph)
    V_{max} ~ 1.54 ΔAbs units per 2.5 min

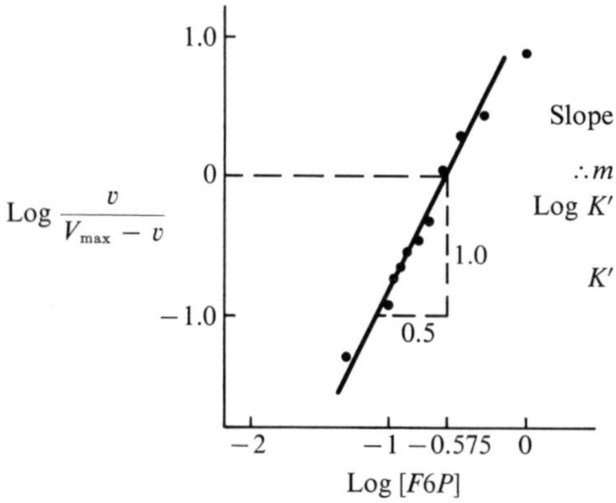

$$\text{Slope} = \frac{1.0}{0.5}$$
$$\therefore m = 2.0$$
$$\text{Log } K' = 2.0 \times (-0.575) = -1.15$$
$$K' = \text{antilog } \bar{2}.85 = 7.1 \times 10^{-2} \text{ m}M = 7.1 \times 10^{-5}\ M$$

15. Computer fit for sequential mechanism

V_{max}	0.48 μmole min^{-1}
$K_{a_{(DHF)}}$	10.11 μM
$K_{b_{(NADPH)}}$	5.02 μM
K_{ia}	3.24 μM

16. V_{max} 0.38 unit min^{-1}
    K_m 6.2×10^{-2} M (see graph)

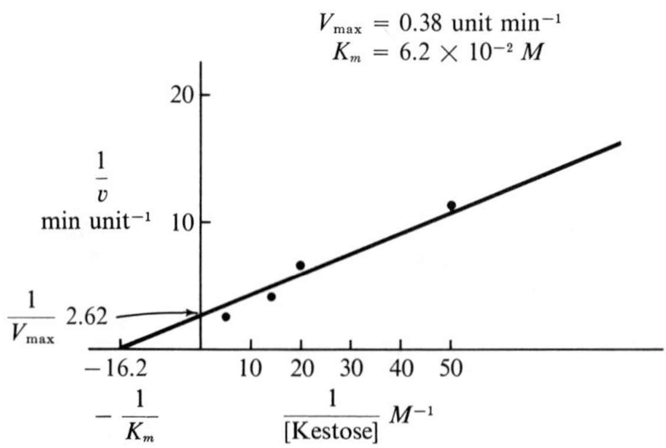

17. Sigmoidal kinetics; m 3.02 K' 1.0×10^{-4} M.

Chapter 12

1. 77.23 kcal $mole^{-1}$

2. (a) 50,000 cm^{-1}
   (b) 14,286 cm^{-1}
   (c) 26,667 cm^{-1}
   (d) 4000 cm^{-1}

3. 84,000 cal $mole^{-1}$; 110,000 cal $mole^{-1}$

4. Absorbance in order: 0, 0.100, 0.196, 0.299, 0.386, 0.474, 0.554, 0.626, 0.685, 0.742, 0.785 (see graph)
   No
   4.4 mg $liter^{-1}$

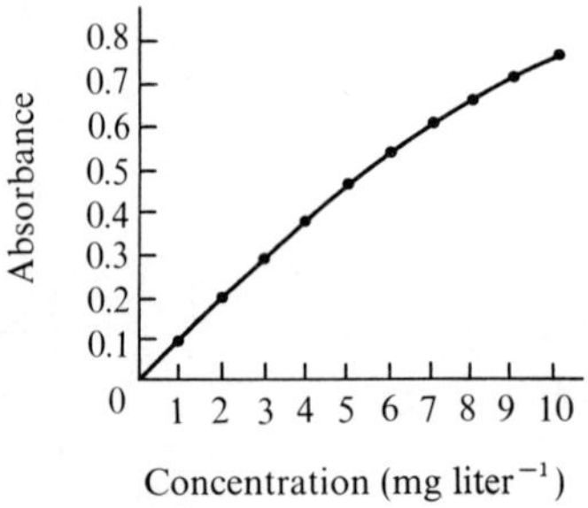

5. (a) 49.5%
   (b) 5 mm
   (c) 85%

6. Yes

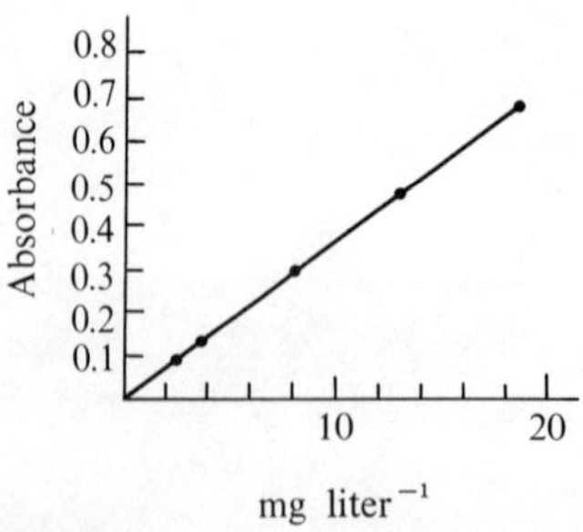

7. 0.347; mM^{-1} cm^{-1}

8. 13.4 mg/liter; Beer's law must be obeyed

9. Filter A; the absorbance is greater per unit concentration

10. 2.418

11. ϵ_{261} 3.46×10^4; ϵ_{340} 1.20×10^4; ϵ_{375} 1.03×10^4; ϵ_{524} 7.77×10^3

Chapter 13

1. Specific activity 1.15×10^{-2} μC_i per g

2. 95%

3. 14.2 hours

4. 2×10^{-3} μmole

5.

t/T	0.10	0.50	0.80	1.00	2.00	3.00	10.00
% remaining	93.3	70.7	57.4	50.0	25.0	12.5	0.1

6. Specific activity 0.02 μC_i per mg
   Counting efficiency 15.9%

7. 15.9%

8. 0.34

9. 1530 days

10. Original activities of $Ba^{14}CO_3$ and $Ba^{35}SO_4$ were 4.50×10^{-2} μC_i and 1.41×10^{-2} μC_i, respectively. Final count at 8% efficiency, 1210 cpm

11. 277 ± 6.6 counts

12. 4.05×10^{-2} mole I_2 per mole ovalbumin; 66–67 days

13. 7361 years

14. 126 days; ^{123}Sn has half-life of 125 days

15. 25% of the label was lost upon oxidation which removes one of the four ^{3}H. The ^{3}H at carbon 2 was therefore proportionately equal to the three ^{3}H at carbon 3

16.

Intestine	4.0×10^{-4} μmole/g
Spleen	3.4×10^{-4} μmole/g
Skin	0.2×10^{-4} μmole/g
Blood	0.1×10^{-4} μmole/g

Index